# Cognitive Technologies

T0181411

Vicenç Torra
Yasuo Narukawa

# Modeling Decisions

Information Fusion
and Aggregation Operators

With 55 Figures and 35 Tables

 Springer

*Authors:*

Vicenç Torra
IIIA-CSIC
Campus Universitat Autonòma de Barcelona
s/n
08193 Bellaterra, Catalonia, Spain
vtorra@iiia.csic.es

Yasuo Narukawa
Tohogakuen
3-1-10, Naka, Kunitachi
186-0004 Tokyo, Japan
narukawa@d4.dion.ne.jp

*Managing Editors:*

Prof. Dov M. Gabbay
Augustus De Morgan Professor of Logic
Department of Computer Science, King's College London
Strand, London WC2R 2LS, UK

Prof. Dr. Jörg Siekmann
Forschungsbereich Deduktions- und Multiagentensysteme, DFKI
Stuhlsatzenweg 3, Geb. 43, 66123 Saarbrücken, Germany

ACM Computing Classification (1998): G.1.6, G.3, H.1, I.2, I.5, I.6.

ISSN 1611-2482
ISBN  978-3-642-08833-9          e-ISBN  978-3-540-68791-7

Springer is a part of Springer Science+Business Media
springer.com

© Springer-Verlag Berlin Heidelberg 2007
Softcover reprint of the hardcover 1st edition 2007

Cover Design: KünkelLopka, Heidelberg

Printed on acid-free paper     45/3100/YL     5 4 3 2 1 0

To our wives, Mònica and Atsuko

and to our children,
Martí, Aina, Meritxell
Masaaki, Yoshifumi, Naohiro

# Preface

Information fusion is a broad area that studies methods to combine data or information supplied by multiple sources. Aggregation operators are some of the functions that can be used for combining data.

This book is intended for those interested in methods for aggregating information and, specially, for those who need to embed such methods in applications. It constitutes an introduction to the field. The main focus is on functions that deal with numerical information although other kinds of functions (specially ones for ordinal scales) are considered as well. It is aimed at senior undergraduate and beginning graduate students of computer science, engineering, and mathematics.

This is an introductory book in the field of aggregation operators, focused on practical applications; we have tried, on the one hand, to limit the operators and results to a set of manageable size and, on the other hand, to include some descriptions and examples of such operators at work.

We have also included a few computational issues. It has to be said that although for most operators no implementation details are given, their implementation is usually straightforward. Most of the operators and methods appearing in the book have been implemented by the authors (in Java).

Due to our objective, results with a mainly mathematical interest are not included in the text. For example, only aggregation operators that combine a finite number of inputs have been studied in detail. Some definitions and results that can be useful for further study but are not relevant for real appli-

---

[1] Rock, Paper, Scissors

cations have been included in separate figures. This is the case for definitions of fuzzy integrals of continuous functions.

## Organization

The book contains an introductory chapter, two chapters presenting some other introductory topics, and the main chapters.

The Introduction describes information integration at large, and locates aggregation operators in this setting.

Chapter 2 describes some of the tools that are needed later in the book. In particular, it focuses on measurement theory, probability and statistics, and fuzzy sets.

Chapter 3 gives an introduction to functional equations. Some well-known equations are reviewed, and a few notes on how to solve them are given.

Chapter 4 is devoted to the synthesis of judgements. It mainly reviews aggregation operators related to separability and quasi-arithmetic means, first without weights and then with them. At this point, the Bajraktarević's mean is defined. A few operators for ordinal scales are also presented.

Chapter 5 gives an overview of fuzzy measures. The most well-known families are studied: belief and plausibility and $\perp$-decomposable and distorted probabilities. Such fuzzy measures are later used in conjunction with fuzzy integrals.

Chapter 6 describes aggregation operators that can be expressed as particular cases of fuzzy integrals. Such operators include weighted means, OWA operators and weighted minimum and maximum. Fuzzy integrals, such as Choquet, Sugeno, t-conorm, and twofold integrals, are also defined and compared.

Chapter 7 is devoted to a few indices to evaluate aggregation operators and their parameters. This section includes descriptions for the Shapley and Banzhaf indices, interactions, average values and orness.

We finish, in Chapter 8, by considering the process of parameter determination for some particular operators, for example, for learning weights for the weighted mean and fuzzy measures for Choquet integrals. Two cases are considered, parameter determination with the help of an expert and parameter determination from examples.

To ease the reading, references have been grouped in bibliographical sections (Bibliographical Notes, at the end of each chapter). The full listing of the references is given at the end of the book. Examples have been given to illustrate the operators, and figures and tables have been included for the same purpose. In some cases, figures have been added to include some definitions or properties that have less interest for practical application (e.g., definitions of some fuzzy integrals in continuous domains). The book finishes with an Appendix where the main properties and some aggregation operators are listed. The lists are not exhaustive.

# How to Use This Book

The book does not assume specific previous knowledge of aggregation operators, and Chapters 2 and 3 give some preliminaries to make it self-contained. Although the chapters have been written to avoid dependences as much as possible, there are some dependences between chapters. The most important relationships are enumerated here. Chapter 4 uses functional equations reviewed in Chapter 3, and Chapter 6 defines fuzzy integrals that use the fuzzy measures described in Chapter 5. Evaluation methods (Chapter 7) are based on the particular operators and the particular parameters explained in previous chapters (e.g., Shapley value for a fuzzy measure). The problem of parameter determination for a given operator (Chapter 8) naturally needs the operator under consideration (described in previous chapters). Nevertheless, to prevent the reader from going back and forth, there are minor repetitions in the text.

The following equation is the most repeated one:

$$\min_i a_i \leq \mathbb{C}(a_1, \ldots, a_N) \leq \max_i a_i$$

# Acknowledgements

First of all, we would like to acknowledge the influence of Professors C. Alsina, U. Cortés, M. Sugeno, and T. Murofushi for introducing us to aggregation operators, fuzzy measures, and integrals.

The origins of this work can be traced back to the graduate courses on *Consensus Theory* at the Polytechnic University of Catalonia (1994-1999) and *Approximate Reasoning and Synthesis of Information* (at the Polytechnic University of Catalonia and the Autonomous University of Barcelona) (2000-present). Preliminary versions of this text have been used in the later courses. We are thankful to the students, specially Jordi Nin, for their comments on earlier versions of this text.

The actual content of the book has been shaped by our own research and joint work in the last few years. The series of conferences we initiated on Modeling Decisions for Artificial Intelligence (2004-present) [412, 415, 416], has also, through its participants, influenced this work. Special thanks go to Professors J. Dujmović and R. Mesiar.

We are grateful to colleagues in our host institutions: Institut d'Investigació en Intel·ligència Artificial IIIA-CSIC (Bellaterra, Catalonia) and Toho Gakuen (Kunitachi, Tokyo, Japan). Special thanks go also to Professors F. Esteva (IIIA director) and L. Godo. Part of this book was written at the University of Tsukuba. The help and support of Professor S. Miyamoto (U. Tsukuba) is gratefully acknowledged.

This book would not have been possible without the libraries of IIIA-CSIC (and the IIIA librarian, Carol Ruiz), Autonomous University of Barcelona,

University of Tsukuba, Rovira i Virgili University at Tarragona, and Tokyo Institute of Technology. We thank Professor J. Medina for providing us with a Catalan translation of the Latin text in Section 6.7. The help of the URV CRISES team (specially Professor J. Domingo and Drs. J. Castellà and J. M. Mateo) and of Drs. K. Fujimoto, J. Castro, H. Imai, A. Giovannucci and S. Yamazaki is also gratefully acknowledged. We thank Ronan Nugent (Springer) for his assistance during the whole production process of the book.

Financial support in the last few years also contributed to making this book possible. The authors acknowledge the support of Catalan, Spanish, and Japanese projects: Research networks XTIC (2002XT 00111, 2004XT 00004), project PROPRIETAS (SEG2004-04352-C04-02), grant PR2005-0337, and JSPS Grant-in-Aid for Scientific Research (16300065).

This book was written in LaTeX and most figures were generated with xfig. The programs required for the book have been implemented in java running on the Linux operating system. The work of free software authors is acknowledged here.

Last but not least, we would like to acknowledge the encouragement of our wives, as well as the help of our children.

Ultimately, the authors are fully responsible for all errors and omissions in this book.

Sabadell (Catalonia) and Kunitachi (Japan),                    Vicenç Torra
April 23rd, 2007 (St. Jordi's Day)                              Yasuo Narukawa

# Contents

# 1

# Introduction

Information fusion techniques, in general, and aggregation operators (or aggregation functions), in particular, are extensively used in several fields of human knowledge. They are used to produce the most comprehensive and specific datum about an entity from data supplied by several information sources (or the same source at different periods of time). They are used in systems to reduce some type of noise, increase accuracy, summarize information, extract information, make decisions, and so on. To illustrate this, we consider below some examples in different fields. Some of the typical applications are also included.

**Economics:** Aggregation techniques are used to define indices about prices such as the *Retail Price Index (RPI)* and, in general, to summarize any kind of economic information. Listings of countries or companies, where individuals are ordered according to their ranking with respect to several criteria, are frequently published in journals and newspapers. Examples are the *Human Development Index (HDI)*, which is an average of the life

---

[1] Natana was asked by all the sisters to describe the method according to which, with the system, one can find and elect the sister who is suited best to be abbess. (...) "By this method," said Natana, "is found the truth; by this truth we will be able to find the sister who is most suitable and best to be our abbess." Translation from [176].

expectancy index, the educational attainment index, and the adjusted real *gross domestic product (GDP)* per capita.

**Biology:** Methods to fuse sequences of DNA and RNA are used in several applications. Aggregation operators have also been developed to combine information about taxonomies (classifications of species). More specifically, methods exist to combine dendrograms (tree-like structures) and partitions.

**Education:** Aggregation operators are extensively used in education for assessing students' knowledge in a given subject or to assign them an overall rating for several subjects. Different methods are used in different countries, according to tradition and to the scale used when giving grades (both numerical and ordinal). Scores for evaluating educational institutions (e.g., universities) are another example of the use of aggregation operators.

**Computer Science:** Aggregation operators are used for different purposes. On the one hand, we have artificial intelligence applications, which are commented on in more detail below. On the other hand, we have decision making procedures that are applied, for example, to evaluate and select hardware and software.

Within artificial intelligence, information fusion is also widely applied, and its use is rapidly increasing as more complex systems are being developed. For example, its uses in robotics (e.g., fusion of data provided by sensors), vision (e.g., fusion of images), knowledge based systems (e.g., decision making in a multicriteria framework, integration of different kinds of knowledge, and verification of knowledge-based systems correctness) and data mining (e.g., ensemble methods) are well known. Recent advances in multiagent systems extend the range of information fusion applications in systems where an agent needs to consider the behavior of other agents to make decisions on the basis of distributed information.

Although the number of information fusion applications in artificial intelligence is large, it can be said that there are only two ultimate goals. They are (i) to make decisions and (ii) to have a better understanding of the application domain. We describe them in more detail below:

**Decision making:** This consists either of selecting the best alternative (*alternative selection*) or building one new alternative (or solution) from a set of them (*alternative construction*).

- In *alternative selection*, fusion is used to evaluate the alternatives. A typical situation is one where there is a set of alternatives and each is evaluated against several criteria (this situation corresponds to the multicriteria decision making – MCDM – problem). For example, when a buying agent has received several offers and wants to select the best one, it needs to consider the best price, the best quality, and so on. This situation can be modeled in terms of several preferences (or utility functions) or by using a single but multivalued preference. That is, for

| alt | Criteria<br>Satisfaction on:<br>Price | Quality | Comfort | | alt | Consensus | | alt | Ranking |
|-----|-------|---------|---------|---|------|-----------|---|------|---------|
| FordT | 0.2 | 0.8 | 0.3 | | FordT | 0.35 | | 206 | 0.72 |
| 206 | 0.7 | 0.7 | 0.8 | | 206 | 0.72 | | FordT | 0.35 |
| ... | ... | | | $\longrightarrow$ | ... | ... | $\longrightarrow$ | ... | ... |

**Fig. 1.1.** Decision making: (a) multicriteria or multivalued preferences; (b) aggregation of degrees of satisfaction (aggregation of preferences) and construction of the global degree of satisfaction; (c) ranking of the alternatives according to the global degree of satisfaction (preferences)

each offer, we consider the degree of satisfaction in terms of price, quality, and so on. Figure 1.1 illustrates the case in point. The figure includes several criteria $c_1, \ldots, c_N$ for each alternative.

The alternative selection problem is usually solved in a two stage process:

(i) For each decision alternative, aggregate the degrees of satisfaction of all criteria. In this way, we obtain for each alternative a single aggregated value that corresponds to a global degree of satisfaction.

(ii) Rank the alternatives with respect to the global degree of satisfaction.

It is clear that the cornerstone of the process is the aggregation method used in the first stage. Figure 1.1 illustrates the whole process.

Systems modeling group decisions also fit in this class of alternative selection problems. In this case, different experts in a group have different opinions and the goal is to obtain some consensus. This field of study is known as group decision making (GDM).

• In *alternative construction*, fusion corresponds to the whole process of building a new alternative from the original ones. It is important to underline that it is often the case that the alternatives correspond to partial solutions and that different alternatives might be incomparable or mutually incompatible. This process has to consider the importance and the reliability of the alternatives, their constraints, and the approaches used when building them. Algorithms for *plan merging* and *ensemble methods* in machine learning can be studied from this perspective.

Plan merging consists of integrating partial plans to build a more complex one. In the integration process the preconditions and the effects of each partial plan have to be considered, as they define constraints on the order in which the partial plans can be executed. For

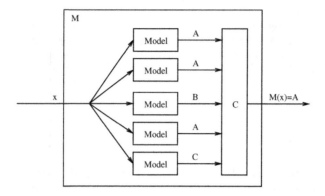

**Fig. 1.2.** Ensemble methods for classification: **x** represents the instance to be classified, and $\mathbb{C}$ represents a method for aggregating partial solutions

example, the plan for tightening a nut cannot be applied after assembling.

Ensemble methods consist of building several models from examples and then combining them to define a new one. The new model is intended to be more reliable and with less error than each of the original ones. Figure 1.2 illustrates this case: in a classification problem, several classifiers or *models $M_i$* are constructed from a set of examples using, for example, different supervised machine learning techniques. Then, for a particular instance (or situation) **x**, all models are applied each giving a solution (or class) $M_i(\mathbf{x})$. In the figure, each model leads to $A$, $B$, or $C$. Then, the solution of the whole system that is denoted by $M(\mathbf{x})$ estimates the class of the instance **x**. This is computed from the classes $M_i(\mathbf{x})$ applying some consensus procedure $\mathbb{C}$. This procedure strongly depends on how $M_i(\mathbf{x})$ are represented. In the problem represented in Figure 1.2, we can use the voting procedure as the consensus procedure $\mathbb{C}$.

**Improving the understanding of the application domain:**  A system solely working with data obtained from a single source of information faces several inconveniences caused by insufficient data quality. In particular, we underline the following difficulties: (i) lack of accuracy of the supplied data due to errors caused by the information source (either intentional or accidental) or due to errors in transmission; (ii) lack of reliability of the sources; (iii) too narrow information supplied in relation to the working domain (the information only describes a part of the application's domain).

To deal with these problems, information fusion techniques can be used. The techniques can increase the reliability of the system, improving their data quality and extending their domain of application. In fact, in some circumstances, such techniques permit the extraction of features that are impossible to perceive from individual sources. Extraction of 3D representation of objects from several images corresponds to this case.

Note that in the setting of improving the quality of the data, information fusion can be applied at the time the system is built or at runtime (for example, by combining the newly acquired information with the previously established one). Knowledge revision can be seen from this perspective.

Although information fusion is a useful tool appropriate for improving the capabilities of intelligent systems, it is important to underline that difficulties arise in their use because such data are frequently not comparable and sometimes inconsistent. Therefore, systems have to embed simple fusion techniques in larger software tools so that results are consistent. These issues are described in more detail in the next section.

## 1.1 Fusion and Integration

This section defines some of the terms in the field of information fusion and integration.

In Section 1.2 we present a general architecture for information integration based on the processes commonly admitted in multisensor fusion and integration. Information integration is considered here as a general framework that embeds information fusion. This follows the approach in the sensor field, where multisensor fusion and multisensor integration are also differentiated. Additionally, we shall use the term *aggregation operators* to refer to concrete mathematical functions. According to this, we describe the terms *information integration*, *information fusion*, and *aggregation operators* as follows.

**Information integration:** This corresponds to the use of information from several sources (or from the same source but obtained at different times) to accomplish a particular task.

**Information fusion:** Information integration requires particular techniques for combining the information. Information fusion is the actual process of combining these different data into one single datum. Therefore, information fusion refers to particular mathematical functions, algorithms, methods, and procedures for data combination. According to this, information fusion is one of the processes embedded in an information integration architecture. In the following, we will use combination as a synonym of fusion.

**Aggregation operators:** These operators (also referred to as *means* or *mean operators*) correspond to particular mathematical functions used for information fusion. Generally, we consider mathematical functions that combine $N$ values in a given domain $D$ (e.g., $N$ real numbers) and return a value in the same domain (e.g., another real number). Denoting these functions by $\mathbb{C}$ (from *Consensus*), aggregation operators are functions of the form:

Unanimity or idempotency: $\mathbb{C}(a, \ldots, a) = a$ for all $a$
Monotonicity: $\mathbb{C}(a_1, \ldots, a_N) \geq \mathbb{C}(a'_1, \ldots, a'_N)$ when $a_i \geq a'_i$
Symmetry: For any permutation $\pi$ on $\{1, \ldots, N\}$ it holds that

$$\mathbb{C}(a_1, \ldots, a_N) = \mathbb{C}(a_{\pi(1)}, \ldots, a_{\pi(N)})$$

**Fig. 1.3.** Main properties of aggregation operators

$$\mathbb{C} : D^N \to D$$

Usually, operators fuse input values taking into account some information about the sources (data suppliers). That is, operators are parametric so that additional knowledge (background knowledge, following artificial intelligence jargon) on the sources can be considered in the fusion process. We express this by $\mathbb{C}_P$, where $P$ represents the parameters of $\mathbb{C}$.

As an example, we can consider the arithmetic mean as one such aggregation operator:

$$\mathbb{C}(a_1, \ldots, a_N) = \sum_{i=1}^{N} a_i / N$$

This expression does not include any information on the data suppliers. Instead, the weighted mean is another aggregation operator that includes a weight for each data supplier:

$$\mathbb{C}_{\mathbf{p}}(a_1, \ldots, a_N) = \sum_{i=1}^{N} p_i \cdot a_i / N$$

Here, $p_i$ is the weight/relevance for the source supplying datum $a_i$.

Aggregation operators are usually required to satisfy unanimity (defined in Figure 1.3) and, when $D$ is an ordinal scale, monotonicity. The two properties imply that aggregation operators are functions that yield a value between the minimum and the maximum of the input values. Formally, they are operators $\mathbb{C}$ that satisfy internality:

$$\min_i a_i \leq \mathbb{C}(a_1, \ldots, a_N) \leq \max_i a_i \tag{1.1}$$

Moreover, in some circumstances symmetry is also required. Here, symmetry stands for the fact that the order of the arguments is not relevant. In other words, there is no source distinguishable.

From this point of view, it is clear that all aggregation operators are information fusion methods. However, only information fusion methods with a *straightforward* mathematical definition are considered here as aggregation operators. Therefore, not all information fusion methods are

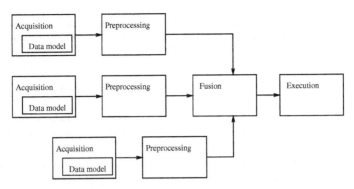

**Fig. 1.4.** General architecture for data fusion

aggregation operators. In particular, methods with a complex operational definition (e.g., complex computer programs) are not considered in this book as such. Naturally, the division between both terms is rather fuzzy.

## 1.2 An Architecture for Information Integration

Now, we turn to a general architecture for information integration. This architecture (see Figure 1.4 for a graphical representation) distinguishes the following stages.

1. **Acquisition:** The first stage corresponds to the process of gathering information from the information sources. This stage is also called detection. In order to have good data quality, a good source model is required, that is, a model of the uncertainty and error of the sources, so that it is possible to have a measure of the quality of the information. This measure can then be used in the fusion process so that it takes into account the reliability of the sources. The requirement of a source model is needed when combining sensory information (*sensor model* to determine the reliability of a particular sensor) or human (symbolic) knowledge (to determine, for example, if the supplied information is within the expert's domain of expertise or belongs to some general knowledge).

   Following the analogy with multisensor fusion, acquisition can be passive (when the information recorded is already present in the surroundings of the system) or active (when the information recorded is a consequence of an action initiated by the system).

2. **Preprocessing:** This second stage consists of preparing the data for the fusion process (i.e., of making data computationally appropriate). Several procedures are encompassed in this stage and they range from simple (like noise reduction and sensor recalibration procedures) to complex (edge detection and filtering methods). Procedures are considered preprocessing

as long as they only use the information of a single information source. This stage also includes procedures for making data commensurable and for solving the registration problem. Several pieces of data are commensurable when they refer to the same position in space and instant in time. The registration problem corresponds to the determination of information from each sensor that refers to the same features in the environment.

Aspects of both data commensurability and registration are relevant not only when we are considering numerical data from sensors but also for other kinds of data and applications, e.g., symbolic data in knowledge elicitation. In this latter setting, knowledge either elicited from experts or (automatically) extracted from databases has to be commensurate with other information before being integrated. Otherwise, results will not be meaningful.

3. **Fusion:** Once data are preprocessed and, thus, are commensurable, they can be fused. At this stage, aggregation operators or more complex fusion methods are applied to obtain a new datum. Typically, all input data uses the same representation formalism, which is also used to represent the outcome of the system. For example, the outcome of a set of images is another image. Nevertheless, some systems differ from the approach. For example, different input data use different formalisms. Then, instead of direct fusion, one source can be used to guide or cue the data from other sources. This is referred to as *guiding* or *cueing* and consists of indirect fusion. The case of visual information guiding the operation of a tactile array mounted on the end of a manipulator is an example of this situation.

4. **Execution:** Appropriate procedures are applied using the datum obtained in the fusion stage. Two kinds of procedures can be distinguished: action application and data interpretation. Control systems correspond to the first case. They use the outcome of the fusion process to decide what action to take. Exploratory robots might correspond to the second case, as they will analyze the new data and add them to their knowledge base. This case corresponds to a world model revision because the system modifies the state of its own model for the operating environment. Again, this classification is rather fuzzy, as the analysis of the data can change the behavior of the robot.

Note that all the procedures and functions that participate in this architecture are task-specific and, thus, change according to the application. For example, a decision making process in a multicriteria environment requires first the acquisition of the values for each criteria. Next, the preprocessing stage consists of the normalization of the data or data translation into a uniform space (e.g., the $[0, 1]$ interval). Then, the fusion is applied using a particular aggregation operator and, finally, a decision is made selecting the alternative that is best rated. In contrast, fusion for obstacle detection in a robot navigation system requires different procedures: gathering sensor data, making them commensurable, fusion, and, finally, raising an alarm if an obstacle is found. Nevertheless, although different procedures are used, the stages above apply to all such cases.

## 1.3 Information Fusion Methods

In the previous section we have focused on fusion processes and their role in an information integration architecture. We now turn to the fusion methods themselves.

Information fusion methods can be studied from different perspectives. In the rest of this section, we describe some of the dimensions in use for classifying them. To some extent, this classification is independent of the type of information source used (sensor or expert) and whether all the information is acquired at the same instant or at different times.

**Type of information:** Two main categories are distinguished. They correspond to redundant and complementary information.
- Redundant information occurs when several information sources describe the same features in the environment. Differences in the data, expected to be small, are due to the lack of the source's reliability. Redundant data are fused to reduce uncertainty and increase data accuracy.
- Complementary information corresponds to the case of sources describing different features of the environment (different subspaces). Different data describes different characteristics that are not similar. Fusion is applied so that the system model cover all subspaces.

**Type of data representation:** A basic consideration for any aggregation operator or fusion method is the type of data it is going to fuse. At present, there exists a large number of aggregation operators applicable to a broad range of data representation formalisms. For example, aggregation operators on the following formalisms have been considered in the literature: *numerical data, ordinal scales, fuzzy sets, belief functions, dendrograms, DNA sequences*, among others. In fact, any kind of data representation formalism is adequate for applying fusion techniques because the plurality rule (mode or voting) can be applied to data of almost any type.

**Level of abstraction:** Due to the information flow within systems (low-level data is transformed into high-level information), fusion techniques can often be applied at different levels of abstraction. For example, in a multisensor fusion system for tank detection, the following levels can be distinguished: signal, pixel, feature, and symbol. Similarly, in a knowledge elicitation problem using data from multiple experts, fusion can be performed either at the matrix level (directly on the raw data supplied by the expert) or at the similarity level (using similarities extracted from experts' raw data).

Let us consider the expression

$$\mathbb{C}(a_1, a_2, \ldots, a_N) = argmin_c \{\sum_{a_i} d(c, a_i)\},$$

where $a_i$ are numbers in $\mathbb{R}$ and where $d$ is a distance defined over $D$. Then, the following hold:

1. When $d(a, b) = (a - b)^2$, $\mathbb{C}$ is the arithmetic mean. That is, $\mathbb{C}(a_1, a_2, \ldots, a_N) = \sum_{i=1}^{N} a_i/N$.
2. When $d(a, b) = |a - b|$, $\mathbb{C}$ is the median. The median of $a_1, a_2, \ldots, a_N$ is the element that occupies the central position when the elements $a_i$ are ordered. The median is formally defined in Definition 6.7.
3. When $d(a, b) = 1$ if and only if $a = b$, $\mathbb{C}$ is the plurality rule (mode or voting). That is, $\mathbb{C}(a_1, a_2, \ldots, a_N)$ selects the element in $\mathbb{R}$ that appears most often in $(a_1, a_2, \ldots, a_N)$.

**Fig. 1.5.** Aggregation as the object that is located at the minimum distance of the objects being aggregated

When several levels can be considered, the selection of the appropriate level depends on the information available. It is usually the case that redundant information is fused at low levels because two pieces of redundant information are usually similar in structure. In contrast, complementary information is usually fused at higher levels of abstraction, as pieces of information are not so similar. For example, in the case of the tank detection system, data from two radars will be fused at the signal level (low level) if both measure the same property at the same time. In contrast, the data from a radar and a radio signal detector should be fused at the symbol level (high level), as in this case the data gathered by the two data suppliers are of a completely different nature and, thus, only the elaborated conclusions (for example, whether data seem to indicate the presence or absence of a tank) can be combined. Nevertheless, there are situations in which two information sources can only be fused at a single level.

### 1.3.1 Function Construction

A pivotal consideration in any information fusion system is the actual method used for combining information. Its definition is the cornerstone of any integration system. Two methods can be distinguished. They roughly correspond to *a priori* and *a posteriori* analyses of the method's properties.

**Definition from properties:**  The starting point for defining the method is a set of properties considered as a requirement for the method. From these properties the function is derived using mathematical tools. This is the approach used when applying functional equations (see Chapter 3). The definition of aggregation as the object that minimizes a given expression follows the same idea.

This approach is formulated as follows: the aggregation of the values $a_1, a_2, \ldots, a_N \in D$, denoted by $\mathbb{C}(a_1, a_2, \ldots, a_N)$, is the object $c$ located at the minimum distance of the objects being aggregated. That is,

$$\mathbb{C}(a_1, a_2, \ldots, a_N) = argmin_c\{\sum_{a_i} d(c, a_i)\}, \qquad (1.2)$$

where $d$ is a distance defined over $D$. The approach is valid in any domain $D$ where a distance $d$ is defined. Figure 1.5 gives an example of it when $D$ is the set of real numbers.

**Heuristic definition:**  In this case, the function is selected or defined because it seems to satisfy user requirements or expectations. The function is studied and its properties analyzed later.

An alternative method has also been proposed for function construction. It can be considered as an intermediate approach between a heuristic definition and a definition from properties.

**Definition from examples:**  This manner of definition follows classical statistical estimation theory and supervised machine learning methods. The function is built as an estimator of some available examples. Therefore, the function approximates example outcomes given example inputs. A typical method is to use neural networks for such approximations.

## 1.4 Goals of Information Fusion

Now that we have introduced information fusion and outlined some of its relevant aspects, we focus on its goals.

We have said that information fusion deals with all the aspects of the fusion process, and its main task is to deal with fusion methods. Due to the development of new representation formalisms, the consideration of new applications, and the growth of computational power, information fusion is a dynamic field, and new methods are constantly being defined. At the same time, existing methods are being analyzed to determine their properties. The two main goals of the field are (i) formalization of aggregation processes and (ii) study of existing methods. The goals are described in more detail below.

**Formalization of the aggregation process:**  This is to find formal descriptions for processes (sometimes, intuitive processes) that are used for decision making and information fusion. Formal descriptions are needed so

Arrow's impossibility theorem applies to aggregation of preferences (over a set of alternatives). It proves that when there are at least three alternatives and at least two preferences, there is no aggregation function that, for all sets of preferences satisfies the following properties:

1. Any preference can be obtained as the result of the function.
2. The function does not imply dictatorship (i.e., the function is not just one of the preferences).
3. The function is monotone, i.e., if one preference is modified so that one alternative is *promoted*, the function should at least avoid *demoting* such alternative.
4. The function satisfies the *independence of irrelevant alternatives*. That is, the final preference of $x$ over $y$ should be independent of preferences for other alternatives.

**Fig. 1.6.** Arrow's impossibility theorem

that problems can be solved in an effective and sound way. Nevertheless, model building (the procedure of building a formal description) is not an easy task. For help, the development of methodologies for function selection and tools for parameter determination (e.g., algorithms) are required. Moreover, in some situations, we need to consider the definition of new aggregation operators, as existing methods are not appropriate because they do not satisfy the desired properties or, worse, do not fit with the current representation formalism in use. The goal can be decomposed as follows:

1. *Function definition:* The construction of new functions on the basis of new properties or when considering new knowledge representation formalisms has been studied for a long time. For example, in the framework of aggregation of preferences (or of alternative selection based on preferences), Llull (thirteenth century) and Nicholas of Cusa (Nicholas Cusanus) (fifteenth century) proposed methods that were later rediscovered by Condorcet and Borda (eighteenth century). They are the Condorcet rule (with the Copeland method for solving ties) and the Borda count. A related approach, important in real-world applications, is to study when no function exists that satisfies a set of properties. Arrow's impossibility (or incompatibility) theorem is a result of this kind. We recall that it applies to functions to aggregate preferences and that Arrow proved that there is no aggregation function that satisfies a set of *natural* axioms. The theorem is reproduced in Figure 1.6.
2. *Function selection:* This corresponds to methods for deciding the most appropriate function in a given situation. At present, this can be done,

as pointed out in Section 1.3.1, heuristically, on the basis of properties or from examples.

3. *Parameter determination:* This stands for algorithms and mechanisms for finding the best parameterization of a given aggregation operator. Methods are mainly based on expert interviews or are example based.

**Study of existing methods:** For most knowledge representation systems there exists a large set of aggregation methods that can be applied. To apply them properly we need to know their intrinsic differences. Three categories can be distinguished in relation to the properties:

1. *Function characterization:* This is to know, on the one hand, which properties a particular operator satisfies and, on the other hand, which operators satisfy a set of properties. Functional equations are basic tools for function characterization.

2. *Determination of function's modeling capabilities:* The selection of an aggregation function corresponds to a tradeoff between expressivity and simplicity. In this respect, we know that aggregation operators can be used to build universal approximators (to approximate an arbitrary function at the desired level of detail). There exist some general models based on quasi-arithmetic means and Choquet integrals. However, to use such general models in practice is a difficult task, because on the one hand they require a large number of parameters and on the other hand they are difficult to interpret. In contrast, the arithmetic mean does not use any parameter, while its modeling capability is very limited (it corresponds to a completely determined hyperplane). In this framework, the determination of a function's modeling capability corresponds to locating it in the broad range of operators between the arithmetic mean and the general model.

3. *Relationship between operators and parameters:* Most aggregation operators are parametric and, therefore, their behavior strongly depends on the parameters. It is important to know how parameters can affect the result. For example, to know whether there exists a parameterization that implies the dictatorship property to one of the information sources (dictatorship can be represented with the weighted mean but not with the OWA operator; see Section 6.1), it is important to know how sensitive the operator is to changes in the data (according to parameterization) or how much the output is changed when the parameters change (needed when parameters are extracted from examples). To help in this analysis, some indices have been defined. Some of them (e.g., *orness*) will be reviewed in Chapter 7.

In this section, we have given a classification of current goals of information fusion and its research. Nevertheless, this classification is not crisp, as there are some research topics that can be found across several different areas. One of them is parameter determination according to the bias-variance trade-off. This is somehow equivalent to the selection of a model that sufficiently fits the data, but does not overfit it. This requires approaches from function selection, parameter determination, and also approaches related to functions' modeling capabilities.

## 1.5 Bibliographical Notes

1. **Information fusion:** Information fusion and integration is a broad field, with applications in several fields of the human knowledge. Due to this, aside from its pure mathematical research, work on it is published in journals and conferences on a wide range of topics. Our bias, and, therefore, the bibliographical references used and consulted for preparing this book, is towards artificial intelligence, mathematics, economics, remote sensing, and multisensor fusion applications.

2. **Information integration and architectures:** The way we have structured this chapter and our vision of the field are mainly based on sensor fusion and integration. Reference books in this field include [1] and [47]. See also the review paper by Hall and Llinas [179]. The chapter by Luo and Kay [241] in [1] gives a nice state-of-the-art description (from the 1990s) of data fusion and sensor integration. Most of the concepts reviewed can be easily translated to other fields, such as artificial intelligence. Several reference papers on fusion and related issues (e.g., sensor and data fusion, decision making) have been collected by Sadjadi in [345].

   Differences between information integration and information fusion explained in this chapter mainly correspond to the ones in [241], while our definition of information fusion is based on [435] and [166].

   Sensor fusion has devoted much effort to research on architectures. The architecture presented here is based on [47], with elements of [241]. In particular, the definition of preprocessing as "putting the data in a form that is computationally appropriate" is from [47]. Additionally, the difference betwen active and passive acquisition can be found in [193].

3. **Aggregation operators:** There is no standard definition of aggregation operators. For example, Cauchy [66] and more recently Ovchinnikov [306] only require a function returning a value between the minimum and the maximum, while [138] and [449] also require symmetry. In this book, we follow the first approach initially and then add some consideration on the background knowledge later.

   As stated, internality (Equation 1.1) means that an operator leads to a value between the minimum and the maximum. This property is used by Cauchy (1821) in [66]. Ovchinnikov in [306] refers to operators that satisfy this property as compensative functions. [246] refers to them as internal functions.

   Additional references on aggregation operators and related topics are given in the bibliographical notes of other chapters of this book, specially Chapters 4 and 6.

Aggregation operators defined as the minimization of a distance (as in Equation 1.2 in Section 1.3.1) have been extensively used in the literature. For example, Fodor and Roubens in their book [146] (p. 143) use this approach to define the aggregation of relations. Similar results focused on biology can be found in [36] and [87]. They correspond, respectively, to methods for aggregating dendrograms and sequences. See also [141] for a recent application of these aggregation methods to bioinformatics. In this setting, the resulting aggregation function is known as the *median rule*. The examples given in Figure 1.5 are proved in Gini's book (1958) [163]. In particular, the result about the arithmetic mean is proved on p. 168, and the one about the median on p. 176. The property concerning the plurality rule is given on p. 185.

Jackson (1921) [201] includes some results for the same problem when the distance equals $d_p(a, b) = |a - b|^p$ for some $p > 1$. It shows that for $p > 1$ there is a single solution. It also studies the case for $p = 1$, which corresponds to the mode. It shows that there is a unique solution when $N$ is either of the form $N = 2k+1$ or of the form $N = 2k$ with $a_{s(k)} = a_{s(k+1)}$ where $s$ is an order statistics (a permutation such that $a_{s(i)} \leq a_{s(i+1)}$). In the case with $N = 2k$ and $a_{s(k)} \neq a_{s(k+1)}$, any value $a$ in the interval $[a_{s(k)}, a_{s(k+1)}]$ is a valid solution. Nevertheless, the paper also shows that the following holds for the limit of $p \to 1$:

$$\lim_{p \to 1} \arg\min_c \left\{ \sum_{i=1}^{N} |c - a_i|^p \right\} = m,$$

where $m$ is characterized by

$$(m - a_{s(1)}) \cdots (m - a_{s(k)}) = (a_{s(k+1)} - m) \cdots (a_{s(N)} - m).$$

From this, it can be shown that for $N = 2$, $m$ should be $m = (a_1 + a_2)/2$ and that for $N = 4$, $m$ corresponds to

$$m = \frac{a_4 a_3 - a_2 a_1}{(a_4 + a_3) - (a_2 + a_1)}.$$

Note that the standard definition of median for $N = 2k$, $(a_{s(k)} + a_{s(k+1)})/2$, does not correspond, in general, to this limit (see Definition 6.7).

The cases with $p = 1$ and $p = 2$ (corresponds to the arithmetic mean) were already studied by different authors. For example, it was known by Laplace [221] (supplement 1812-1818) and Svanberg (attributed) [20] p. 194-195 (1821). The case of $p \to \infty$ corresponds to the midrange of $\{a_1, \ldots, a_N\}$. That is, $(a_{s(1)} + a_{s(N)})/2$. Foster (1922) [149] studied the case of $p \to 0$, showing that it corresponds to the mode. The case of weighted distances was studied in [35] (1938). It leads to the mode ($p \to 0$), weighted median ($p = 1$), weighted mean ($p = 2$), and, again, the midrange for $p \to \infty$.

4. **Applications and examples:** The cited chapter by Luo and Kay [241] describes several systems in some detail. They are examples of sensor fusion. Among them, we underline the example of the tank detection system, where fusion is performed at several levels. This example was outlined in Section 1.3. The other example in the same section on knowledge elicitation is taken from [409].

   Luo and Kay also give an example that corresponds to the indirect fusion (guiding and cueing) described in Section 1.2. It is the description of a robotic object recognition system that uses vision to guide tactile sensing. Other examples of aggregation operators for either numerical or ordinal scales are given in [43]. In particular, [43] includes a description of the Human Development Index and several methods for aggregating grades. Some fusion methods in biology are described in [244], [67], and [86]. [244] deals with fusion of taxonomies ([318] is an application of such aggregation methods for comparing phylogenetic trees), while [67] and [86] deal with fusion of sequences. Methods for the aggregation of partitions, also used to aggregate nonhierarchical classifications in biology, can be found in [143] and [266]. Examples of fusion techniques for computer science can be found in [104] and [105].

   Decision making is described in several books. See [340] for a state-of-the-art (1996) description of the field. Other examples briefly pointed out in this chapter include plan merging and ensemble methods. Methods for plan merging are described in [96, 150]. Ensemble methods are a successful technique applied in machine learning and are nowadays described in most machine-learning books. See [182] and [436].

5. **Goals of information fusion:** Section 1.4 is basically based on our own research. Ramon Llull (thirteenth century) findings on electoral systems can be found in [234] (Chapter XXIV), [176], and also on a Web page [235]. [176] and [235] include English translations as well as transcripts of Llull's original works in either Catalan (for example, the novel Blanquerna [234] written c. 1283 [369]) or Latin (*Artifitium electionis personarum* and *De arte eleccionis*). Llull's election method anticipated Condorcet (eighteenth century) (he uses Copeland's method for solving ties). Nicholas of Cusa (or Cusanus) introduced an alternative method to Llull's in 1431 (in his work *De concordantia catholica*) that corresponds to Borda's account. Ramon Llull and Cusanus were motivated by a need to find a method for honest elections in the Church.

   The papers by McLean [257] and McLean and London [258] are also of interest here. They discuss Ramon Llull's contributions in the context of medieval voting, and the influence of Ramon Llull in Cusanus. Chapter 37 of Book III of *De concordantia Catholica* by Cusanus is reproduced in [257] and [258]. This book was written while Cusanus was attending the Council of Basel (1431-1434). [258] argues that the method proposed by Llull in Blanquerna corresponds to the Borda count. In this respect, we agree with the later interpretation by Hägele and Pukelsheim [176], rather than with the one by McLean and London.

The papers *De arte eleccionis* and *Artifitium electionis personarum* were rediscovered, respectively, by Honecker [191] in 1937 and by Perez Martínez [320] in 1959. The first work was found in the library of the Sankt Nikolaus-Hospital/Cusanusstift in Bernkastel-Kues and seems to have been copied by Cusanus himself (see [176] p. 6).

Arrow's impossibility theorem was given in [23]. For a history of voting procedures, see [426] or [259]. Arrow's theorem is described in several books on preference, choice, and decision. See [332], and also the handbook edited by Arrow, Sen, and Suzumura [24].

The definition of models based on aggregation operators that are universal approximators can be found in [399] and [277, 290, 413]. The former work defines a model based on quasi-weighted means and the latters define models based on Choquet integrals.

The bias-variance tradeoff is described in most machine learning and statistical learning books. See [182] and [296].

# 2

# Basic Notions

In this chapter we will review some of the concepts that are needed later in the book. In particular, we focus on *measurement theory* and some basic elements of *probability theory* and *fuzzy sets theory*.

## 2.1 Measurement Theory

*Per levar los molts y notables inconvenients que resúltan de haver-hi diversitat de pesos, midas y mesuras en las ciutats, vilas y locs del Principat de Cathalunya y Comtats de Rosselló y Cerdanya, (...) statuïm e ordenam, (...) que en part alguna de dits Principat y Comtats no·s puga tenir, rebre, ni usar altre pes, mida ni mesura, sinó la que se usa, y és approbada, en la ciutat de Barcelona (...)*[2]

――――――――――――――

Corts de Montsó, Chapter 89 (1585), reproduced from [16], p. 99

A working definition of measurement reads "the process of assigning numbers to characteristics of objects or persons according to rules." Then, aggregation operators are applied to such measurements so that the numbers are *improved*. As Roberts [334], after Hays [183], points out, one can always perform mathematical operations on numbers (add them, average them, take logarithms,

――――――――――――――

[1] Begin with something closest to you

[2] To eliminate the large and remarkable inconveniences caused by the diversity of weights, sizes and measures used in the cities, towns and villages of the Principality of Catalonia and the Counties of Rosselló and Cerdanya, (...) we enact and order (...) that in no part of the mentioned Principality and Counties one could have, receive or use other weight, size nor measure different than the one that is used, and accepted, in the town of Barcelona

and so on). However, the question is whether, after having performed such operations, one can still deduce true (or, better, meaningful) statements about objects.

This statement is of special relevance in the field of data fusion, as the way in which aggregation operators operate can distort the real meaning of the values, and the outcome can be meaningless. Two examples are considered for illustrating this process.

*Example 2.1.* Let $\mathbf{v_1} = (1,1)$ and $\mathbf{v_2} = (-1,1)$ be two vectors to be fused. One alternative is to aggregate componentwise (using an arithmetic mean) and, thus, define the aggregated vector $\mathbf{v}_C$ by $\mathbf{v}_C = ((1-1)/2, (1+1)/2)$. So, the outcome is $\mathbf{v}_C = (0,1)$ using the arithmetic mean.

Nevertheless, this approach can be useless if $\mathbf{v_1}$ and $\mathbf{v_2}$ are the outcome of two planning systems for a robot and correspond to the direction the robot should take. If outcomes $\mathbf{v_1}$ and $\mathbf{v_2}$ are due to the fact that the planners want to avoid a collision with an object just in front of the robot (precisely, in the direction $\mathbf{v}_C = (0\ 1)$), the aggregation is completely inappropriate.

An alternative aggregation method that is more appropriate in this case is to fuse the angles between the vectors and the $(0,1)$ vector.

*Example 2.2.* Let us consider the values $(1, 1, 4, 4, 5)$ to be aggregated. One approach is to apply the arithmetic mean to them. Then, the output would be $(1+1+4+4+5)/5 = 3$. Three alternative scenarios for this computation are presented below.

(i) Let the values $(1, 2, 3, 4, 5)$ correspond to the identifiers of some search engines and let the selected values $(1, 1, 4, 4, 5)$ be the best engines with respect to the performance, according to five different criteria. In this case, the arithmetic mean is not appropriate because it causes the selection of a nonoptimal search engine. An alternative approach suitable for this kind of problem is majority voting (i.e., select either search engine 1 or 4).

(ii) Let the values represent grades of satisfaction in the set {*very low, low, medium, large, very large*} (i.e., 1 is *very low*, 2 is *low*, and so on). In this case, the average seems satisfactory. The outcome would correspond to *low*.

(iii) Let the values represent grades of satisfaction in the set {*low, medium, large, very large, optimum*}. In this case, a value equal to 1 corresponds to low and a value equal to 2 corresponds to *medium*. The average obtained with the arithmetic mean corresponds to *large*. Nevertheless, if we consider that the two values 1 (*low*) in $(1, 1, 4, 4, 5)$ are compensated by the two values 4 (*very large*), it seems that the outcome should be *larger* than *medium*. The use of the median operator in the aggregation process would lead, with the data given above, to 4 (*very large*). So, in this context, the median seems a more suitable operator than the average. Moreover, when ordered categories are considered, the median is sounder than the arithmetic mean.

These examples show that an essential matter to be considered in information fusion is the kind of operator meaningful in a given domain. This is related to *Measurement Theory*. Measurement Theory gives a sound foundation of all matters related to measurement and scale.

### 2.1.1 Measurement

Measurement may be regarded as the construction of homomorphisms (or scales) from empirical relational structures into numerical relational structures. Informally, the empirical relational structures correspond to structures found in the real world and the numerical ones correspond to ones in the framework we build to measure.

An example related to Example 2.2 is the Mohs scale of hardness. Hardness is a measure of a mineral's resistance to abrasion and it is well known that it is defined in terms of the following ten minerals ($A :=\{1.$ Talc, 2. Gypsum, 3. Calcite, 4. Fluorite, 5. Apatite, 6. Orthoclase, 7. Quartz, 8. Topaz, 9. Corundum, 10. Diamond$\}$). In this set, the $i$th mineral can scratch all minerals $j$ such that $j < i$. Then, the hardness of any other mineral is determined from the standard set of minerals by checking which ones scratch it.

In relation to the Mohs scale, the empirical relational structure is the set of minerals and their order in the real world with respect to hardness. The numerical relational structure is the set of natural numbers in $\{1, 10\}$ with the $<$ relation. Measurement is, thus, the process of constructing the homomorphism between the two relational structures. In other words: measurement is the process of assigning numbers that preserve certain conditions (such as being able to scratch).

By the way, it has to be said that it is not always required that homomorphism be defined as a *numerical* relational structure. Measurement without numbers is also possible. One example is the way students are graded in some countries ($\{A, B, C, D, F\}$ in American institutions).

Formally, a relational structure is a set $A$ with one or more relations $R_i$ (not necessarily binary) on $A$ and, possibly, some operations $o_i$ on $A$ ($o_i : A \times A \to A$). Structures are expressed by tuples, as in $\langle A, R_1, R_2, o_1 \rangle$. For example, in the case of the Mohs scale of hardness, we would have a set $A$ and a relation $\succ$. That is, $R :=\succ$. Therefore, the relational structure is of the form: $\langle A, \succ \rangle$. Here, $a_1 \succ a_2$ for $a_1, a_2 \in A$ holds when $a_1$ is harder than $a_2$.

Another example is the measurement of long objects (i.e., measurement of length). In this case, the measurement requires a set (say $A$), a relation (say $\succ$), and an operation (say $\circ$). Thus, the relational structure is of the form: $\langle A, \succ, \circ \rangle$. In this case, the relation $\succ$ has a meaning similar to the case above: $a_1 \succ a_2$ for $a_1, a_2 \in A$ holds when $a_1$ is longer than $a_2$. The operation $\circ$ stands for the concatenation of long objects; $a_1 \circ a_2$, stands for putting $a_2$ after $a_1$. The consideration of this operation is appropriate for dealing with addition of lengths.

Numerical relational structures correspond to the case in which we consider a set of numbers (e.g., $\mathbb{R}$), and, therefore, relations and operations correspond to relations and operations over the set of numbers. In the case of the Mohs scale, we would have $\langle \mathbb{N}_{10}, > \rangle$, and in the case of the measurement of length, we would have $\langle \mathbb{R}^+, >, + \rangle$. Here $\mathbb{N}_{10}$ stands for integers between 1 and 10, and $\mathbb{R}^+$ is the set of positive real numbers.

Given two relational structures, a homomorphism is a mapping from one relational structure to the other in such a way that relations and operations are preserved. Formally, given the relational structures

$$\langle A, R_1, \ldots, R_r, \circ_1, \ldots, \circ_s \rangle$$

and

$$\langle B, R'_1, \ldots, R'_r, \circ'_1, \ldots, \circ'_s \rangle,$$

$\phi : A \to B$ is a homomorphism from the first structure into the second if, for all $a_1, \ldots, a_r \in A$, the following two conditions hold:

- $R_i(a_1, \ldots, a_{r_i})$ if and only if $R'_i(\phi(a_1), \ldots, \phi(a_{r_i}))$ for all $i = 1, \ldots, r$.
- $\phi(a_1 \circ_i a_2) = \phi(a_1) \circ'_i \phi(a_2)$ for all $i = 1, \ldots, s$.

Measurement is based on homomorphism instead of isomorphism because the mapping is not usually one-to-one. In other words, $\phi(a_1) = \phi(a_2)$ does not imply that $a_1$ and $a_2$ are the same. For example, two objects may have the same hardness or two students may have the same mark.

### 2.1.2 Representation and Uniqueness Theorems

As has been stated, the purpose of measurement is to establish a homomorphism between the empirical relational structure and the numerical relational structure. This homomorphism is built once a set of axioms over the structure are established. Then, the representation theorems and uniqueness theorems are proved.

Representation theorems establish the existence of the homomorphism (say $\phi$) into the numerical relational structure. Uniqueness theorems establish the permissible transformations over $\phi$ that also yield to homomorphisms into the same numerical relational structure. For example, in the case of measuring length, it is not possible to replace values in $\mathbb{R}$ with their logarithms (i.e., change lengths $a$ to $\log(a)$) because the addition, $+$, will not be consistent with the $\circ$ operation. However, multiplying all values by a positive constant will keep $+$ and $\circ$ consistent.

An example of a representation and uniqueness theorem is given below for illustration. It corresponds to the establishment of an ordinal scale on finite sets. We start with the definition of a relational structure with the set and the weak order $\succ$. Note that this example relates to the Mohs scale considered above.

| Name | Permissible transformations | Examples |
|------|------------------------------|----------|
| Absolute | $\psi(x) = x$ | counting, numbers |
| Ratio scale | $\psi(x) = \alpha x$ (for $\alpha > 0$) | mass (kg to pounds: $\psi(k) = 0.4536k$) |
| | | length (miles to km: $\psi(m) = 1.6093m$) |
| Interval scale | $\psi(x) = \alpha x + \beta$ | time (calendar) |
| | | temperature (Celsius/Fahrenheit) |
| Ordinal scale | $\psi(x)$ such that | preferences |
| | $x > y$ implies $\psi(x) > \psi(y)$ | Mohs' scale to measure hardness |
| | $x = y$ implies $\psi(x) = \psi(y)$ | |
| Nominal | $\psi(x)$ one-to-one | subjects in a school |
| | | brands of products |

**Table 2.1.** Major scale types

**Definition 2.3.** *Let $A$ be a set and $\succeq$ be a binary relation on $A$. The relational structure $\langle A, \succeq \rangle$ is a weak order if and only if, for all $a_1, a_2, a_3 \in A$, the following two axioms are satisfied:*

*Connectedness: Either $a_1 \succeq a_2$ or $a_2 \succeq a_1$*
*Transitivity: If $a_1 \succeq a_2$ and $a_2 \succeq a_3$, then $a_1 \succeq a_3$.*

Now, we give a representation theorem for this relational structure studied by Cantor:

**Theorem 2.4.** *(Cantor's representation theorem) Let $\langle A, \succeq \rangle$ be a weak order with $A$, a finite nonempty set; then, there exists a real-valued function $\phi$ on $A$ such that, for all $a_1, a_2 \in A$,*

$$a_1 \succeq a_2 \text{ if and only if } \phi(a_1) \geq \phi(a_2).$$

Therefore, when connectedness and transitivity are satisfied in the relational structure $\langle A, \succeq \rangle$, there exists a homomorphism $\phi$ into the numerical relational structure $\langle \mathbb{R}, \geq \rangle$. The homomorphism $\phi$ is said to be an order homomorphism.

**Theorem 2.5.** *Let $\langle A, \succeq \rangle$ be a weak order with $A$, a finite nonempty set, and let $\phi$ be an order homomorphism. Then $\phi'$ is another order homomorphism on $A$ if and only if there exists a strictly increasing function $f$, with domain and range equal to $\mathbb{R}$, such that, for all $a \in A$,*

$$\phi'(a) = f(\phi(a)).$$

This uniqueness theorem establishes that the only valid transformations are the ones generated by strictly increasing functions.

### 2.1.3 Uniqueness Theorems and Scale Type

Uniqueness theorems establish permissible transformations for $\phi$. We say that a transformation $\psi$ is permissible when, applied to $\phi$, yields to a homomorphism from the empirical relational structure into the numerical one. For example, we have in Theorem 2.5 that strictly increasing functions are the only valid permissible transformations for $\phi$.

Table 2.1 displays the most common scales. The name of the scale, permissible transformations, and some examples are given for each scale. We briefly review them here.

1. **Absolute scale** (first row in Table 2.1): This corresponds to the case where no transformations are possible. Counting and numbers are examples of this scale.

2. **Ratio scale** (second row): This corresponds to measures in which there is an absolute zero and only the unit of measurement can be changed. Thus, $\psi(x) = \alpha x$ for positive $\alpha$. This is the case for length (meters vs. miles) and mass (grams vs. pounds). Note that zero is an absolute value (no length or no mass) and that one measure in one unit can be changed into a measure in the other unit by multiplying it by a constant. For example, 1 mile = 1.609344 km and 1 pound = 0.4536 kg.

   In ratio scales, the ratio between two scale values are independent of the actual scale used. For example, the ratio of the distances between *Barcelona* and *València* (349 km, 216.85855 miles) and *Barcelona* and *Tarragona* (99 km, 61.515747 miles) does not depend on the unit. In other words, $\phi(a_1)/\phi(a_2) = \psi(\phi(a_1))/\psi(\phi(a_2))$. In the particular case of the example considered, this is 349 / 99 = 216.85855 / 61.515747 = 3.5252526.

3. **Interval scale** (third row): In this case, affine transformations are allowed (i.e., $\psi(x) = \alpha x + \beta$). Temperatures are an example of interval scales. Fahrenheit temperatures $(F)$ are computed from Centigrade ones $(C)$ by $F = 1.8C + 32$. In interval scales, the ratios of intervals are invariant. This is formally expressed by

$$\frac{\phi(a_1) - \phi(a_2)}{\phi(b_1) - \phi(b_2)} = \frac{\psi(\phi(a_1)) - \psi(\phi(a_2))}{\psi(\phi(b_1)) - \psi(\phi(b_2))}.$$

4. **Ordinal scale** (fourth row): This example has been considered in Theorems 2.4 and 2.5. Any monotone increasing function is a permissible transformation. The Mohs scale of hardness is an example of this scale. Preferences are also often expressed using ordinal scales. Any ordered set of values is equally appropriate to express the ordering. In the case of preferences, it is equally valid to use the set $\{1, 2, 3, 4, 5\}$, the set $\{A, B, C, D, E\}$, or the set $\{very\ low, low, medium, large, very\ large\}$ to express which alternative we prefer. In this case, $\phi(1) = A$, $\phi(2) = B$, ..., $\phi(5) = E$, or, similarly, $\phi'(1) = very\ low$, $\phi'(2) = low$, ..., $\phi'(5) = very\ large$.

| Name | Permissible transformations |
|---|---|
| Difference scale | $\psi(x) = x + \beta$ |
| Log-interval scale | $\psi(x) = \alpha x^{\beta}$, for $\alpha, \beta > 0$ |

**Table 2.2.** Other scale types

5. **Nominal scale** (fifth row): In this case, any transformation is appropriate, as numbers do not have any intrinsic value (they do not codify information). Any (numerical/nonnumerical) coding system is appropriate for representing nominal scales. Subjects in schools and brands of products are examples of nominal scales.

The consideration of the transformations on scales is of great importance in aggregation. This is so because when data is supplied in a given scale, the outcome of the method is expected to be in the same scale. In Chapter 4 (Section 4.3), we will consider again the use of scales and their implications in aggregation.

**Other scales**

The scales we have just described are the major scale types found in the literature. Other scales have also been defined and used in various applications. Some of them are included in Table 2.2. The difference scales might be used after logarithmic transformations have been applied to ratio scales. In the same way, log-interval scales correspond to exponential transformations of interval scales.

## 2.2 Probability and Statistics

*It is possible to develop a theory of measure with the countable additivity requirement replaced by the weaker condition of finite additivity. The disadvantage of doing this is that the resulting mathematical equipment is much less powerful. However, a convincing physical justification of countable additivity has yet to be given.*

R. B. Ash, p. 6 [26]

In this section we review some basic facts about probability measures. Such measures are the basic tool in probability theory to model randomness. In a random experiment the outcome cannot be predicted in advance, as different executions of the same experiment often lead to different outcomes. Dice and coins are classical examples, as each time we toss them we might get a different output.

Probability theory is the mathematical theory to model these kinds of situations. Probability measures assign values to possible outcomes. Formally, probability measures are defined taking into account the set of all possible outcomes of the process, i.e., the state space, the sample space, or the reference set, often denoted by $X$, $\Omega$, or $\mathcal{X}$. In this book, we will use $X$.

When tossing a coin, we have $X = \{head, tail\}$, and when throwing a die, we have $X = \{1, 2, 3, 4, 5, 6\}$. In these examples, the space $X$ is finite. Nevertheless, in general, $X$ can be infinite, and either countable or uncountable. For example, we can consider as $X$ the set of integers or the $[0, 1]$ interval.

Another basic concept is *event*. This corresponds to a property that can be checked after an experiment has been done. For example, after tossing the coin we can check whether it is tail or not, and after throwing the die we can determine whether it has an odd number or not. Formally, an event is a subset of the set $X$.

**Definition 2.6.** *In probability theory, we consider the concepts of state space and event:*

1. *The state space or sample space is the set of all possible outcomes.*
2. *An event is a subset of the state space.*

Probability measures are functions that assign a number to an event. Given an event $A$ ($A \subseteq X$), the value $P(A)$ measures the likelihood of the event $A$ before performing the experiment. It is well known that the higher $P(A)$, the higher the likelihood that $A$ occurs. As events are subsets of $X$, probability measures are set functions.

When finite sets $X$ are considered, probability measures can be defined on all subsets of $X$. That is, $P$ is a function on the set $\wp(X)$ into $[0, 1]$. Nevertheless, in general, it is not possible to consider all subsets of $X$. This is the case, for example, when $X$ is not finite. In such situation, measures are defined over $\sigma$-algebras. They are subsets $\mathcal{A}$ of $\wp(X)$ with some particular properties. These properties and the definitions of algebra and $\sigma$-algebra are recalled below.

**Definition 2.7.** *Let $X$ be a reference set, and let $\mathcal{A}$ be a subset of $\wp(X)$. Let us consider the following properties:*

*Property 1: $\emptyset \in \mathcal{A}$ and $X \in \mathcal{A}$*
*Property 2: if $A \in \mathcal{A}$ then $X \setminus A \in \mathcal{A}$*
*Property 3: $\mathcal{A}$ is closed under finite unions and finite intersections:*

$$\text{if } A_1, \ldots, A_n \in \mathcal{A}, \text{ then } \cup_{i=1}^n A_i \in \mathcal{A} \text{ and } \cap_{i=1}^n A_i \in \mathcal{A},$$

*Property 4: $\mathcal{A}$ is closed under countable unions and intersections:*

$$\text{if } A_1, A_2, \cdots \in \mathcal{A}, \text{ then } \cup_{i=1}^\infty A_i \in \mathcal{A} \text{ and } \cap_{i=1}^\infty A_i \in \mathcal{A},$$

1. *$\mathcal{A}$ is an algebra (or a field) if it satisfies Properties 1, 2, and 3.*

2. $\mathcal{A}$ *is a $\sigma$-algebra (or a $\sigma$-field) if it satisfies Properties 1, 2, and 4.*

Note that Properties 1 and 4 imply Property 3. Therefore, any $\sigma$-algebra is an algebra. Nevertheless, the reverse is not true.

When $\mathcal{A}$ is defined as $\mathcal{A} := \wp(X)$, we have that $\mathcal{A}$ is a $\sigma$-algebra, and, therefore, $\mathcal{A}$ is also an algebra. In general, when $\mathcal{A}$ is an algebra on the reference set $X$, a pair $(X, \mathcal{A})$ is known as a measurable space. Therefore, $(X, \wp(X))$ is a measurable space.

Now, let us define probability measures.

**Definition 2.8.** *Let $X$ be a reference set and let $\mathcal{A}$ be a $\sigma$-algebra on $X$; then, a set function $\mathcal{P}$ is a probability measure if it satisfies the following conditions:*

*(i) $P(A) \geq 0$ for all $A \in \mathcal{A}$,*
*(ii) $P(X) = 1$, and*
*(iii) $P(\cup_{i=1}^{\infty} A_i) = \sum_{i=1}^{\infty} P(A_i)$ for every countable sequence $A_i$ ($i \geq 1$) of $\mathcal{A}$ that is pairwise disjoint (i.e., $A_i \cap A_j = \emptyset$ when $i \neq j$).*

Condition $(iii)$ in this definition is known as *countable additivity*. The axiom $(iii')$ given below might be used to replace Condition $(iii)$ when $X$ is finite. This alternative condition is known as the *finite additivity*:

$(iii')$ $P(A \cup B) = P(A) + P(B)$ for all $A, B \in \mathcal{A}$ when $A \cap B = \emptyset$.

Nevertheless, in general, $(iii)$ and $(iii')$ are not equivalent, as $(iii')$ only implies $(iii)$ for a finite number of pairwise disjoint sets $A_i, A_j$. That is, $(iii')$ implies the following equality instead of implying (iii):

$(iii'')$ $P(\cup_{i=1}^{n} A_i) = \sum_{i=1}^{n} P(A_i)$ for every countable sequence $A_i$ ($i \geq 1$) of $\mathcal{A}$ that is pairwise disjoint (i.e., $A_i \cap A_j = \emptyset$ when $i \neq j$).

So, countable additivity is not implied by finite additivity. For the purpose of this book, the difference is not very relevant, as we focus on finite sets. Therefore, in practice, condition $(iii')$ suffices.

Conditions $(i)$, $(ii)$ and $(iii')$ on an algebra correspond to the Kolmogorov axioms.

The following properties can be deduced for probability measures from Definition 2.8.

**Proposition 2.9.** *Let $(X, \mathcal{A})$ be a measurable space, and let $P$ be a probability measure. Then, for all $A, A_1, \ldots, A_n, B$ in $\mathcal{A}$, the following holds:*

1. $P(\emptyset) = 0$
2. $P(A) \leq 1$
3. $P$ *is additive*
4. $P(A) \leq P(B)$ *if* $A \subset B$
5. $P(B \setminus A) = P(B) - P(A)$ *if* $A \subseteq B$
6. $P(X \setminus A) = 1 - P(A)$
7. $P(A_1 \cup \cdots \cup A_n) \leq P(A_1) + \cdots + P(A_n)$ *for* $A_i \subset X$

Let $S$ be subsets of $X$; then, the $\sigma$-algebra generated by $S$ is the set of subsets of $X$ that

1. contains $S$
2. is a $\sigma$-algebra
3. is as small as possible

This $\sigma$-algebra generated by $S$ is denoted by $\sigma(S)$.
Let $\mathcal{O}$ be the open subsets of $\mathbb{R}$; then, $\mathcal{B} = \sigma(\mathcal{O})$ is the *Borel $\sigma$-algebra of* $\mathbb{R}$.

**Fig. 2.1.** Definition of Borel $\sigma$-algebra

8. $P(A \cup B) = P(A) + P(B) - P(A \cap B)$
9. $P(\cup_{i=1}^{n} A_i) = \sum_i P(A_i) - \sum_{i<j} P(A_i \cap A_j) + \sum_{i<j<k} P(A_i \cap A_j \cap A_k) + \cdots + (-1)^{(n+1)} P(A_1 \cap \cdots \cap A_n)$

*The last equality is known as the inclusion-exclusion formula.*

Taking into account what has been explained above, a model for a random experiment is defined in terms of the reference set $X$, the algebra $\mathcal{A}$, and the probability measure $P$. This is commonly referred as a probabilistic space, and is denoted by $(X, \mathcal{A}, P)$.

### 2.2.1 Random Variables

A random variable is a function that assigns values to the outcomes in the reference set. That is, a random variable is a function from $X$ into a space $S$. For example, if $X$ is the outcome of throwing two dice, then we can consider the random variable $f$ that assigns to each outcome the sum of the values obtained by the two dice. Thus, in this case, we have $X = \{(1,1), (1,2), \ldots, (6,5), (6,6)\}$ and $f((a,b)) := a + b$.

Given a probability space $(X, \mathcal{A}, P)$ and a random variable $f : X \rightarrow S$, we can define a new probability measure on the space $S$ from $P$. This is shown below. Note that in this definition we need a $\sigma$-algebra $S$ on the space $S$, as the new probability should be defined over subsets of $S$.

Often, as $S$ is the set of real numbers, we define the $\sigma$-algebra $S$ to be the Borel $\sigma$-algebra. This algebra (defined in Figure 2.1) contains all subintervals in $S$.

**Definition 2.10.** *Let $(X, \mathcal{A}, P)$ be a probability space and let $f : X \rightarrow S$ be a random variable; then, the probability measure $P_f$ induced by $f$ on a $\sigma$-algebra $S$ on $S$ is defined as*

$$P_f(A) := P(\{w : f(w) \in A, w \in X\})$$

*for all $A$ in $S$.*

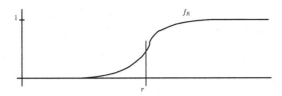

**Fig. 2.2.** Cumulative distribution function as the integral of a probability density function

This probability is also known as *distribution of the random variable f* or as *the law of f*. The probability $P_f$ can be alternatively expressed by $P(f^{-1}(A))$ or $P(f \in A)$, where $f^{-1}(A)$ is the inverse of $f$ and corresponds to the set $\{w : f(w) \in A, w \in X\}$.

**Definition 2.11.** *Let* $(X, \mathcal{A}, P)$ *be a probability space and let* $f : X \to S$ *be a random variable. Then, the cumulative distribution function of the random variable f is defined:*

$$G_f(r) := P(\{x : f(x) \leq r\})$$

*or, using the probability measure* $P_f$ *on* $(S, \mathcal{S})$, *by*

$$G_f(r) := P_f((-\infty, r])$$

This function is often referred to as *distribution function, probability distribution* or *cdf* (for *cumulative distribution function*).

When the cumulative distribution function $G_f$ can be expressed by

$$G_f(r) = \int_{-\infty}^{r} g_f(s)ds$$

for all $r$ in $\mathbb{R}$, we say that $g_f$ is the *probability density function* (or *pdf*), or, simply, *density of the random variable f*. Figure 2.2 represents this expression.

### 2.2.2 Expectation and Moments

If $f$ is a random variable of $X$, then the following definitions are of interest:

**Definition 2.12.** *Let* $(X, \mathcal{A}, P)$ *be a probability space and let* $f : X \to S$ *be a random variable. Then, the expectation of f is defined by*

$$E[f] := \int_X f dP$$

*provided the integral exists.*

When a random variable $f$ takes only a finite number of values, it is said that $f$ is simple. In this case, if $f$ takes values $\{r_1, \ldots, r_s\}$, the expectation is

$$E[f] = \sum_{i=1}^{s} r_i P(\{f = r_i\})$$

When $f$ is a random variable on $(X, \mathcal{A}, P)$, the following expectations are often considered for $k > 0$

1. $E[f^k]$, the $k$th moment of $f$.
2. $E[|f|^k]$, the $k$th absolute moment of $f$.
3. $E[(f - E[f])^k]$, the $k$th central moment of $f$.

These moments are only defined when $E[f]$ is finite.

The mean (denoted by $mean(f)$ or $\bar{f}$) and the variance (denoted by $\sigma^2(f)$, $\sigma_f^2$, or $Var(f)$) are terms often used to refer to the first moment of $E[f]$ and to the second central moment of $f$. The positive square root of the variance is known as the standard deviation of $f$ (denoted by $\sigma$). That is, $mean(f) := E[f]$ and $\sigma^2(f) := E[(f - E[f])^2]$.

**Proposition 2.13.** *If $E[f^2]$ is finite, then $E[|f|]$ is also finite, and we have*

$$\sigma^2(f) = E[f^2] - E[f]^2.$$

*Therefore, the following equation also holds:*

$$E[|f|]^2 \leq E[f^2].$$

### 2.2.3 Independence

An important concept that appears in probability theory is independence. We review some definitions and results in this section. The main idea behind independence is that when two events are independent, some knowledge about the occurrence (or nonoccurrence) of one of the events does not change the odds of the occurrence of the other.

Let $(X, \mathcal{A}, P)$ be a probability space. Then, if we consider the event $B$, we have that its probability will be $P(B)$. Next, let us consider the event $A \cap B$. Then, we have that when $A$ and $B$ are independent, the probability that $A$ occurs does not depend on the occurrence of $B$. Informally, this means that the proportion of $A$ in $X$ is the same as the proportion of $A$ in $B$. This is established by the equality $P(A)/P(X) = P(A \cap B)/P(B)$.

In mathematical terms, independence is defined as follows:

**Definition 2.14.** *Two events are independent if and only if $P(A \cap B) = P(A) \cdot P(B)$*

If two events are independent, knowing something about the occurrence (or nonoccurrence) of one of the events does not change the probability of the other. When independence does not hold, it is relevant to consider the conditional probability to measure the change of a probability. The conditional probability of $A$ relative to $B$ is defined as

$$P(A|B) = \frac{P(A \cap B)}{P(B)} \tag{2.1}$$

provided $P(B) > 0$.

We consider now a few results for independent events.

**Theorem 2.15.** *Let $P$ be a probability on a probability space $(X, \mathcal{A}, P)$. Then, the following holds:*

1. *If $P(A) > 0$, then $A$ and $B$ are independent events if and only if*

$$P(A|B) = P(A).$$

2. *If, for arbitrary events $A_1$, $A_2$, ..., $A_n$, we have $P(A_1 \cap A_2 \cap \cdots \cap A_n) > 0$, then the following holds:*

$$P(A_1 \cap A_2 \cap \cdots \cap A_n) = P(A_1)P(A_2|A_1)P(A_3|A_2 \cap A_1) \ldots P(A_n|A_{n-1} \cap \cdots \cap A_1).$$

3. *Let $A_1, \ldots, A_n$ be a partition of the sample space ($A_i \cap A_j = \emptyset$ for $i \neq j$, and $\cup A_i = X$); then, for any $B \subseteq X$, we have*

$$P(B) = \sum_{i=1}^{n} P(A_i)P(B|A_i).$$

4. *Let $A_1, \ldots, A_n$ be a partition of the sample space; then, for any $B \subseteq X$ such that $P(B) > 0$, we have*

$$P(A_i|B) = \frac{P(A_i)P(B|A_i)}{\sum_{i=1}^{n} P(A_i)P(B|A_i)}.$$

*The last property is Bayes' theorem.*

Now we turn to the independence of random variables. Let $(X, \mathcal{X}, P)$ be a probability space and let $f_1, \ldots, f_n$ be random variables into $(S, \mathcal{S})$; then, independence means that knowledge about one or more $f_i$ does not change the probability of the other. Knowledge about a random variable means knowledge about one event of the form $A_i = \{f_i \in \mathcal{B}\}$, where $B$ is in $\mathcal{S}$. Therefore, independence of the random variables $f_1, \ldots, f_n$ means that the events $A_1, \ldots, A_n$ should be independent.

This is formally defined as follows:

**Definition 2.16.** *Let* $(X, \mathcal{A}, P)$ *be a probability space and let* $f_1, \ldots, f_n$ *be random variables into* $(S, \mathcal{S})$; *then,* $f_1, \ldots, f_n$ *are independent if and only if, for all sets* $B_1, \ldots, B_n \in \mathcal{S}$, *we have*

$$P(\{f_1 \in B_1, \ldots, f_n \in B_n\}) = P(\{f_1 \in B_1\}) \cdots \cdot P(\{f_n \in B_n\}).$$

The following results can be proved for independent variables.

**Theorem 2.17.** *Let* $f_1, \ldots, f_n$ *be random variables on* $(X, \mathcal{A}, P)$. *Let* $G_i$ *be the distribution functions of* $f_i$ *for* $i = 1, \ldots, n$, *and let* $G$ *be the distribution function of* $f = \{f_1, \ldots, f_n\}$. *Then,* $f_1, \ldots, f_n$ *are independent if and only if*

$$G(x_1, \ldots, x_n) = G_1(x_1) \cdot G_2(x_2) \cdots \cdot G_n(x_n)$$

*for all real* $x_1, \ldots, x_n$.

**Corollary 2.18.** *If* $f_1, \ldots, f_n$ *are independent and* $f_i$ *has density* $g_i$ *for* $i = 1, \ldots, n$, *then* $f$ *has density* $g$ *given by*

$$g(x_1, \ldots, x_n) = g_1(x_1) \cdots \cdot g_n(x_n).$$

**Theorem 2.19.** *Let* $f_1, \ldots, f_n$ *be independent random variables on* $(X, \mathcal{A}, P)$. *Then, if* $E[f_i]$ *is finite for all* $i = 1, \ldots, n$, $E[f_1, \ldots, f_n]$ *exists and the following equation holds:*

$$E[f_1, \ldots, f_n] = E[f_1] \cdot E[f_2] \cdots \cdot E[f_n].$$

When random variables are not independent, it is meaningful to consider their covariance and correlation coefficients. The correlation coefficient (or Pearson's correlation coefficient) is also known as the product moment correlation. We define it below.

**Definition 2.20.** *Let* $f_1, f_2$ *be two random variables with finite expectation, and assume* $E[f_1 f_2]$ *is finite. Then, the covariance of* $f_1$ *and* $f_2$ *is defined by*

$$Cov(f_1, f_2) := E[(f_1 - E[f_1])(f_2 - E[f_2])] = E[f_1 f_2] - E[f_1]E[f_2]$$

*Given two sets of random variables* $f = \{f_1, \ldots, f_s\}$ *and* $f' = \{f'_1, \ldots, f'_t\}$, *the matrix defined by*

$$Cov(f, f') := \begin{pmatrix} Cov(f_1, f'_1) & Cov(f_1, f'_2) & \cdots & Cov(f_1, f'_t) \\ Cov(f_2, f'_1) & Cov(f_2, f'_2) & \cdots & Cov(f_2, f'_t) \\ \vdots & \vdots & & \vdots \\ Cov(f_s, f'_1) & Cov(f_s, f'_2) & \cdots & Cov(f_s, f'_t) \end{pmatrix}$$

*is the covariance matrix. For* $f' = \{f\}$, *this definition generalizes the variance:*

$$Var(f) = Cov(f, f).$$

*Note that* $\sigma^2(f) = Var(f) = E[(f - E[f])^2]$.

If $f_1$ and $f_2$ are independent, then the covariance is zero. Nevertheless, the converse is not true. For example, $f_1 = cos(\theta)$ and $f_2 = sin(\theta)$, where $\theta$ is uniformly distributed between 0 and $2\pi$, has covariance zero.

**Definition 2.21.** *Let $f_1$ and $f_2$ be two random variables such that $\sigma^2(f_1)$ and $\sigma^2(f_2)$ are finite and greater than zero; then, the correlation coefficient between $f_1$ and $f_2$ is defined by*

$$\rho(f_1, f_2) = \frac{Cov(f_1, f_2)}{\sigma(f_1)\sigma(f_2)}.$$

*Equivalently, we have*

$$\rho^2(f_1, f_2) = \frac{E[(f_1 - E(f_1))(f_2 - E(f_2))]}{E[(f_1 - E(f_1))^2]E[(f_2 - E(f_2))^2]}.$$

**Proposition 2.22.** *The correlation coefficient satisfies the following properties:*

1. $-1 \leq \rho(f_1, f_2) \leq 1$
2. $|\rho(f_1, f_2)| = 1$ *if and only if $f_1' = f_1 - E[f_1]$ and $f_2' = f_2 - E[f_2]$ are linearly independent. That is, if $P(\{af_1' + bf_2' = 0\}) = 1$ for some real numbers $a$ and $b$, not both zero.*

So, in general, the nearer $\rho$ is to $-1$ or to 1, the more linear is the model.

### 2.2.4 Parametric Models and Nonparametric Methods

*Tout le monde y croit cependant, me disait un jour M. Lippmann, car les expérimentateurs s'imaginent que c'est un théorème de mathématiques, et les mathématiciens que c'est un fait expérimental.*[3]

H. Poincaré, [324]

Given a set of observations, parametric models permit us to represent data in a compact way. Once a particular model is properly selected, a huge amount of data can be reduced into the few parameters that the model requires. Normal (either univariate or multivariate normal) distributions and $\chi^2$ are examples of parametric models. Then, parametric techniques rely on the properties of parametric distributions. Alternatively, nonparametric methods have been developed so that they can be applied when no parametric model can be made to fit the data. So, these methods do not require strong assumptions on the data distribution.

Normal distribution is one of the parametric models. In it, a whole data set is reduced to two values: mean and variance. Normal distributions are defined in the next example.

---

[3] Everyone believes in it – Mr. Lippmann said to me one day – because the experimenters think it is a theorem of mathematics, and the mathematicians think it is an experimental fact.

*Example 2.23.* A random variable $f$ follows a (univariate) normal distribution with mean $\mu$ and variance $\sigma^2$ if the probability density function of $x$ for $x \in (-\infty, \infty)$ is of the form

$$g_f(x) = \frac{1}{\sigma\sqrt{2\pi}}e^{-\frac{1}{2}(x-\mu)^2/\sigma^2}.$$

*Example 2.24.* A random variable $f$ follows a multivariate normal distribution of dimension $n$ with mean vector $\mu$ and variance-covariance matrix $\Sigma$ if the probability density function of $\mathbf{x}$ is of the form

$$g_f(\mathbf{x}) = \frac{1}{(2\pi)^{n/2}|\Sigma|^{1/2}}e^{-\frac{1}{2}(\mathbf{x}-\mu)'\Sigma^{-1}(\mathbf{x}-\mu)},$$

where $|\Sigma|$ denotes the determinant of the variance-covariance matrix $\Sigma$.

Note that, in the multivariate case, $\mu, \mathbf{x} \in \mathbb{R}^n$ for a given $n$.

When $f$ follows a normal distribution, we write $x \sim N(\mu, \sigma^2)$ or $\mathbf{x} \sim N(\mu, \Sigma)$.

### 2.2.5 Regression

> *Pour cet effet, la méthode qui me paroit la plus simple et la plus générale, consiste à rendre* minimum *la somme des quarrés des erreurs. On obtient ainsi autant d'équations qu'il y a de coëfficiens inconnus; ce qui achève de déterminer tous les élémens de l'orbite.*[4]

Legendre, A. M., p. viii [224]

In this section we present an overview of regression, in which we construct a model of one variable in terms of some other variables. We restrict ourselves to the case of linear models. So, the outcome is a linear combination of the input variables.

Let us start considering a simple model concerning two random variables $f_1$ and $f_2$. In this case, if $f_2$ is expressable in terms of $f_1$ following a linear model $\beta_0 + \beta_1 f_1$, we have that the following equation holds when there is some error involved in the process:

$$f_2 = \beta_0 + \beta_1 f_1 + \epsilon.$$

Here, $f_1$ is the explanatory variable (or carrier, regressor or the independent variable), $f_2$ is the response variable (or the dependent variable), $\epsilon$ is the error (in fact, another random variable), and $\beta_0$ and $\beta_1$ are the parameters of the model.

---

[4] For this effect, it seems to me that the simplest and more general method consists of making minimum the sum of the squares of the errors. We obtain in this way as many equations as unknown coefficients; which serves to determine all the elements of the orbit.

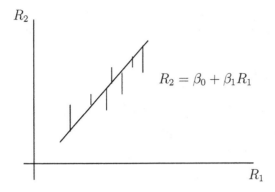

$$R_2 = \beta_0 + \beta_1 R_1$$

**Fig. 2.3.** Graphical representation of least sum of squares of the vertical distances

| Index | $x_i$ | $y_i$ |
|-------|-------|-------|
|       | Year  | Export revenues |
| 1 | 1970 | 59 |
| 2 | 1975 | 111 |
| 3 | 1980 | 314 |
| 4 | 1985 | 653 |
| 5 | 1990 | 903 |
| 6 | 1995 | 1209 |
| 7 | 2000 | 1425 |

**Table 2.3.** Japanese annual sales revenue from exports to U.S., in hundred million dollars

Now, given some particular data (such as in Table 2.3), we might be able to estimate $\beta_0$ and $\beta_1$ so that the model fits the data. This is classically solved by minimizing the sum of the squares of the vertical distances (see Figure 2.3). This method is known as *least sum of squares* (LSS) (or least squares). Formally, we consider the error between the observed value of $f_2$ and the linear model computed for $f_1$:

$$f_2 - (\beta_0 + \beta_1 f_1).$$

Thus, when the particular data for variables $f_1$ and $f_2$ corresponds to the pairs $\{(x_i, y_i)\}_i$, we consider minimization of the following error:

$$ERR(\beta_0, \beta_1) = \sum_{i=1}^{n}(y_i - (\beta_0 + \beta_1 x_i))^2 \tag{2.2}$$

To solve this problem, we consider the partial derivatives of the expression $ERR$ with respect to the two parameters $\beta_0$ and $\beta_1$, and then set these derivatives to zero. That is,

$$\frac{\partial ERR(\beta_0, \beta_1)}{\partial \beta_0} = 2 \sum_{i=1}^{n} (y_i - \beta_0 - \beta_1 x_i)(-1) = 0$$

and

$$\frac{\partial ERR(\beta_0, \beta_1)}{\partial \beta_1} = 2 \sum_{i=1}^{n} (y_i - \beta_0 - \beta_1 x_i)(-x_i) = 0.$$

The solution of these expressions is:

$$\hat{\beta}_1 = \frac{\sum_{i=1}^{n} x_i y_i - (\sum_{i=1}^{n} x_i)(\sum_{i=1}^{n} y_i)/n}{(\sum_{i=1}^{n} x_i^2) - (\sum_{i=1}^{n} x_i)^2/n} = \frac{\sum_{i=1}^{n}(x_i - \bar{x})(y_i - \bar{y})}{\sum_{i=1}^{n}(x_i - \bar{x})^2}, \quad (2.3)$$

$$\hat{\beta}_0 = \bar{y} - \hat{\beta}_1 \bar{x}. \quad (2.4)$$

As the determinant of the matrix of second-order partial derivatives of $ERR$ is positive, the solutions correspond to the minimum of $ERR$. $\hat{\beta}_0$ and $\hat{\beta}_1$ are estimators of $\beta_0$ and $\beta_1$, respectively.

*Example 2.25.* Let us consider the data in Table 2.3. This data is represented in Figure 2.4. It can be observed that this data follows a linear model, except for the first observation.

Accordingly, we consider the regression using a linear model of the form

$$revenue = \beta_0 + \beta_1 year.$$

Using all observations in Table 2.3, the linear regression model we obtain (using Expressions 2.4 and 2.3) has the following estimates for $\hat{\beta}_0$ and $\hat{\beta}_1$:

- $\beta_0 = -96923.393$
- $\beta_1 = 49.164$

So, the model is

$$revenue = -96923.393 + 49.164 year.$$

In contrast, when the first observation is not included in the computations, we obtain the following model:

$$revenue = -107180.476 + 54.314 year.$$

The models are represented in Figure 2.4, together with the original observations. It can be seen that the second model is more adjusted to the original data while the first one is slightly displaced to accommodate the first observation.

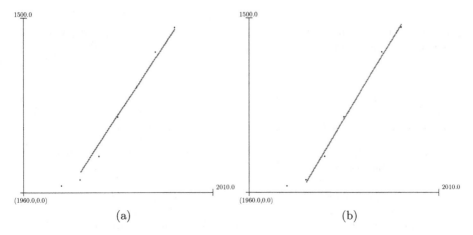

**Fig. 2.4.** Representation of the data in Table 2.3 and regression models: (a) regression using all observations; (b) regression using all observations except the first one, corresponding to the year 1970.

In a more general case, we can consider $r$ explanatory variables $\{X_1, \ldots, X_r\}$, and one response variable $Y$. In this case, when a linear model is considered, we have

$$Y = \beta_0 + \beta_1 X_1 + \cdots + \beta_r X_r + \epsilon. \tag{2.5}$$

When $n$ observations of the form $\{(x_{i1}, \ldots, x_{ir}, y_i)\}$ are given, we should consider $n$ equations of the form:

$$y_i = \beta_0 + \beta_1 x_{i1} + \cdots + \beta_r x_{ir} + \epsilon_i$$

for $i = 1, \ldots, n$, which, considering $x_{i0} = 0$ for all $i$ in $1, \ldots, n$, can be put in matrix form as follows:

$$\begin{pmatrix} y_1 \\ y_2 \\ \vdots \\ y_n \end{pmatrix} = \begin{pmatrix} x_{10} & x_{11} & \cdots & x_{1r} \\ x_{20} & x_{21} & \cdots & x_{2r} \\ \vdots & \vdots & & \vdots \\ x_{n0} & x_{n1} & \cdots & x_{nr} \end{pmatrix} \begin{pmatrix} \beta_0 \\ \vdots \\ \beta_r \end{pmatrix} + \begin{pmatrix} \epsilon_1 \\ \epsilon_2 \\ \vdots \\ \epsilon_n \end{pmatrix}.$$

That is,

$$Y = X\beta + \epsilon.$$

As with the simplest case of one explanatory variable, in the general case we also use the least sum of squares. Again, this is to minimize the error with respect to $\beta$. In matrix form, this is expressed as the minimization of the following expression:

Let $X$ be a matrix with dimensions $m \times n$; then, its generalized inverse $X^-$ is another matrix with dimensions $n \times m$ such that the following equation holds:

$$XX^-X = X$$

For each $X$, there is at least one generalized inverse, but it is not necessarily unique. One method for computing $X^-$ when $X$ is a symmetric square matrix of dimension $n$ and rank $r$ is as follows:

(i) Take $X$ and delete $n - r$ rows and the corresponding columns to obtain an $r \times r$ nonsingular principal matrix. Let $M = \{m_{ij}\}$ denote the new matrix. Note that $M$ exists because the rank of $X$ is $r$.

(ii) Invert $M$ and obtain $M^{-1}$. Let us denote the elements in $M^{-1}$ by $m'_{ij}$.

(iii) Take $X$ again and replace the elements $m_{ij}$ of $M$ by the elements $m'_{ij}$ in $M^{-1}$. Replace the other elements of $X$ not in $M$ by zero. The matrix obtained is a generalized inverse of $X$.

---

**Fig. 2.5.** Generalized inverse of a matrix $X$

$$ERR(\beta) = ||Y - X\beta||^2.$$

When the columns of $X$ are linearly independent, there exists a unique vector $\hat{\beta}$ that minimizes this error (i.e., there is a unique solution of the minimization problem). This solution is

$$\hat{\beta} = (X'X)^{-1}X'Y.$$

When the columns of $X$ are not linearly independent, a solution is given by

$$\hat{\beta} = (X'X)^- X'Y.$$

Here, $(X'X)^-$ is a generalized inverse of $(X'X)$. The computation of a generalized inverse is given in Figure 2.5.

Once a model is built, it is necessary to evaluate to what extent the model fits the data, i.e., to measure the goodness of fit of the model. This can be done using $\rho^2$, i.e., the square of the correlation coefficient between the observed variable $Y$ and the predicted $\hat{Y}$. This is formally defined by (see Definition 2.21 for details)

$$\rho^2 = \frac{[\sum(y_i - \bar{y})(\hat{y}_i - \bar{\hat{y}})]^2}{\sum(y_i - \bar{y})^2 \sum(\hat{y}_i - \bar{\hat{y}})^2}. \tag{2.6}$$

An equivalent expression can be used for $\rho^2$. It is defined in terms of the sum of squares due to regression $SSR$ ($SSR = \sum(\hat{y} - \bar{\hat{y}})^2$) and the total sum of squares $SST$ ($SST = \sum(y - \bar{y})^2$). This expression is as follows:

$$\rho^2 = \frac{SSR}{SST} = \frac{\sum(\hat{y}_i - \bar{\hat{y}})^2}{\sum(y_i - \bar{y})^2},\qquad(2.7)$$

with $\hat{y}_i = \beta_0 + \beta_1 x_i$.

Note that the equivalence of Expressions 2.6 and 2.7 uses the fact that we are considering a linear model (represented in Equation 2.5).

Denoting the sum of squares of errors by $SSE$ (i.e., $SSE = \sum(y_i - \hat{y})^2$), we have that the following equation holds:

$$SST = SSE + SSR.$$

That is,

$$\sum(y - \bar{y})^2 = \sum(y - \hat{y})^2 + \sum(\hat{y} - \bar{y})^2.$$

Here, as $\sum(y - \bar{y})^2$ measures the variability in $Y$, Expression 2.7 can be interpreted as the percentage of the variability in $Y$ that is explained by the model. Naturally, the higher the percentage, the better the model. Recall that $\rho^2$ is between 0 and 1.

*Example 2.26.* The goodness of fit of the regression models in Example 2.25 are as follows:

- $\rho^2 = 0.97937$ (regression using all observations)
- $\rho^2 = 0.99637$ (regression using all observations except the first one)

The linear model developed so far is based on several assumptions. They are highlighted below:

1. The linear model is an adequate approximation. That is, Equation 2.5 is appropriate to represent $Y$ as a function of $X$.
2. The variance of the error $\epsilon_i$ is independent of the observation. That is, $Var(\epsilon_i) = K$ for all $i$.
3. The errors are uncorrelated: $\epsilon_i$ is correlated neither with $\epsilon_j$ for $i \neq j$ nor with $X$.
4. $\epsilon$ should follow a normal distribution with mean equal to zero.

Other regression methods have been developed for situations in which these assumptions do not hold. For example, weigthed regression can be applied when the variance is not constant for all $\epsilon_i$. Robust regression methods (see Section 2.2.8) are some other tools available for these situations.

## 2.2.6 Robust Statistics

*The point of robust statistics is that one may keep a parametric model although the latter is known to be wrong*

Hampel et al., p. 403 [181]

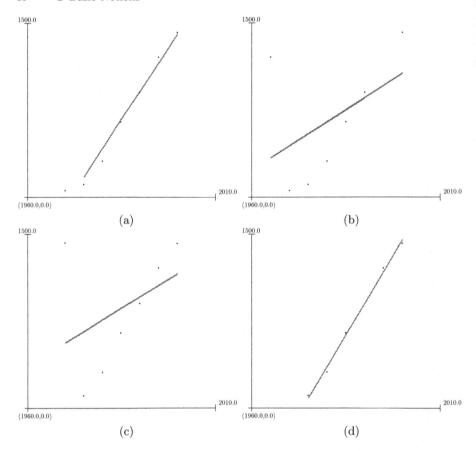

**Fig. 2.6.** Representation of the data in Table 2.3, and the same data with some perturbation: (a) data and regression using the original observations; (b) data and regression with a perturbation on the $x$-axis (1995 changed into 1965 for observation 6); (c) data and regression with a perturbation on the $y$-axis (59 changed into 1429 in observation 1); (d) data and regression with a perturbation on the $x$- and $y$-axis (pair (1970, 59) changed into (1999, 1400) in observation 1). The corresponding data sets are presented in Table 2.4

The actual practice of statistical methods with real data poses several problems. An important one is due to the fact that data usually contain errors and do not always fit the assumed data model. The differences between model and real data can cause the results obtained to deviate from the ones that would be obtained if errors were not present, i.e., to deviate from the results obtained in an ideal situation. Robust statistics has been developed to deal with this situation.

Differences between model and real data can be due to different reasons. Some of them are the following:

| Index | $x_i^a$ | $y_i^a$ | $x_i^b$ | $y_i^b$ | $x_i^c$ | $y_i^c$ | $x_i^d$ | $y_i^d$ |
|-------|------|----------|------|----------|------|----------|------|----------|
|       | Year | Revenues | Year | Revenues | Year | Revenues | Year | Revenues |
| 1 | 1970 | 59 | 1970 | 59 | 1970 | 1429 | 1999 | 1400 |
| 2 | 1975 | 111 | 1975 | 111 | 1975 | 111 | 1975 | 111 |
| 3 | 1980 | 314 | 1980 | 314 | 1980 | 314 | 1980 | 314 |
| 4 | 1985 | 653 | 1985 | 653 | 1985 | 653 | 1985 | 653 |
| 5 | 1990 | 903 | 1990 | 903 | 1990 | 903 | 1990 | 903 |
| 6 | 1995 | 1209 | 1965 | 1209 | 1995 | 1209 | 1995 | 1209 |
| 7 | 2000 | 1425 | 2000 | 1425 | 2000 | 1425 | 2000 | 1425 |

**Table 2.4.** Japanese annual sales revenue from exports to U.S., in hundred million dollars: (a) original data; (b) data with error in the year; (c) data with error in the revenues; (d) data with errors in both year and revenue

- Errors in the data, either intentional or accidental. Intentional errors include rounding (or grouping) or censoring (e.g., applying masking methods for data protection). Some of the errors can be hidden in the data while others can cause important damage to the conclusions of an analysis. Table 2.4 and Figure 2.6 represent four data sets, an original one (considered in Example 2.25), and three others obtained from it with some perturbations on the $x$ and/or $y$ axes. The figure also represents the corresponding linear regression model using the least sum of squares method. It can be observed that the perturbation provokes changes in the linear model. In fact, extreme perturbation of a single value might even change the sign of the $\beta_1$ coefficient.
- Assumptions are violated; some of the assumptions of the model do not fit. For example, assumptions on independence do not hold for some variables.

Tools have been developed so that data diverting from the model does not affect the result in a significant way. In other words, robust statistics develops procedures in a way that their behavior in the neighborhood of parametric models is similar to the one obtained with the data completely fulfilling the model. So, in the case of the data represented in Figure 2.6, the goal would be to obtain in all cases a result as similar as possible to the one in Figure 2.4 (a).

In order to characterize to what extent a procedure is robust, some concepts have been defined. The breakdown point is one of these concepts. Informally, the breakdown point corresponds to the minimal proportion of bad data that can ruin the output of the procedure.

As an example we can consider the mean and the median of a sample $\{x_1, \ldots, x_N\}$. If we replace any of the $x_i$ for $x_i'$, it is clear that for $x_i' \to \infty$ the mean $(\sum_{i \neq j} x_j + x_i')/N$ is unbounded while the median is bounded.

As will be seen later, the influence functions and the breakdown point of mean and median show that the latter is more robust. We review below some concepts related to robust statistics.

First, we consider the influence function. This notion, which is a local measure, stands for a measure of the influence of infinitessimal perturbations on a statistic. This notion is formalized as follows.

**Definition 2.27.** *Let $G$ be a probability distribution and let $T$ be an estimator; then, the influence function (IF) of the estimator $T$ at the distribution $G$ is given by*

$$IF(x; T, G) := \lim_{\epsilon \to 0} \frac{T((1 - \epsilon)G + \epsilon \Delta_x) - T(G)}{\epsilon}$$

*in the $x$ where this limit exists.*

$\Delta_x$ denotes the probability distribution that puts all its mass in $x$.

The influence function is a linear approximation of the estimator for a distribution contaminated by amount $\epsilon$. That is,

$$T((1 - \epsilon)G + \epsilon \Delta_x) \approx T(G) + \epsilon IF(x; T, G).$$

For finite samples, several alternative expressions exist. The sensitivity curve and the empirical influence functions are two of the existing expressions.

**Definition 2.28.** *Let $S = \{x_1, \ldots, x_{n-1}\}$ be a sample and let $T_n$ for $n \geq 1$ be an estimator; then, the sensitivity curve of $T$ for the sample $S$ at $x$ is*

$$SC(x; T_n, S) := n\big(T_n(x_1, \ldots, x_{n-1}, x) - T_{n-1}(x_1, \ldots, x_{n-1})\big)$$

**Definition 2.29.** *Let $S = \{x_1, \ldots, x_{n-1}\}$ be a sample and let $T_n$ for $n \geq 1$ be an estimator; then, the empirical influence function of $T$ for the sample $S$ at $x$ is*

$$EIF(x; T_n, S) := T_n(x_1, \ldots, x_{n-1}, x).$$

*Example 2.30.* Let $S = \{0.5, 0.25, 0.8, 0.75\}$; then, the EIF of the arithmetic mean (AM) at the sample $S$ is

$$EIF(x; AM, S) = \frac{0.5 + 0.25 + 0.8 + 0.75 + x}{5}.$$

The EIF of the median at the sample $S$ is

$$EIF(x; median, S) = \begin{cases} 0.5 & \text{if } x \leq 0.5 \\ x & \text{if } 0.5 \leq x \leq 0.75 \\ 0.75 & \text{if } 0.75 \geq x. \end{cases}$$

So, the EIF of the mean is unbounded while that of the median is bounded. Note that the presence of an erroneous measurement affects the outcome of the median as it does in the case of the mean: in both cases the outcome shifts in the direction of the error. Nevertheless, the influence of the error is limited in the case of the median, but not in the case of the mean, as EIF indicates.

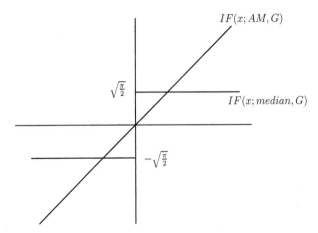

**Fig. 2.7.** Influence functions (IF) for the arithmetic mean (AM) and the median

The influence function can be used for computing the gross-error sensitivity and the local-shift sensitivity. They are defined as follows:

**Definition 2.31.** *Let G be a probability distribution and let T be an estimator. Then,*

*1. The gross-error sensitivity of T at G is defined as*

$$\gamma^*(T,G) := sup_x |IF(x;T,G)|.$$

*x is taken where IF(x; T, G) exists*
*2. The local-shift sensitivity of T at G is defined by*

$$\lambda^*(T,G) := sup_{x\neq y}\frac{|IF(y;T,G) - IF(x;T,G)|}{|y-x|}$$

Here, the gross-error sensitivity gives information about the worst case situation. It measures the largest influence as computed using $IF$. In robust procedures, this sensitivity is expected to be finite. This is not the case for the arithmetic mean.

The local-shift sensitivity measures the influence of replacing the observation $x$ by some $y$.

*Example 2.32.* Let $G$ be the normal distribution $N(0,1)$; then, the influence functions of the arithmetic mean (AM) and the median at the distribution $G$ correspond to the following expressions (see Figure 2.7):

$$IF(x; AM, G) = x$$

$$IF(x; median, G) = \begin{cases} -\sqrt{\pi/2} & \text{if } x < 0 \\ \sqrt{\pi/2} & \text{if } x > 0 \end{cases}$$

Their gross-error sensitivity is

- $\gamma^*(AM, G) = \infty$
- $\gamma^*(median, G) = \sqrt{\pi/2}$

and the local-shift sensitivity corresponds to

- $\lambda^*(AM, G) = 1$
- $\lambda^*(median, G) = \infty$

Note that $\lambda^*(AM, G) = 1$ because $IF(x; AM, G) = x$, and thus we have $\lambda^*(AM, G) = sup_{x \neq y} \frac{|y-x|}{|y-x|}$. In contrast, in the case of the median, we have, for $x < 0$, $IF(x; median, G) = -\sqrt{\pi/2}$, but, for $x > 0$, $IF(x; median, G) = \sqrt{\pi/2}$.

The concepts defined above are local ones, as they only refer to the variation of the output for a single observation. In contrast, the breakdown point is a global concept, as it takes into account the effect of changes in the whole distribution. More specifically, the breakdown establishes the fraction of the data that can cause the estimator to lead to a meaningless result. There exist several definitions for the breakdown point on finite samples. One of them is given below.

**Definition 2.33.** *Let $S = \{x_1, \ldots, x_n\}$ be a sample, and let $T_n$ be an estimator; then, the breakdown point of the estimator $T$ at the sample $S$ is defined as*

$$\epsilon_n^*(T, S) := \frac{1}{n} \min\{m; bias(m; T, S) = \infty\},$$

*with $bias(m; T, S)$ defined by*

$$bias(m; T, S) := \sup_{S' \in R(S, m)} ||T(S') - T(S)||,$$

*where $R(S, m)$ represents all samples obtained from $S$ with $m$ original observations replaced by arbitrary values.*

Note that, here, $bias(m; T, S) < \infty$ means that the effect of $m$ perturbations is bounded, while $bias(m; T, S) = \infty$ means that it is not. Therefore, $\epsilon_n^*(T, S)$ corresponds to the smallest fraction of contamination with unbounded effect. Naturally, the larger the breakdown point, the better.

In this definition, the smallest possible breakdown point is $1/n$, and the largest one is around $1/2$. In fact, the breakdown point of the arithmetic mean is $1/n$ and the one of the median is 50%.

### 2.2.7 M- and L-Estimators

Robust statistics has studied several families of estimators. Here, we review two families: M-estimators and L-estimators. The first take their name from generalized maximum likelihood and the second from linear combination of order statistics.

**Definition 2.34.** *Given a sample $x_1, \ldots, x_n$, an estimator $T_n$ is an M-estimator if $T_n$ is the minimum of*

$$\sum_{i=1}^{n} \rho(x_i, T_n)$$

*for an arbitrary function $\rho$.*

The following is an alternative definition.

**Definition 2.35.** *Let $\phi$ be equal to $\phi(x, \Theta) = (\partial/\partial\Theta)\rho(x, \Theta)$ when $\rho$ has a derivative. Then, $T_n$ is an M-estimator if it satisfies*

$$\sum_{i=1}^{n} \phi(x_i, T_n) = 0$$

Next, we define L-estimators.

**Definition 2.36.** *Given a sample $x_1, \ldots, x_n$, an estimator $T_n$ is an L-estimator if $T_n$ is of the form*

$$T_n(x_1, \ldots, x_n) = \sum_{i=1}^{n} c_i x_{s(i)},$$

*where $x_{s(1)}, \ldots, x_{s(n)}$ is the ordered sample (e.g., $x_{s(1)} \leq x_{s(2)} \leq \cdots \leq x_{s(n)}$) and the $c_i$ are coefficients. As $x_{s(i)}$ corresponds to the ith order statistic, $T_n$ is a linear combination of order statistics.*

Recall how the order statistics are defined.

**Definition 2.37.** *Let $i$ be an index $i \in \{1, \ldots, N\}$; then, a mapping OS: $\mathbb{R}^N \to \mathbb{R}$ is the ith order statistic of dimension $N$ if and only if*

$$OS_i(a_1, \ldots, a_N) = a_{s(i)},$$

*where $s$ is a permutation of $\{1, \ldots, N\}$ such that $a_{s(i)} \leq a_{s(i+1)}$ for $i \in \{1, \ldots, N-1\}$.*

## 2.2.8 Robust Regression

Standard regression using the least sum of squares has a breakpoint of $1/n$. Therefore, the model is highly affected by errors in the data. The goal of robust regression is to build models that are resilient to errors in the data, so that the regression model obtained from data with errors is not much different from the one obtained without erroneous data. To illustrate, let us consider again the data in Figure 2.4. A robust regression method would obtain for all data sets a model similar to the one of Figure 2.4 (a).

We consider below two methods for robust regression: the *Least Median of Squares* (LMS) and the *Least Trimmed Squares* (LTS).

Given the set of observations $\{(x_i, y_i)\}_{i=1,...,n}$, we have that LMS corresponds to finding the parameters $\beta = (\beta_0\beta_1)$ that minimize the following expression:

$$ERRLMS(\beta_0, \beta_1) := r_{s(h)},$$

where $r_i = |y_i - (\beta_0 + \beta_1 x_i)|$, and $r_{s(j)}$ is the $j$th element in $r_i$ when the residuals are ordered in increasing order. That is, $r_{s(1)} \leq r_{s(2)} \leq \cdots \leq r_{s(n)}$. When $h = n/2$, this approach corresponds to Equation 2.2, replacing the sum by the median. Note that, as seen in Example 2.32, the median is more robust than the arithmetic mean.

LTS corresponds to finding the parameters $\beta = (\beta_0\beta_1)$ that minimize

$$ERRLTS(\beta_0, \beta_1) := \sum_{j=1}^{h} d_{s(j)}, \tag{2.8}$$

where $d_i = |y_i - (\beta_0 + \beta_1 x_i)|$, and $d_{s(j)}$ corresponds to the $j$th element in $d_i$ when the residuals are ordered in increasing order.

The minimization of Equation 2.8 ignores all residuals $d_{s(j)}$ for $j > h$, which are the largest ones. In this way, the influence of $n - h$ erroneuous data in the model is reduced.

Note that the solution of the LMS problem (minimization of $r_{s(h)}$) is equivalent to the solution of minimizing $d_{s(h)}$, because the indices $i$ for the ordering $r_{s(i)}$ corresponds to the indices for the ordering $d_{s(i)}$. The value of $h$ is given and should be larger than $n/2$. For linear models with $p$ parameters, an optimal value for $h$ is the integer part of $(n + p + 1)/2$. Thus, in our case, with $\beta_0$ and $\beta_1$, it corresponds to $h$ being equal to the integer part of $(n + 2 + 1)/2$. That is, $h = \lfloor (n + 3)/2 \rfloor$. The breakdown point for such linear regression models and with such a value for $h$ is equal to 50%.

To compute the optimal $\beta = (\beta_0, \beta_1)$, the following steps are required:

1. Consider all pairs of two observations $o = (x, y)$ and $o' = (x', y')$ drawn from the original set of data.
   a) Determine $\beta_0^{o,o'}$ and $\beta_1^{o,o'}$ so that the line $y = \beta_0^{o,o'} + \beta_1^{o,o'} x$ goes through observations $o$ and $o'$. This problem, solved with a system of

two unknowns and two equations, results in: $\beta_1^{o,o'} = (y - y')/(x - x')$
and $\beta_0^{o,o'} = y - \beta_1^{o,o'} x$.

b) Ignore $\beta_0^{o,o'}$, and define $\tilde{\beta}_0^{o,o'}$ as the value that minimizes the corresponding error given $\beta_1$. That is, either

$$\tilde{\beta}_0^{o,o'} = argmin_{\beta_0} ERRLMS(\beta_0, \beta_1^{o,o'})$$

or

$$\tilde{\beta}_0^{o,o'} = argmin_{\beta_0} ERRLTS(\beta_0, \beta_1^{o,o'}).$$

The minimization problem is solved as follows.

For the LMS problem, (i) define $r_i = y_i - \beta_1^{o,o'} x_i$; (ii) order $r_i$ and obtain $r_{s(i)}$ so that $r_{s(i)} \leq r_{s(i+1)}$; (iii) determine the length of the contiguous intervals containing $h$ points and the midpoint of such intervals (that is, $l_i = |r_{s(i+h-1)} - r_{s(i)}|$ and $m_i = (r_{s(i+h-1)} + r_{s(i)})/2$, for $1 \leq i \leq n-h+1$): (iv) find the minima of the lengths, and then return the median of the corresponding midpoints (that is, if $I = \{i | l_i = \min_j l_j\}$, then return $median_{i \in I} m_i$).

For the LTS problem, (i) define $r_i = y_i - \beta_1^{o,o'} x_i$; (ii) order $r_i$ and obtain $r_{s(i)}$ so that $r_{s(i)} \leq r_{s(i+1)}$; (iii) for each contiguous interval containing $h$ points, determine the mean and the sum of the squared deviations from the mean (that is, $\bar{r}_i = \sum_{j=i}^{i+h-1} r_{s(j)}/h$ and $d_i = \sum_{j=i}^{i+h-1}(r_{s(j)} - \bar{r}_i)^2$); (iv) find the minima of the sums of square deviations, and return the corresponding mean (that is, if $I = \{i | d_i = \min_j d_j\}$, then return $\bar{r}_i$).

c) Define $e^{o,o'}$ as follows.
For the LMS problem,
$$e^{o,o'} := r_{s(h)} \text{ where } r_i := |y_i - (\tilde{\beta}_0^{o,o'} + \beta_1^{o,o'} x_i)|.$$
For the LTS problem,
$$e^{o,o'} := \sum_{j=1}^{h} d_{s(j)} \text{ where } d_i := |y_i - (\tilde{\beta}_0^{o,o'} + \beta_1^{o,o'} x_i)|.$$

2. Return the pair of parameters $\tilde{\beta}_0^{o,o'}$ and $\beta_1^{o,o'}$ with a minimum error $e^{o,o'}$.

We have restricted ourselves to the case of one explanatory variable and one response variable. In the more general case of $r$ explanatory variables, the same approach is applied. To properly determine the $r$ parameters, $r$ observations, instead of two observations, should be considered. Then, instead of a line, an hyperplane is determined. Nevertheless, in such a general case it is not always possible to consider all possible subsets of $r$ observations as the number of possible subsets is $\binom{n}{r} = n!/(r!(n-r)!)$. Random subsets are considered, but this implies that the method does not always find the optimum of the robust regression problem.

*Example 2.38.* Let us consider the data sets represented in Table 2.4 and Figure 2.6, the linear regression models of these data using the least sum of squares (LSS), least median of squares (LMS), and least trimmed squares

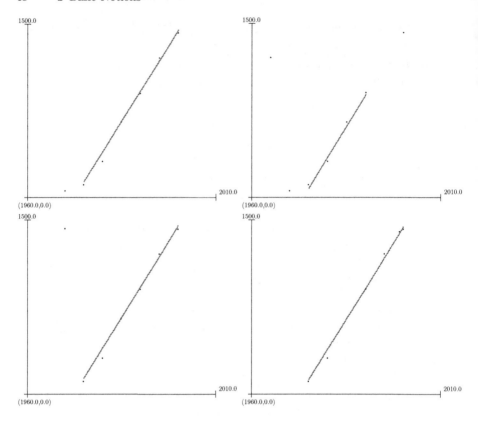

**Fig. 2.8.** Data in Figure 2.6 with robust regression (LMS)

(LTS) are represented in Figures 2.6, 2.8, and 2.9. They correspond to the following models:

- Original data with no perturbation.
  LSS: $revenue = -96923.393 + 49.164 * year$
  LMS: $revenue = -103671.6 + 52.56 * year$
  LTS: $revenue = -104159.0 + 52.8 * year$
- Data with perturbation on the year.
  LSS: $revenue = -40646.413 + 20.858 * year$
  LMS: $revenue = -106968.0 + 54.2 * year$
  LTS: $revenue = -106960.0 + 54.2 * year$
- Data with perturbation on the revenues.
  LSS: $revenue = -38453.75 + 19.807 * year$
  LMS: $revenue = -103671.6 + 52.56 * year$
  LTS: $revenue = -104159.0 + 52.8 * year$
- Data with perturbation on both year and revenue.
  LSS: $revenue = -107400.949 + 54.426 * year$

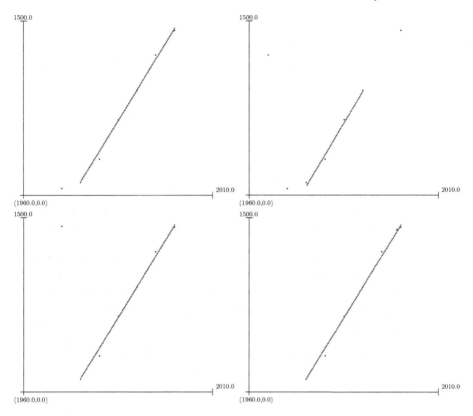

**Fig. 2.9.** Data in Figure 2.6 with robust regression (LTS)

LMS: $revenue = -103681.22 + 52.56 * year$
LTS: $revenue = -104163.040 + 52.8 * year$

## 2.3 Fuzzy Sets

> *On another day, when a visitor came and inquired,*
> *"Is your honorable father in?"*
> *The son replied,*
> *"To a certain extent, yes; to a certain extent, no"*

Z. Zhuang, p. 284 [464]

One of the basic properties of standard sets is that elements either belong completely to the set or do not belong to the set at all. In this way, given a

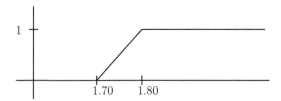

**Fig. 2.10.** Graphical representation of the membership function for tall: $\mu_{tall}$

property, for example, the property of "being odd," and a number, we can check whether or not the number is odd.

The main characteristic of fuzzy sets is that membership is no longer a boolean property. Instead, membership is graded, and, accordingly, there are different degrees of membership.

Formally, standard sets, known as *crisp sets*, can be defined in different ways. One of them is in terms of characteristic functions. Given a reference set $X$, a characteristic function $\chi$ is a function that labels each element in $X$ as either belonging to the set or not. When we denote membership in a set by 1, and nonmembership by 0, we have that $\chi$ is a function from $X$ into $\{0,1\}$.

*Example 2.39.* Let $X$ be the set of values on a die, and let $A$ denote the odd values in $X$, then $\chi_A : \{1,2,3,4,5,6\} \rightarrow \{0,1\}$ is defined as follows:

$$\chi_A(x) = 1 \ \ \text{if and only if } x = 1, 3, 5.$$

As stated, fuzzy sets permit degrees of membership. This is modeled using membership functions with range $[0,1]$ instead of $\{0,1\}$. In this way, the membership permits a smooth transition from nonmembership (membership value equal to zero) to complete membership (membership value equal to one). Then, the larger the value, the larger the membership in a set. Accordingly, a membership function of a concept $A$ on a reference set $X$ is a function $\mu_A$ from $X$ into $[0,1]$.

A typical example of fuzzy set is the set of tall heights (for people). It is clear that someone with a height of 1.20 m is not tall, and that someone with a height of 2.00 m is very tall. Besides, it is clear that the degree of membership in the fuzzy set tall of someone with height equal to 1.75 m should be larger than the degree of membership of someone with height equal to 1.70 m. A possible definition for this fuzzy set is given below.

**Definition 2.40.** *Let $X$ be the set of possible heights ($X \subseteq \mathbb{R}$); then, we define the membership function $\mu_{tall} : X \rightarrow [0,1]$ as*

$$\mu_{tall}(x) = \begin{cases} 0 & \text{if } x \leq 1.70 \\ (x - 1.70)/0.1 & \text{if } 1.70 \leq x \leq 1.80 \\ 1 & \text{if } x \geq 1.80. \end{cases}$$

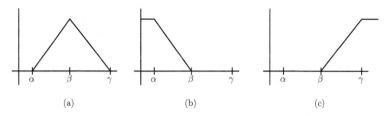

**Fig. 2.11.** Some typical membership functions: (a) Triangular; (b) $L$-shape; (c) $\Gamma$-shape

Figure 2.10 shows a graphical representation of this membership function.

Fuzzy sets are used to model a large variety of concepts. Especially, they are used for representing the meanings of graded adjectives and adverbs, and for computing with them. So, in some sense, they are used for computing with words, and thus having a practical application. For example, they can be used to model concepts related to distance (e.g., a value $x$ is *near* a point, such as *near zero*, *near Barcelona*), temperature (e.g., a room is *warm* or the temperature of a device is *high*), cost (e.g., the trip from Tokyo to Shinjuku is not *cheap*), time (e.g., the trip from Tokyo to Osaka takes *around 2.5 hours*, the train arrived *almost on time*), and so on.

The use of fuzzy sets for describing concepts in real applications makes them context dependent. For example, in the case of *tall*, some countries would define the membership function of $\mu_{tall}$ by

$$\mu_{tall}(x) = \begin{cases} 0 & \text{if } x \leq 1.60 \\ (x - 1.60)/0.1 & \text{if } 1.60 \leq x \leq 1.70 \\ 1 & \text{if } x \geq 1.70. \end{cases}$$

Similarly, the concept of being *late* for a date would be also context dependent.

To represent fuzzy concepts in a simple way, triangular membership functions are often considered. They are defined in terms of three parameters $(\alpha, \beta, \gamma)$, as follows.

**Definition 2.41.** *Let $X$ be a reference set ($X \subseteq \mathbb{R}$), and let $\alpha < \beta < \gamma$ in $X$; then, the triangular membership function $\mu_{\alpha,\beta,\gamma}^t(x)$ is defined as follows:*

$$\mu_{\alpha,\beta,\gamma}^t(x) = \begin{cases} 0 & \text{if } x \leq \alpha \text{ or } x \geq \gamma \\ (x - \alpha)/(\beta - \alpha) & \text{if } \alpha < x < \beta \\ (\gamma - x)/(\gamma - \beta) & \text{if } \beta \geq x < \gamma \end{cases}$$

Figure 2.11 represents a triangular membership function as well as two other functions ($L$-shape and $\Gamma$-shape) that are often used in practical applications.

We have defined the membership of a set to take values in $[0, 1]$. Alternatively, to express partial membership it would be enough to evaluate the elements in an arbitrary (partially) ordered set $L$ (e.g., a finite ordinal scale).

In the next example we consider the definition of a few quantifiers. Fuzzy quantifiers are fuzzy sets $\mu : [0, 1] \to [0, 1]$ whose domain corresponds to the proportion of the elements that satisfy a property. So, if $Q_{>50}$ represents the quantifier "more than 50%," we have that $Q_{>50}(x)$ corresponds to the membership of the proportion $x$ in the concept "more than 50%."

*Example 2.42.* We consider the definition of four quantifiers, giving for each of them their membership functions. They correspond to the concepts "there exists," "for all," and "more than 50%." For the last quantifier, a crisp and a fuzzy definition are given.

Figure 2.12 gives a graphical representation of the quantifiers.

1. Quantifier "there exists":

$$Q_\exists(x) = \begin{cases} 0 \text{ if } x = 0 \\ 1 \text{ if } x \in (0, 1]. \end{cases}$$

Note that when the proportion of elements is not zero, we have that the quantifier is completely satisfied.

2. Quantifier "for all":

$$Q_\forall(x) = \begin{cases} 0 \text{ if } x \in [0, 1) \\ 1 \text{ if } x = 1. \end{cases}$$

In this case, we need that all elements $x$ are included in the proportion. Therefore, $Q_\forall(x) = 1$ if and only if $x = 1$.

3. Quantifier "more than 50%":

$$Q_{>50}(x) = \begin{cases} 0 \text{ if } x \in [0, 0.5) \\ 1 \text{ if } x \in [0.5, 1]. \end{cases}$$

4. Quantifier "more than 50%" (fuzzy definition):

$$Q_{>\tilde{50}}(x) = \begin{cases} 0 & \text{if } x \in [0, 1/3) \\ 3(x - 1/3) & \text{if } x \in [1/3, 2/3] \\ 1 & \text{if } x \in (2/3, 1] \end{cases}$$

Note that $Q_{>50}$ is satisfied when $x > 1/2$. In fact, we have a crisp transition for $x = 1/2$. In contrast, in the case of the fuzzy quantifier $Q_{>\tilde{50}}$, there is a smooth transition from nonmembership to full membership. The transition starts with $x = 1/3$ and finishes with $x = 2/3$.

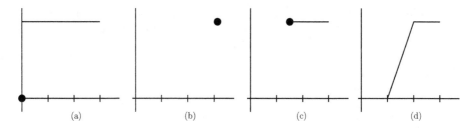

**Fig. 2.12.** Fuzzy quantifiers: (a) "for all"; (b) "there exists"; (c) "more than 50%"; (d) "more than 50%" (fuzzy definition)

### 2.3.1 Operations on Fuzzy Sets

The basic set theoretic operations defined on crisp sets (intersection, union, and complement) have counterparts in fuzzy sets. These functions are the fuzzy intersection, fuzzy union and fuzzy complement. For union and the intersection we will consider two fuzzy sets $A$ and $B$ represented by the membership functions $\mu_A$ and $\mu_B$. For the sake of simplicity, it is assumed that both functions are defined on the same domain. That is, $\mu_A : X \to [0,1]$ and $\mu_B : X \to [0,1]$.

We start considering the definition of the fuzzy intersection, tThat is, the function that computes the intersection of two fuzzy sets. The output of this function is naturally another fuzzy set, and this fuzzy set is represented by its membership function. Thus, given two fuzzy sets $A$ and $B$ represented by their membership functions $\mu_A$ and $\mu_B$, the fuzzy intersection permits us to construct a new fuzzy set $A \cap B$ represented by its membership function $\mu_{A \cap B}$. More specifically, $\mu_{A \cap B}$ is defined for each $x$ as a function $\top$ of $\mu_A$ and $\mu_B$ on $x$. Formally,

$$\mu_{A \cap B}(x) = \top(\mu_A(x), \mu_B(x)).$$

This function $\top$ is the t-norm. Note that $\top$ takes two values in $[0,1]$ and returns another one in the same interval.

In order to have an intersection consistent with the one on crisp sets, the function $\top$ should satisfy $\top(0,0) = \top(0,1) = \top(1,0) = 0$ and $\top(1,1) = 1$. All t-norm functions satisfy these properties, as will be seen below. Nevertheless, other properties are also required.

The formal definition of a t-norm is given below:

**Definition 2.43.** *A function* $\top : [0,1] \times [0,1] \to [0,1]$ *is a t-norm if and only if it satisfies the following properties:*

*(i)* $\top(x,y) = \top(y,x)$ *(symmetry or commutativity)*
*(ii)* $\top(\top(x,y),z) = \top(x,\top(y,z))$ *(associativity)*
*(iii)* $\top(x,y) \leq \top(x',y')$ *if* $x \leq x'$ *and* $y \leq y'$ *(monotonicity)*
*(iv)* $\top(x,1) = x$ *for all* $x$ *(neutral element 1)*

Additionally, t-norms are often required to satisfy continuity and subidempotency $(\top(x, x) < x$ for $x \neq 0)$. Such t-norms are called Archimedean t-norms.

**Definition 2.44.** *A continuous t-norm satisfying subidempotency ($\top(x, x) < x$) is an Archimedean t-norm.*

While in crisp sets there is only a single function for intersection, this is not the case for fuzzy sets. We consider below some examples.

*Example 2.45.* The following functions are t-norms.

Minimum: $\top(x, y) = \min(x, y)$. The minimum is often denoted by $\wedge$. That is, $x \wedge y = \min(x, y)$. We will use this notation in this book.
Algebraic product: $\top(x, y) = xy$.
Bounded difference/Lukasiewicz: $\top(x, y) = \max(0, x + y - 1)$.
Yager family: $\top_w(x, y) = 1 - \min\left(1, ((1 - x)^w + (1 - y)^w)^{1/w}\right)$ for $w \geq 0$.

All the t-norms are proper generalizations for conjunctions on crisp sets, as, for all of them, $\top(0, 0) = \top(0, 1) = \top(1, 0) = 0$ and $\top(1, 1) = 1$. In fact, these equalities follow from Definition 2.43. The study of t-norms has led to several characterizations. One of them is as follows.

**Theorem 2.46.** $\top : [0, 1] \times [0, 1] \to [0, 1]$ *is an Archimedean t-norm if and only if there exists a continuous strictly decreasing function $f$ from $[0, 1]$ to $\mathbb{R}$ with $f(1) = 0$ such that*

$$\top(x, y) = f^{(-1)}(f(x) + f(y)) \tag{2.9}$$

*for all $x, y \in [0, 1]$*

This function $f$ is called a decreasing generator, and $f^{(-1)} : \mathbb{R} \to [0, 1]$ corresponds to its quasi-inverse (or pseudo-inverse). Such a quasi-inverse is defined as

$$f^{(-1)} = \begin{cases} 1 & \text{if } x \in (-\infty, 0) \\ f^{-1} & \text{if } x \in [0, f(0)] \\ 0 & \text{if } x \in (f(0), \infty). \end{cases}$$

The definition of fuzzy union, the function to compute the union of fuzzy sets and to model disjunction, follows a pattern similar to the one of fuzzy intersection. Here, we will use a function $\perp$ that is known as t-conorm. The same properties considered for the t-norm apply here, except for the neutral element. In the case of union, the neutral element is zero. The requirement of t-conorms are formalized below.

**Definition 2.47.** *A function $\perp : [0, 1] \times [0, 1] \to [0, 1]$ is a t-conorm if and only if it satisfies the following properties:*

*(i)* $\perp(x, y) = \perp(y, x)$ *(symmetry or commutativity)*
*(ii)* $\perp(\perp(x, y), z) = \perp(x, \perp(y, z))$ *(associativity)*
*(iii)* $\perp(x, y) \leq \perp(x', y')$ *if $x \leq x'$ and $y \leq y'$ (monotonicity)*
*(iv)* $\perp(x, 0) = x$ *for all $x$ (neutral element 0)*

So, given two fuzzy sets $A$ and $B$, their union $A \cup B$ is represented by the membership function $\mu_{A \cup B}(x) = \perp(\mu_A(x), \mu_B(y))$.

*Example 2.48.* The following functions are t-cornorms.

Maximum: $\perp(x, y) = \max(x, y)$. The maximum is often denoted by $\vee$. That
    is, $x \vee y = \max(x, y)$. We will use this notation in this book.
Algebraic sum: $\perp(x, y) = x + y - xy$.
Bounded sum/Lukasiewicz: $\perp(x, y) = \min(1, x + y)$.
Yager: $\perp_w(x, y) = \min(1, (x^w + y^w)^{1/w})$ for $w \geq 0$.
Sugeno: $\perp_\lambda(x, y) = min(1, x + y + \lambda xy)$ for $\lambda > -1$.

For the sake of completeness, we give below a result concerning the representation of Archimedean t-conorms. A t-conorm is Archimedean when it is a continuous superidempotent t-conorm, where superidempotency means that $\perp(x, x) > x$ for all $x$. The result is analogous to the one given for t-norms.

**Theorem 2.49.** $\perp : [0, 1] \times [0, 1] \rightarrow [0, 1]$ *is an Archimedean t-conorm if and only if there exists a continuous strictly increasing function $g : [0, 1] \rightarrow \mathbb{R}$ with $g(0) = 0$ such that*

$$\perp(x, y) = g^{(-1)}(g(x) + g(y)) \tag{2.10}$$

*for all $x, y \in [0, 1]$.*

In this case, $g$ is known as an increasing generator, and its quasi-inverse is defined by (here, $g^{-1}$ is the inverse of $g$)

$$g^{(-1)} = \begin{cases} 0 & \text{if } x \in (-\infty, 0) \\ g^{-1} & \text{if } x \in [0, g(1)] \\ 1 & \text{if } x \in (g(1), \infty). \end{cases}$$

To illustrate, we give below a generator for the Sugeno t-conorm.

*Example 2.50.* The increasing generator of $\perp_\lambda(x, y) = min(1, x + y + \lambda xy)$ (for $\lambda > -1$) is

$$g_\lambda(x) = ln(1 + \lambda x)/ln(1 + \lambda). \tag{2.11}$$

The inverse of $g_\lambda$ is the function

$$g_\lambda^{-1}(x) = \lambda^{-1}(e^{x ln(1+\lambda)} - 1). \tag{2.12}$$

Now, we define a substraction operator $-_\perp$ for a t-conorm $\perp$: .

**Definition 2.51.** *Let $\perp$ be a t-conorm; then, the operation $-_\perp$ on $[0, 1]^2$ is defined by*

$$x -_\perp y := inf\{z | y \perp z \geq x\}.$$

For Archimedean t-conorms and for the maximum, this equation can be rewritten as follows.

(i) If $\perp$ is an Archimedean t-conorm with generator $g$, then

$$x -_\perp y = g^{(-1)}(g(x) - g(y)).$$

(ii) If $\perp$ is equal to the maximum, then

$$x -_{\max} y = \begin{cases} x \text{ if } x \geq y \\ 0 \text{ if } x < y. \end{cases}$$

Next, we define in a similar way the operator $\mathcal{I}_\top(x,y)$ for a t-norm $\top$.

**Definition 2.52.** *Let $\top$ be a t-norm; then, the operation $\mathcal{I}_\top(x,y)$ on $[0,1]^2$ is defined by*

$$\mathcal{I}_\top(x,y) := \sup\{z \in [0,1] | x\top z \leq y\}. \tag{2.13}$$

This operation is known as $\top$-residuum.

Fuzzy complements for fuzzy sets follow a similar pattern. In this case, however, the operation is unary. Therefore, the complement of a fuzzy set $A$ is defined in terms of a function $neg : [0,1] \to [0,1]$, as follows:

$$\mu_{\neg A}(x) = neg(\mu_A(x))$$

This function, known as negation, is defined as follows.

**Definition 2.53.** *A function $neg : [0,1] \to [0,1]$ is a negation if it satisfies:*

*(i) $neg(0) = 1$ and $neg(1) = 0$ (boundary conditions)*
*(ii) $x < y$ imply $neg(x) \geq neg(y)$ (order reversing)*
*(iii) $neg(neg(x)) = x$ for all $x$ (negation is involutive)*

We now present some examples:

*Example 2.54.* The functions listed below satisfy the requirements of complement

Standard: $neg(x) = 1 - x$.
Yager: $neg_w(x) = (1 - x^w)^{1/w}$ for $w > 0$.
Sugeno: $neg_\lambda(x) = (1 - x)/(1 + \lambda x)$ for $\lambda > -1$.

Now, we give a characterization of negations

**Proposition 2.55.** *A function $neg : [0,1] \to [0,1]$ is a negation if and only if $neg$ is of the form*

$$neg(x) = h^{-1}(1 - h(x))$$

*for all $x$ in $[0,1]$ for a strictly increasing function from $[0,1]$ to $[0,1]$, with $h(0) = 1$ and $h(1) = 0$.*

Although the operations over fuzzy sets satisfy the properties we have detailed above, it is not true that they satisfy all properties satisfied by union, intersection, and complement in classical set theory. For some properties, only a few families satisfy them. This is the case with the law of excluded middle, law of contradiction, and the law of De Morgan. These laws are formulated for fuzzy sets as $\perp(a, neg(a)) = 1$, $\top(a, neg(a)) = 0$, and $neg(\perp(a, b)) = \top(neg(a), neg(b))$ (for all $a, b \in [0, 1]$). Among the existing operators, we have $\perp(x, y) = min(1, x + y)$, $\top = max(0, x + y - 1)$, and $neg(x) = 1 - x$ satisfy the exluded middle and the law of contradiction but not idempotency ($\top(x, x) = x$, $\perp(x, x) = x$). In contrast, $\perp = max$, $\top = min$, and $neg(x) = 1 - x$, which satisfy idempotency, do not satisfy these properties.

Negation functions permit us to define the dual of any binary operator.

**Definition 2.56.** *Let $B$ be a binary operator on $[0, 1] \times [0, 1]$; then, its dual operator with respect to a negation function neg is defined, for all $x, y \in [0, 1]$, as follows:*

$$\tilde{B}(x, y) := neg(B(neg(x), neg(y))).$$

An order can be defined on binary operators. This order, denoted by $\preceq$, permits us to classify the operators. The definition is based on a pointwise ordering.

**Definition 2.57.** *Let $B_1$ and $B_2$ be two binary operators on $[0, 1] \times [0, 1] \rightarrow [0, 1]$; then, we say that $B_1 \preceq B_2$, if for all $x, y \in [0, 1]$, we have $B_1(x, y) \leq B_2(x, y)$.*

When, for two operators $B_1$ and $B_2$, either $B_1 \preceq B_2$ or $B_2 \preceq B_1$, the two operators are said to be comparable. Although this definition is given for binary operators, it can be applied to $N$-dimensional operators.

From this definition, it follows that min $\preceq$ max. This definition can be used to characterize the parameters of some families of norms. For example, in Yager's family of t-conorms, we have that if we consider two values $\alpha$ and $\beta$ such that $\alpha \leq \beta$, then $\perp_\beta \preceq \perp_\alpha$.

Additionally, we have the following proposition:

**Proposition 2.58.** *For all t-norms $\top$ and all t-conorms $\perp$, we have:*

$$\top_{min} \preceq \top \preceq min$$

$$max \preceq \perp \preceq \perp_{max}$$

Here, $\perp_{max}$ and $\top_{min}$ stand for the following operations:

$$\perp_{max}(x, y) = \begin{cases} x \text{ if } y = 0 \\ y \text{ if } x = 0 \\ 1 \text{ otherwise.} \end{cases} \tag{2.14}$$

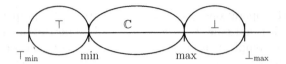

**Fig. 2.13.** t-norms, binary aggregation operators, and t-conorms

$$\mathsf{T}_{\min}(x,y) = \begin{cases} x \text{ if } y = 1 \\ y \text{ if } x = 1 \\ 0 \text{ otherwise.} \end{cases} \tag{2.15}$$

In the light of the previous definition, Equation 1.1 (which establishes that aggregation operators $\mathbb{C}$ are functions that yield a value between the minimum and the maximum of the input values) can be expressed equivalently for two inputs as

$$\min \preceq \mathbb{C} \preceq \max.$$

Therefore, putting all these results together, we have the following property:

$$\mathsf{T}_{\min} \preceq \mathsf{T} \preceq \min \preceq \mathbb{C} \preceq \max \preceq \bot \preceq \bot_{\max}.$$

That is, aggregation operators, t-norms, and t-conorms define different regions in the space of binary operators. t-norms are conjunctive, t-conorms are disjunctive, and aggregation operators are compensatory operators. These regions are illustrated in Figure 2.13.

### 2.3.2 Implications

Among other existing operations on fuzzy sets, we underline fuzzy implications. There exist several approaches to define them, based on different definitions of classical logic. One of them is based on the equivalence of the implication $a \rightarrow b$, with $\neg a \vee b$. Implications of this form are known as $\bot$-implications, as they are based on a t-conorm $\bot$ (the one used to model the disjunction in $\neg a \vee b$).

**Definition 2.59.** *A binary operator* $\mathcal{I} : [0,1] \times [0,1] \rightarrow [0,1]$ *is a fuzzy $\bot$-implication if it can be expressed as*

$$\mathcal{I}(x,y) = \bot(neg(x), y). \tag{2.16}$$

The Kleene-Dienes implication is one of the $\bot$-implications. Its definition is based on $\bot = \max$ and $neg(x) = 1 - x$. Therefore, the Kleene-Dienes implication corresponds to

$$\mathcal{I}(x,y) = \max(1 - x, y). \tag{2.17}$$

The Lukasiewicz implication is another example (using bounded sum and $neg(x) = 1 - x$):

$$\mathcal{I}(x, y) = \min(1, 1 - x + y).$$

R-implications (R for residuated) represents another family of implications. They extend to the fuzzy setting the definition in classical logic that an implication can be defined by

$$\mathcal{I}(x, y) = \max\{z \in \{0, 1\} | x \wedge z \leq y\}. \tag{2.18}$$

**Definition 2.60.** *A binary operator* $\mathcal{I} : [0, 1] \times [0, 1] \rightarrow [0, 1]$ *is a fuzzy R-implication if it can be expressed using Equation 2.13 for a t-norm* $\top$:

$$\mathcal{I}_\top(x, y) := \sup\{z \in [0, 1] | x \top z \leq y\}.$$

Although in the crisp setting Equations 2.16 and 2.18 are equivalent, in the fuzzy setting $R$-implications and $\perp$-implications do not define the same set. Nevertheless, there are some implications, such as the Lukasiewicz one, that are both R- and $\perp$-implications.

### 2.3.3 Fuzzy Relations

Fuzzy relations are a generalization of crisp relations. We will consider a few definitions regarding finite sets of reference.

**Definition 2.61.** *Let* $X_1, X_2, \cdots, X_N$ *be reference sets; then,* $R \subset X_1 \times X_2 \times \cdots \times X_N$ *is a crisp relation and a fuzzy set on* $X_1 \times X_2 \times \cdots \times X_N$ *is a fuzzy relation.*

Among the existing operations on fuzzy relations, we are interested in the composition of binary relations. Such relations corresponds to Definition 2.61 with $N = 2$. Composition is defined below.

**Definition 2.62.** *Let* $R$ *be a fuzzy relation on* $X_1 \times X_2$, *and let* $S$ *be a fuzzy relation on* $X_2 \times X_3$; *then, the* $\perp$-$\top$ *composition of* $R$ *and* $S$ *is a new relation on* $X_1 \times X_3$, *denoted by* $T := R \circ S$ *and defined by*

$$T(x_1, x_3) := \perp_{x \in X_2} \top(R(x_1, x), S(x, x_3))$$

*for all* $x_1$ *in* $X_1$ *and all* $x_3$ *in* $X_3$.

In the particular case of $\perp = \max$ and $\top = \min$, we get the max-min composition. This is defined as follows.

**Definition 2.63.** *The standard composition for fuzzy relations corresponds to the max-min composition. That is, the max-min composition of* $R$ *and* $S$ *is* $T = R \circ S$, *with*

$$T(x_1, x_3) = \max_{x \in X_2} \min(R(x_1, x), S(x, x_3))$$

*for all* $x_1$ *in* $X_1$ *and all* $x_3$ *in* $X_3$.

Let us consider an example of the max-min composition.

*Example 2.64.* Let $R_1 : X_1 \times X_2 \to [0,1]$ and $R_2 : X_2 \times X_3 \to [0,1]$ be two fuzzy relations defined as follows:

$$R_1 := (0 \quad 0.2 \quad 0.8 \quad 1 \quad 0.5 \quad 0.2 \quad 0 \quad 0)$$

$$R_2 := \begin{pmatrix} 0 & 0 & 0.1 & 0.1 \\ 0 & 0 & 0.2 & 0.1 \\ 0 & 0.1 & 0.3 & 0.1 \\ 0 & 0.2 & 0.8 & 0.2 \\ 0 & 0.3 & 1.0 & 0.1 \\ 0.1 & 0.4 & 0.5 & 0 \\ 0 & 0.1 & 0.2 & 0 \\ 0 & 0 & 0 & 0 \end{pmatrix}$$

Then, if $R$ is the max-min composition of $R_1$ and $R_2$, we have

$$R = R_1 \circ R_2 = (0.1 \quad 0.3 \quad 0.8 \quad 0.2)$$

### 2.3.4 Truth Degrees

It is known that, in the crisp setting, there is a tight relationship between sets and classical logics so that intersection corresponds to conjunction, union to disjunction, and complement to negation. This also occurs in the fuzzy setting. Here, fuzzy intersection, fuzzy union, and fuzzy complement can be used to model conjunction, disjunction, and negation for fuzzy predicates. In particular, when we have truth degrees $\tau(p_1)$ and $\tau(p_2)$ to denote the degree of truth of predicates $p_1$ and $p_2$, then we can combine them by means of t-norms, t-conorms, and negations to denote the degree of truth of composite predicates.

*Example 2.65.* Let us consider the following two rules:

1) If Barcelona is near and the ticket is cheap, then we visit La Sagrada Familia.

2) If Tokyo is near and we are not tired, then we visit Miraikan.

Let $a, b, c$, and $d$ be the truth degrees for "Barcelona is near," "the ticket to Barcelona is cheap," "Tokyo is near," and "we are tired." Then, the truth degree of the antecedent of the first rule would be $\top(a, b)$ and of the second rule would be $\top(c, neg(d))$.

When the predicates include a vague term, as in the case of "Barcelona is near," fuzzy sets can be used to determine the truth degree of the predicate. This is illustrated in the following example.

*Example 2.66.* Let us consider the concept "near" described by the following fuzzy set:

$$\mu_{near}(x) = \begin{cases} 0 & \text{if } |x| \geq 50km \\ (50 - x)/50 & \text{if } |x| \leq 50km. \end{cases}$$

Then, if our actual position is Bellaterra and the distance between this town and Barcelona (Plaça Catalunya) is 24 km, we have that the truth degree of "Barcelona is near" can be computed as follows:

$$\tau(\text{Barcelona is near}) = \mu_{near}(distance(Barcelona, actual position)) =$$
$$= \mu_{near}(distance(Barcelona, Bellaterra)) =$$
$$= \mu_{near}(24km) = (50 - 24)/50 = 0.52.$$

In general, if we have a predicate "x is A" and a membership $\mu_A$ to represent the concept $A$, then the truth degree of $\tau(x$ is $A)$, given that $x$ equals $x_0$ is defined as

$$\tau(x \text{ is } A) = \mu_A(x_0).$$

Truth degrees have also been defined for quantified predicates. In this case, we have expressions of the form "Q A's are B's," where $Q$ is a quantifier, such as *for all*, *most*, or *some*. The computation of the truth degree of such an expression assumes that there is an interpretation of $Q$ in terms of a fuzzy set, and we evaluate the proportion of $B$ in $A$ with such a fuzzy set. This is formally established below. To do so, we first need some definitions.

**Definition 2.67.** *Let $\mu_A$ be a fuzzy set on a finite set of reference $X$ representing the concept $A$; then, the $\Sigma count$ of $A$ is defined by*

$$\Sigma count(A) = \sum_{x \in X} \mu_A(x).$$

*Let $\mu_A$ and $\mu_B$ be two fuzzy sets on a finite reference set $X$ representing the concepts $A$ and $B$; then, the $\Sigma count(B|A)$ is defined by*

$$\Sigma count(B|A) = \frac{\Sigma count(A \cap B)}{\Sigma count(A)}.$$

$\Sigma count(B|A)$ is a relative measure of the cardinalities of $A$ and $B$. It corresponds to the proportion of $A$ in $B$. Note that, when all elements in $A$ are in $B$, we have $\Sigma count(B|A) = 1$, and, in contrast, when no element in $A$ is in $B$, $\Sigma count(B|A) = 0$. Note that this definition is similar to the one of conditional probabilities (see Equation 2.1).

**Definition 2.68.** *Let $A$ and $B$ be two fuzzy sets and let $Q$ be a fuzzy quantifier. Then, the truth degree of the statement "Q A's are B's" is defined by*

$$\tau(Q \text{ A's are B's}) = Q(\Sigma count(B|A)).$$

*Note that the sentence "Q A's are B's" is interpreted as "$\Sigma count(B|A)$ is Q."*

Now, we consider an example of computing the truth degree of one statement with fuzzy quantifiers.

*Example 2.69.* Let us consider the evaluation of the statement

"More than 50% of the students are supporters of F.C. Barcelona"

in a given class of 25 students, with 15 supporters of the team, and with the fuzzy quantifier "more than 50%" defined as in Example 2.42:

$$Q_{>\tilde{5}0}(x) = \begin{cases} 0 & \text{if } x \in [0, 1/3) \\ 3(x - 1/3) & \text{if } x \in [1/3, 2/3] \\ 1 & \text{if } x \in (2/3, 1] \end{cases}$$

Then, with these definitions, the truth degree of the statement given above is:

$$Q_{>\tilde{5}0}(\Sigma count(B|A)) = Q_{>\tilde{5}0}\left(\frac{\Sigma count(A \cap B)}{\Sigma count(A)}\right) = Q_{>\tilde{5}0}\left(\frac{15}{25}\right) = 0.8$$

In this example, we have used $\Sigma count(A) = 25$ (students in the class) and $\Sigma count(A \cap B) = 15$ (students in the class who are supporters). Instead, of *crisp supporters*, we could envision *fuzzy supporters*. In this case, we would consider a membership function giving the *support* degree for each student, and then $\Sigma count(A \cap B)$ would be computed as the summation of this function for all the students in the class. That is, let $X = A = \{x_1, x_2, \ldots, x_{25}\}$ be the 25 students and let $\mu_B(x_i)$ the degree of $x_i$ supporting the Barcelona team. Then, $\Sigma count(A \cap B)$ is defined by $\sum_{x_i \in X} \mu_B(x_i)$.

### 2.3.5 Fuzzy Inference Systems

Fuzzy systems are a particular type of Knowledge-Based System, where knowledge is represented by rules and concepts are represented by fuzzy sets. They are typically used to describe a function with $m$ inputs and one output. In general, there are $N$ rules of the form

$$R_i: \textbf{IF } x^1 \text{ is } A_i^1 \textbf{ and } \ldots \textbf{ and } x^m \text{ is } A_i^m \textbf{ THEN } y \text{ is } B_i, \quad (2.19)$$

where the $x^j$ correspond to the input variables of the system, $y$ is the output, and $A_i^j$ and $B_i$ denote fuzzy terms, represented by their corresponding membership functions $\mu_{A_i}^j$ and $\mu_{B_i}$. For example, $A_i^j$ might correspond to *near*, as in Example 2.66.

From now on, we will restrict the discussion to a single input - single output system. We will denote the input variable by $x$. Under this restriction, the rules follow the structure:

$$R_i: \textbf{IF} \quad x \text{ is } A_i \textbf{ THEN } \quad y \text{ is } B_i. \tag{2.20}$$

Given a set of fuzzy rules $\{R_i\}_i$, the system computes the value for variable $y$ given a value for variable $x$, say $x_0$. As the terms $A_i$ and $B_i$ are fuzzy sets, the output of the system for input $x_0$ is also a fuzzy set, that is, a function from the range of $y$ into $[0, 1]$ (a possibility distribution). The actual computation of this output value depends on the interpretation of the fuzzy rules. There are two main interpretations: disjunctive and conjunctive rules.

**The case of disjunctive rules**

When a fuzzy system is described by $N$ rules of the form of Equation 2.20, the output of the system for $x = x_0$ is computed using the following steps:

1. Compute the truth degree or satisfaction degree for the antecendent of all rules $R_i$. Let this value be $\alpha_i$. That is, $\alpha_i = \tau(x_0 is A_i)$. In our case, as there is a single condition in the antecedent, $\alpha_i = \mu_{A_i}(x_0)$. For more complex antecedents, we would use the approach described in Section 2.3.4.
2. Compute the conclusion of rule $R_i$. The most common approach is Mamdani's approach. From an operational point of view, Mamdani's approach is as follows: for each rule $R_i$, its output fuzzy set $\mu_{B_i}$ is clipped according to the degree of satisfaction $\alpha_i$. That is, the output of rule $R_i$ is $\mu_{B_i} \wedge \alpha_i = \mu_{B_i} \wedge \mu_{A_i}(x_0)$. This expression means that $\mu'_{B_i}(x) = \mu_{B_i}(x) \wedge \mu_{A_i}(x_0)$. Formally, this computation is equivalent to considering the input as equivalent to the set $A' = \{x_0\}$, and then defining the output as either $\cup_j (A' \circ R_j)$ or $A' \circ (\cup_j R_j)$, with $\circ$ being a max-min composition and $R_j$ being the intersection of $A_j$ and $B_j$. The expressions $\cup_j (A' \circ R_j)$ and $A' \circ (\cup_j R_j)$ are equivalent. The description uses the first expression.
3. Compute the output of the set of rules $\{R_i\}_i$. Once we have the output of each rule $R_i$, denoted by $\mu'_{B_i}$, we define the output for the whole system as the union of the outputs of each rule $R_i$. Using the maximum of the union (the most usual operator), we obtain

$$\tilde{B} = \vee_{i=1}^N \big( B_i \wedge A_i(x_0) \big). \tag{2.21}$$

4. Finally, the output fuzzy set $\tilde{B}$ is usually defuzzified. This corresponds to integrating the whole fuzzy set $\tilde{B}$ into a single number. There are different approaches. One of them is the center of gravity. It corresponds to defining the output as follows:

$$y_0 = \frac{\int x \mu_{\tilde{B}}(x) dx}{\int \mu_{\tilde{B}}(x) dx}$$

Thus, the output of the system is $y = y_0$.

**The case of conjunctive rules**

When rules are interpreted in a conjunctive way, the output of a system can be computed as either $\cap_j(A' \circ R_j)$ or $A' \circ (\cap_j R_j)$. While the two expressions for disjunctive rules are equivalent, this is not true for conjunctive ones. That is, in general, $\cap_j(A' \circ R_j) \neq A' \circ (\cap_j R_j)$. However, the equality holds when $A'$ is a single value.

We will now describe the computation of the output when $A'$ is a single value. For convenience, the description follows the computation of $\cap_j(A' \circ R_j)$. The output of the system $\{R_i\}$ when $x = x_0$ is computed as follows:

1. Compute $A' \circ R_j$ for all rules $R_i$. First, the relation $R_i$ is defined in terms of the implication function as follows: $R_i := \mathcal{I}(A_i, B_i)$. Then, as $A'$ is the singleton $A' = \{x_0\}$, we have that it can be proved that $\tilde{B}_i = A' \circ R_i = A' \circ \mathcal{I}(A_i, B_i)$ is equivalent to $\tilde{B}_i(y) = \mathcal{I}(A_i(x_0), B_i(y))$. The latter expression is the outcome of rule $R_i$.
2. Compute the intersection of all the outcomes of the previous step. This is expressed as follows:

$$\tilde{B}(y_0) = \wedge_{i=1}^{N}\big(\mathcal{I}(A_i(x_0), B_i(y_0))\big). \tag{2.22}$$

## 2.4 Bibliographical Notes

1. **Measurement theory:** The roots of measurement theory are old and diverse (see [238], Chapter 20, on Scale Types, for details). Recent work is strongly based on S. S. Stevens's research. In particular, he defined scale type according to the class of permissible transformations (see [375, 376, 377]). He introduced the terms ordinal, interval, log-interval, ratio, and absolute scales. Section 2.1 is based on [334] and [217]. The definition of measurement given in the first paragraph of this section is based on [417] and [376]. [334] and [217] underline the fact that one can always operate on numbers, but that the outcome might be meaningless. The definition of measurement as the construction of homomorphisms is taken from [217]. For a more complete account on Measurement Theory, see [386] and [238]. They complete [217].
2. **Probability theory:** There exists a large number of books on probability theory. For the first sections of Section 2.2 (random variables, expectation, and moments), the references [202] and [42] are adequate. The book by R. B. Ash [26] also deals with these topics and includes the results on the independence of variables given here in Section 2.2.3. Ash's book is more mathematically oriented.

   Kolmogorov's axioms were originally published in [215]. Nonparametrical methods are not discussed in detail in this book, although they are mentioned in Section 2.2.4. Such methods are described in [188] and [225].

   The normal distribution was discovered by De Moivre in 1733, but later independently rediscovered by Gauss (1809) and Laplace (1812) [314].

3. **Robust statistics and outliers:** Robust statistics is described in [194, 333, 181]. [194] was a seminal work in the area. [181] (Chapter 8) includes some discussion on the interest and usefulness of robust statistics.

   Order-statistics are included in some standard books on statistics, such as [381] (Chapter 14). For more details, the reader can use the book by Arnold, Balakrishan, and Nagaraja [21]. It is a course on order statistics.

   Barnet and Lewis [34] is a book devoted to the study of outliers in statistical data. The book also includes descriptions of L-estimators, M-estimators, and R-estimators. The book by Hawkins [130] can also be used for outliers.

4. **Regression and robust regression:** Regression is described in most books on statistics. It is worth mentioning [342, 352]. [342] includes a chapter on robust regression. For robust regression see also [339]. This book focuses on this topic and describes robust regression methods in detail, including some implementation issues. Nevertheless, [339] should be complemented with [338], which describe some improvements to LMS and LTS methods. The examples for LMS and LTS included in Section 2.2.8 have been solved following this work.

   For regression, [353], which reviews matrix algebra, is also useful. It includes chapters devoted to inverse matrices and generalized inverses, as well as such simpler operations as rank determination and methods for solving linear equations. Additionally, it includes a chapter on regression. Computational issues are also considered. [353] (Chapter 8.6), [352] (Chapter 1.5), and [354] (p. 469) describe algorithms for computing the generalized inverse. The computation of the rank of a matrix is given in [353] (Chapter 7.2).

   In relation to regression, the first description of the least sum of squares found in the literature was due to Legendre [224], but Gauss claimed later its discovery. The interesting papers by Plackett [323] and Stigler [378] discuss this matter. Harter, in his series of articles [123, 124, 125, 126, 127, 128, 129], traces the history of this field (mainly from Galileo Galilei, 1632 to 1974).

5. **Fuzzy sets:** Fuzzy sets were originally defined by L. A. Zadeh in 1965 [459]. [211] is a standard textbook in this field, and the Handbook of Fuzzy Systems [341] gives an account of the main topics. L-fuzzy sets are an alternative to membership functions in $[0, 1]$. L-fuzzy sets were introduced by Goguen [165]. For fuzzy sets and models related to uncertainty, see [219]. The concept "computing with words" and the use of fuzzy sets in this framework is detailed in [462]. See also [463] for some related research in this area.

   The origin of t-norms can be found in probabilistic metric spaces [260, 350, 349]. Original characterizations of t-norms and negations can be found, respectively, in [227] and [418]. Initial results in the field of fuzzy logic establishing t-norms, t-conorms, and negations can be found in [18]. See also the book by Klement, Mesiar and Pap [210] on t-norms.

The book by Alsina, Frank, and Sklar [17] devoted to associative operators include results concerning t-norms and t-conorms. Fuzzy systems, specially for applications in fuzzy control, are explained in several specific books. Fuzzy control is described in [98]. The distinction between disjunctive and conjunctive rules can be found in [211].

Fuzzy quantifiers, that were studied by Zadeh (see *e.g.*, [461]) are reviewed by Liu and Kerre in [232] and [233].

# 3

# Introduction to Functional Equations

> *1r PROBLÈME. Déterminer la fonction $\phi(x)$, de manière qu'elle reste continue entre deux limites réelles quelconques de la variable $x$, et que l'on ait pour toutes les valeurs réelles des variables $x$ et $y$*
>
> (I)      $\phi(x+y) = \phi(x) + \phi(y)^1$
>
> ───────────────────────────
>
> Augustin-Louis Cauchy, [66] (p. 104)

Functional equations are equations where the unknowns are functions. A well-known example of functional equation is the following Cauchy equation:

$$\phi(x+y) = \phi(x) + \phi(y). \tag{3.1}$$

A function $\phi$ is a solution of this equation if, for any two values $x$ and $y$, the application of $\phi$ to $x+y$ equals the addition of the application of $\phi$ to $x$ and to $y$. Therefore, the equation establishes conditions that functions $\phi$ have to satisfy. Typical solutions of this Cauchy equation are the functions $\phi(x) = \alpha x$ for an arbitrary value for $\alpha$.

In information fusion, functional equations can be used in two different contexts.

1. Functional equations can be used when we need to define an aggregation operator and we know which basic properties it has to satisfy. We can express the conditions of such an operator using functional equations. The operator is then derived from the equations.
2. Functional equations can be used to study the properties of information fusion methods. This is so because they can characterize the operators. Here, a characterization consists of finding a minimum set of properties

───────────────

[1] Determine the function $\phi(x)$ so that it remains continuous between two arbitrary real limits of the variable $x$, and that for all real values of the variables $x$ and $y$ one has (I) $\phi(x+y) = \phi(x) + \phi(y)$

(a minimum set of equations) that uniquely implies the operator. It is important to say that the set of properties that imply an operator is usually not unique.

An example about the use of functional equations in the definition of numerical aggregation operators is given below. The theorem establishes that the most general solution of two functional equations (Equations 3.2 and 3.3) is the weighted mean with nonrestricted weights.

**Theorem 3.1.** *The most general function of two variables satisfying the functional equations*

$$\phi(x + t, y + t) = \phi(x, y) + t \tag{3.2}$$

*and*

$$\phi(xu, yu) = \phi(x, y)u \text{ for } u \neq 0 \tag{3.3}$$

*for all $x, y, t$, and $u$ is*

$$\phi(x, y) = (1 - k)x + ky. \tag{3.4}$$

*Proof.* Let $y' = (y - x)$, $x' = 0$, and $t' = x$; then, Equation 3.2 corresponds to

$$\phi(x, y) = \phi(0 + x, (y - x) + x) = \phi(0, (y - x)) + x. \tag{3.5}$$

Now, for $x \neq y$, we can use Equation 3.3 to rewrite $\phi(0, y - x)$ as follows:

$$\phi(0, y - x) = \phi(0(y - x), 1(y - x)) = (y - x)\phi(0, 1) \text{ for } (y - x) \neq 0.$$

This equation means that

$$\phi(0, y - x) = k(y - x) \text{ with } k = \phi(0, 1). \tag{3.6}$$

Thus, for $y \neq x$, Equations 3.5 and 3.6 lead to

$$\phi(x, y) = k(y - x) + x = (1 - k)x + ky.$$

Now, let us consider the case of $y = x$; taking $y = x = 0$, we have that Equation 3.3 implies that $\phi(0, 0) = 0$. Thus, Equation 3.2 implies that $\phi(x, x) = \phi(0, 0) + x = x$. As, in both cases, Equation 3.4 holds, this equation is implied by Equations 3.2 and 3.3.

Equation 3.4 satisfies both Equation 3.2 and Equation 3.3. Therefore, the theorem is proved.  □

Alternatively, the theorem can be seen as a characterization of Equation 3.4. Nevertheless, other characterizations are possible. For example, the following proposition gives another characterization of Equation 3.4.

**Theorem 3.2.** *The most general function of two variables satisfying the functional equations*

$$\phi(x_1 + y_1, x_2 + y_2) = \phi(x_1, x_2) + \phi(y_1, y_2) \tag{3.7}$$

*and*

$$\phi(x, x) = x \tag{3.8}$$

*for all $x_1, x_2, y_1, y_2$ and $x$ is*

$$\phi(x, y) = (1 - k)x + ky. \tag{3.9}$$

## 3.1 Basic Functional Equations

In this section, some examples of basic functional equations are given with their solutions. We start with the Cauchy equation described above, and then present some of Cauchy's other equations: the exponential, the logarithm, and $\phi(xy) = \phi(x)\phi(y)$. We will also present the Jensen equations. For some of them, generalizations are given. Note that, from now on, we will denote $\phi(x + y) = \phi(x) + \phi(y)$ by the first Cauchy equation, as it is extensively used in the rest of this chapter.

**Proposition 3.3.** *If a continuous function $\phi : \mathbb{R} \to \mathbb{R}$ satisfies the* Cauchy *equation*

$$\phi(x + y) = \phi(x) + \phi(y),$$

*then there exists a real constant $\alpha$ such that*

$$\phi(x) = \alpha x$$

*for all real $x$.*

The theorem holds even if the function $\phi$ is continuous at a point or monotone or bounded on one side on an interval of positive length. In the following, we assume that such conditions hold when we come up with the Cauchy equation.

Now, we consider a generalization of the previous equation. This generalization is for functions $\phi$ on $\mathbb{R}^N$.

**Proposition 3.4.** *If a real function $\phi : \mathbb{R}^N \to \mathbb{R}$ satisfies*

$$\phi(x_1+y_1, x_2+y_2, \ldots, x_N+y_N) = \phi(x_1, x_2, \ldots, x_N)+\phi(y_1, y_2, \ldots, y_N), \tag{3.10}$$

*then it is of the form*

$$\phi(x_1, x_2, \ldots, x_N) = \alpha_1 x_1 + \alpha_2 x_2 + \cdots + \alpha_N x_N$$

*for an arbitrary real constant $\alpha_i$.*

*Proof.* To solve this equation, we will reduce it to the first Cauchy equation. To do so, we start by generalizing the equation, by induction, to

$$\phi(\mathbf{x_1} + \mathbf{x_2} + \cdots + \mathbf{x_p}) = \phi(\mathbf{x_1}) + \phi(\mathbf{x_2}) + \cdots + \phi(\mathbf{x_p})$$

$$\text{for all } \mathbf{x_1}, \mathbf{x_2}, \ldots, \mathbf{x_p} \in \mathbb{R}^N \text{ and } p = 2, 3, \ldots$$

Therefore, as any vector $\mathbf{x}$ can be rewritten as

$$\mathbf{x} = (x_1, x_2, \ldots, x_N) = (x_1, 0, \ldots, 0) + (0, x_2, \ldots, 0) + \cdots + (0, 0, \ldots, x_N),$$

we can express $\phi(\mathbf{x})$ as

$$\phi(\mathbf{x}) = \phi((x_1, x_2, \ldots, x_N)) =$$

$$= \phi((x_1, 0, \ldots, 0)) + \phi((0, x_2, \ldots, 0)) + \cdots + \phi((0, 0, \ldots, x_N)),$$

which, defining

$$\psi_i(x_i) := \phi((0, \ldots, 0, x_i, 0, \ldots, 0)), \tag{3.11}$$

where $x_i$ occupies the $i$th position in the vector, can be rewritten as

$$\phi(\mathbf{x}) = \psi_1(x_1) + \psi_2(x_2) + \cdots + \psi_N(x_N) = \sum_{i=1}^{N} \psi_i(x_i) \tag{3.12}$$

for all $x_i \in \mathbb{R}$ and $i = 1, 2, \ldots, N$

Equation 3.10 (and 3.11) implies that $\psi_i$ satisfies the first Cauchy equation for all $i = 1, 2, \ldots, n$ and for all $x$, $y$, and $x + y$ in $\mathbb{R}$. That is,

$$\psi_i(x + y) = \psi_i(x) + \psi_i(y). \tag{3.13}$$

Therefore, applying Proposition 3.3, we have that $\psi_i$ is of the form

$$\psi_i(x) = \alpha_i x.$$

Now, replacing $\psi_i$ with its equivalent expression in Equation 3.12, we have

$$\phi(\mathbf{x}) = \phi(x_1, x_2, \ldots, x_N) = \sum_{i=1}^{N} \alpha_i x_i.$$

As this expression satisfies Equation 3.10, the proposition is proved.  □

**Proposition 3.5.** *If a nonidentically zero function $\phi : \mathbb{R} \to \mathbb{R}$ satisfies the equation*

$$\phi(x + y) = \phi(x)\phi(y), \tag{3.14}$$

*then there exists an arbitrary real constant $\alpha$ such that*

$$\phi(x) = e^{\alpha x}.$$

Equation 3.14 is known as the exponential equation.
A similar proposition applies for the logarithm equation. That is, when

$$\phi(x \cdot y) = \phi(x) + \phi(y) \tag{3.15}$$

holds for all positive $x$ and $y$, then $\phi$ is of the form

$$\phi(x) = \alpha \log(x).$$

The most general solution of $\phi(x \cdot y) = \phi(x) + \phi(y)$ when the equation is valid not only for positive but for all real $x \neq 0$ and $y \neq 0$ is $\phi(x) = \alpha log|x|$.
Finally, the solution of

$$\phi(xy) = \phi(x)\phi(y), \tag{3.16}$$

when it holds for all positive $x$, is $\phi(x) = x^c$ or $\phi(x) = 0$.
The proof of the equations is obtained through the transformation of Equations 3.14 and 3.15 into the first Cauchy equation. For example, the substitutions $x = e^u$, $y = e^v$, and $\phi(e^w) = \psi(w)$ into the logarithm equation (Equation 3.15) yield the equation $\psi(u + v) = \psi(u) + \psi(v)$. Therefore, $\psi(u) = \alpha u$, and, thus, $\phi(e^u) = \alpha u$ or $\phi(x) = \alpha \log(x)$. Similarly, $\phi(x) = e^{\psi(x)}$ also leads to the first Cauchy equation for Equation 3.14. Equation $\phi(xy) = \phi(x)\phi(y)$ is also solved by rewriting it as the first Cauchy equation.

**Proposition 3.6.** *Let $N$ be any fixed $N \geq 2$, let $x_i$ be in $J$ (an open real interval) for all $i = 1, \ldots, N$, and let $\phi$ be a function continuous at a point or monotone or bounded on one side on an interval of positive length. Then, the general solution of the $N$-term Jensen equation*

$$\phi(\frac{1}{N} \sum_{i=1}^{N} x_i) = \frac{1}{N} \sum_{i=1}^{N} \phi(x_i) \tag{3.17}$$

*is*

$$\phi(x) = \alpha x + \beta, \tag{3.18}$$

*with $\alpha$ and $\beta$ arbitrary real constants.*

Note that, in the case of $N = 2$, the above equation reduces to

$$\phi(\frac{x + y}{2}) = \frac{\phi(x) + \phi(y)}{2}. \tag{3.19}$$

The 2-term Jensen equation is known as the *Jensen equation*. It comes from the definition of convex functions. A function which satisfies $\phi((x + y)/2) \leq (\phi(x) + \phi(y))/2$ in a certain interval is said to be convex in that interval. Geometrically, when this inequality holds, any chord lies above or on the curve (see Figure 3.1). The proposition shows that, for continuous functions, the equality can only be satisfied for lines of the form: $\phi(x) = \alpha x + \beta$. A generalization for the $N$-term is given below.

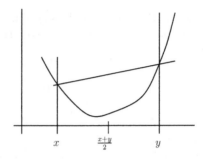

**Fig. 3.1.** Convex functions and Jensen equation

**Proposition 3.7.** *Let $N$ be any fixed $N \geq 2$, let $x_i \in J$ (an open interval), and let $\phi$ and $\psi$ be strictly monotone functions. Then, the general solution of:*

$$\phi^{-1}\left(\frac{1}{N}\sum_{i=1}^{N}\phi(x_i)\right) = \psi^{-1}\left(\frac{1}{N}\sum_{i=1}^{N}\psi(x_i)\right) \tag{3.20}$$

*is*

$$\phi(x) = \alpha\psi(x) + \beta,$$

*with $\alpha$ and $\beta$ arbitrary real constants such that $\alpha \neq 0$.*

A generalization of the Jensen equation (Equation 3.19) is given here.

**Proposition 3.8.** *The general solution for $\phi : \mathbb{R}^2 \to \mathbb{R}$ of*

$$\phi\left(\frac{x_1 + y_1}{2}, \frac{x_2 + y_2}{2}\right) = \frac{\phi(x_1, x_2) + \phi(y_1, y_2)}{2} \tag{3.21}$$

*is*

$$\phi(x, y) = \alpha x + \beta y + c.$$

Now, we present a completely different type of functional equation that will be useful in Section 3.3. In fact, the problem corresponds to a system of equations to define the area of a rectangle. Two equations are established that define the area on the basis of the two sides of the rectangle: $\phi(side_1, side_2)$. The meaning of the equations is represented in Figure 3.2.

**Proposition 3.9.** *The most general positive solution of the system of equations*

$$\phi(x_1 + x_2, y) = \phi(x_1, y) + \phi(x_2, y) \tag{3.22}$$

$$\phi(x, y_1 + y_2) = \phi(x, y_1) + \phi(x, y_2) \tag{3.23}$$

*is*

$$\phi(x, y) = kxy.$$

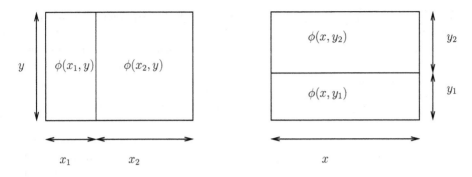

**Fig. 3.2.** The area of a rectangle: graphical representation of Equation 3.22 (left) and Equation 3.23 (right)

*Proof.* Let us start by considering Equation 3.22, and by assuming that $\phi$ is constant on $y$. Then, defining $\psi(x) = \phi(x, y)$, the equation reduces to $\psi(x_1 + x_2) = \psi(x_1) + \psi(x_2)$. As this is the first Cauchy equation, $\psi(x)$ is of the form $\psi(x) = \alpha x$. Nevertheless, although we can conclude that $\phi(x, y)$ is also of the form $\alpha x$, $\alpha$ depends on $y$; i.e., the constant $\alpha$ we have obtained was for a particular $y$, and, thus, $\alpha$ is a function of $y$. Therefore, $\phi(x, y) = \alpha(y) \cdot x$.

Similarly, considering 3.23, we conclude that $\phi(x, y)$ is the product of $y$ and a function of $x$. Formally, $\phi(x, y) = \beta(x) \cdot y$.

Thus, the following equation holds:

$$\phi(x, y) = \alpha(y) \cdot x = \beta(x) \cdot y. \tag{3.24}$$

Now, dividing by the product $x \cdot y$, we obtain

$$\frac{\phi(x, y)}{x \cdot y} = \frac{\alpha(y)}{y} = \frac{\beta(x)}{x}.$$

The only way that $\alpha(y)/y$ is equal to $\beta(x)/x$ for all $x$ and $y$ is with both quotients always equal to a constant. Denoting such a constant by $k$, we have that the following holds:

$$\frac{\phi(x, y)}{xy} = \frac{\alpha(y)}{y} = \frac{\beta(x)}{x} = k. \tag{3.25}$$

Therefore, $\phi(x, y) = kxy$. Finally, as this expression satisfies Equations 3.22 and 3.23, the proposition is proved. $\square$

The value of the constant $k$ depends on the scale we consider for the area of the rectangle. If we add the requirement that the area of a rectangle with $x = 1$ and $y = 1$ is equal to 1, then $k$ should also be 1.

In Section 2.3.1, some other examples of functional equations are given. They establish the t-norms (Definition 2.43), t-conorms (Definition 2.47), and complements (Definition 2.53).

## 3.2 Using Functional Equations for Information Fusion

In this section, we present an example of the application of functional equations to construct an operator that aggregates numerical information. We start by formalizing the problem, and then turn to the development of its solution. The example illustrates the main procedures to prove theorems by means of functional equations.

*Example 3.10.* In a committee, $m$ different projects have been evaluated by $N$ different experts to allocate a total budget of $s$ euros. Then, a decision maker has to aggregate the information of all the experts and give a final allocation of the budget. The problem is how to formalize the aggregation process. We start by formalizing the available data and then the requirements for the aggregation process.

Available data is modeled as follows: $x_j^i$ stands for the quantity that the $i$th expert assigns to the $j$th project. $\mathbf{X}$ stands for the whole matrix with the opinions of all the experts ($\mathbf{X} = \{x_j^i\}$) and $\mathbf{x}^i$ corresponds to the vector with the values of the $i$th expert for all the projects. Similarly, $\mathbf{x}_j$ is the vector with all the assignments to the $j$th project. Table 3.1 illustrates the data supplied by the experts. In this table, the $i$th expert is denoted by $E_i$.

The requirements for the aggregation process will assume

R1: The amount allocated to projects is always positive; i.e., $x_j^i \geq 0$ for all $i$ and $j$.

R2: Each of the experts distributes the whole amount $s$ among all the projects; i.e., $\sum_{j=1}^m x_j^i = s$ for all $i$.

Once the experts have supplied their evaluation, the decision maker (denoted by $DM$ in the last row of Table 3.1) has to make a final assignment. This final assignment is expressed as a function $\mathbf{g}$ of $\mathbf{X}$. That is, $\mathbf{g}(\mathbf{X})$ denotes all the assignments and $g_j(\mathbf{X})$ corresponds to the final assignment to a particular project $j$. Therefore, $\mathbf{g}(\mathbf{X}) = (g_1(\mathbf{X}), \ldots, g_m(\mathbf{X}))$. Now, we consider our basic assumptions about functions $g_j$ for $j = \{1, \cdots, m\}$:

(i) The total amount distributed by $\mathbf{g}$ should be $s$. This condition is only required when all the experts assign the whole quantity $s$, i.e., when requirement R2 above is fulfilled.

(ii) The final quantity that the decision maker assigns to a project only depends on the assignments to that project. This is, instead of considering functions $g_j$ on the whole matrix $\mathbf{X}$, it is enough to consider functions defined on a single column. Denoting such function by $f_j$ and the column by $\mathbf{x}_j$, we have $g_j(\mathbf{X}) = f_j(\mathbf{x}_j)$. These functions appear in the last row of Table 3.1. The definition of $f_i$ in this way satisfies the condition of independence of irrelevant alternatives.

(iii) If all the experts assign 0 to a certain project, then the decision maker will also assign 0 to the project. This is, $f_j(\mathbf{0}) = 0$.

| | Proj 1 | Proj 2 | $\cdots$ | Proj j | $\cdots$ | Proj m |
|---|---|---|---|---|---|---|
| $E_1$ | $x_1^1$ | $x_2^1$ | $\cdots$ | $x_j^1$ | $\cdots$ | $x_m^1$ |
| $E_2$ | $x_1^2$ | $x_2^2$ | $\cdots$ | $x_j^2$ | $\cdots$ | $x_m^2$ |
| $\vdots$ | $\vdots$ | $\vdots$ | | $\vdots$ | | $\vdots$ |
| $E_i$ | $x_1^i$ | $x_2^i$ | $\cdots$ | $x_j^i$ | $\cdots$ | $x_m^i$ |
| $\vdots$ | $\vdots$ | $\vdots$ | | $\vdots$ | | $\vdots$ |
| $E_N$ | $x_1^N$ | $x_2^N$ | $\cdots$ | $x_j^N$ | $\cdots$ | $x_m^N$ |
| $DM$ | $f_1(\mathbf{x_1})$ | $f_2(\mathbf{x_2})$ | $\cdots$ | $f_j(\mathbf{x_j})$ | $\cdots$ | $f_m(\mathbf{x_m})$ |

**Table 3.1.** Assignment of $s$ euros to $m$ projects by $N$ human experts. $\{E_1, \cdots, E_N\}$ stand for experts, $\{Proj_1, \cdots, Proj_m\}$ for projects, $\mathbf{x_j} = (x_j^1 \cdots x_j^N)$ for assignments to the $j$th project, and $f_j(\mathbf{x}_j)$ for the final decision for the $j$th project

All together, we have that the function $g(X)$ has to satisfy the following conditions:

1. $\mathbf{g(X)} = (g_1(\mathbf{X})g_2(\mathbf{X})\ldots g_m(\mathbf{X})) = (f_1(\mathbf{x_1})f_2(\mathbf{x_2})\ldots f_m(\mathbf{x_m}))$, where $f_j : [0,s]^N \to \mathbb{R}^+$ for $j = 1, \cdots, m$.
2. $\sum_{j=1}^m \mathbf{x_j} = \mathbf{s}$ implies that $\sum_{j=1}^m f_j(\mathbf{x_j}) = s$.
3. $f_j(\mathbf{0}) = 0$ for $j = 1, \cdots, m$.

Functions $f_i$ that satisfy these conditions are characterized in the next proposition.

**Proposition 3.11.** *The general solution of the system*

$$f_j : [0,s]^N \to \mathbb{R}^+ \text{ for } j = \{1, \cdots, m\} \tag{3.26}$$

$$\sum_{j=1}^m \mathbf{x_j} = \mathbf{s} \text{ implies that } \sum_{j=1}^m f_j(\mathbf{x_j}) = s \tag{3.27}$$

$$f_j(\mathbf{0}) = 0 \text{ for } j = 1, \cdots, m \tag{3.28}$$

*for a given $m > 2$ is given by*

$$f_1(\mathbf{x}) = f_2(\mathbf{x}) = \cdots = f_m(\mathbf{x}) = f((x_1, x_2, \ldots, x_N)) = \sum_{i=1}^N \alpha_i x_i,$$

*where $\alpha_1, \cdots, \alpha_N$ are nonnegative constants satisfying $\sum_{i=1}^N \alpha_i = 1$, but are otherwise arbitrary.*

*Proof.* We start by considering Equation 3.27. As this equation holds for all $\mathbf{x_j}$, then, in particular, it should hold for the substitutions $\mathbf{x_1} = \mathbf{s}$, $\mathbf{x_2} = \cdots = \mathbf{x_m} = \mathbf{0}$. Then, as $f_j(\mathbf{0}) = 0$, according to Equation 3.28, we get

$$f_1(s) = s. \tag{3.29}$$

Nevertheless, the selection of $x_1$ was arbitrary. Therefore, the equality holds for all $j$. That is,

$$f_j(s) = s \text{ for all } j \in \{1, 2, \ldots, m\}.$$

Let us now consider a different substitution in the same equation: $x_1 = z$, $x_3 = s - z$, $x_2 = x_4 = \cdots = x_m = 0$. With this substitution, and taking advantage of Equation 3.28, we get

$$f_1(z) + f_3(s - z) = s.$$

Therefore, the following holds:

$$f_1(z) = s - f_3(s - z) \text{ for all } z \in [0, s]^N. \tag{3.30}$$

Let us consider again another substitution in the same Equation 3.27. In this case, $x_1 = x$, $x_2 = y$, $x_3 = s - x - y = 0$. Then, we get

$$f_1(x) + f_2(y) + f_3(s - x - y) = s,$$

which is equivalent to

$$f_1(x) + f_2(y) = s - f_3(s - x - y). \tag{3.31}$$

Note that the terms in the right hand side of this equation can be made equal to the ones in Equation 3.30 with $z = x + y$. So, $s - f_3(s - x - y) = f_1(x + y)$. Taking this into account, we obtain the following equation:

$$f_1(x) + f_2(y) = f_1(x + y) \text{ for all } x, y, x + y \in [0, s]^N. \tag{3.32}$$

As this equation holds for all $x$, it also holds for $x = 0$. So, as $f_1(0) = 0$ (by Equation 3.28 for $j = 1$), the following can be established:

$$f_2(y) = f_1(y).$$

As the selection of functions $f_1$ and $f_2$ was arbitrary, the equality can be established for all $f_j$. We will denote this function by $f$:

$$f_1 = f_2 = \cdots = f_m = f$$

Using $f$, Equation 3.32 is rewritten

$$f(x) + f(y) = f(x + y) \text{ for all } x, y, x + y \in [0, s]^N. \tag{3.33}$$

This equation was solved in Proposition 3.4; therefore, $f$ is as follows:

$$f(\mathbf{x}) = \sum_{i=1}^{N} \alpha_i x_i.$$

Nevertheless, as Equation 3.26 implies that $f_i(\mathbf{x}) \geq 0$ for all $i = 1, 2, \ldots, m$, and, thus, $\alpha_i x_i \geq 0$ for all $x \in [0, s]$, we conclude that $\alpha_i \geq 0$.

Moreover, taking into account Equation 3.29, which for this particular form for $f$ is equivalent to $\sum_{i=1}^{N} \alpha_i s = s$, we further constrain the values for $\alpha_i$, requiring

$$\sum_{i=1}^{N} \alpha_i = 1$$

Finally, as the functions

$$f_1(\mathbf{x}) = f_2(\mathbf{x}) = \cdots = f_m(\mathbf{x}) = f((x_1, x_2, \ldots, x_N)) = \sum_{i=1}^{N} \alpha_i x_i,$$

with $\alpha_1, \alpha_2, \ldots, \alpha_N$ such that $\sum_{i=1}^{N} \alpha_i = 1$,

satisfy Equations 3.26, 3.27, and 3.28, the proposition is proved. $\square$

**Corollary 3.12.** *For a system satisfying*

$$f_j : [0, s]^N \to \mathbb{R}^+ \text{ for } j = \{1, \cdots, m\} \tag{3.34}$$

$$\sum_{j=1}^{m} \mathbf{x}_j = \mathbf{s} \text{ implies that } \sum_{j=1}^{m} f_j(\mathbf{x}_j) = s \tag{3.35}$$

$$f_j(\mathbf{0}) = 0 \text{ for } j = 1, \cdots, m \tag{3.36}$$

*for a given $m > 2$, there exists a probability $P$ such that $f$ is represented as an expectation:*

$$f_1(\mathbf{x}) = f_2(\mathbf{x}) = \cdots = f_m(\mathbf{x}) = E_P(\mathbf{x})$$

## 3.3 Solving Functional Equations

Now, we review the main techniques that are commonly in use to solve functional equations. They will be illustrated with some examples and refer to the propositions and proofs given in this section.

**Variables by values:** When a functional equation is satisfied for all values $d$ in a domain $D$, it must also be satisfied by a particular value $d_0$. The substitution of any variable with a particular value might simplify the equation.

The proof of Proposition 3.11 gives several examples of this technique. One of them is the substitution of $\mathbf{x}_1$ with $\mathbf{s}$, and of $\mathbf{x}_2, \ldots, \mathbf{x}_m$ with $\mathbf{0}$ in Equation 3.27, which yields Equation 3.29.

Sometimes, different substitutions are applied to the same equation, obtaining different equations. The proof of Proposition 3.11 also illustrates this case. Equation 3.27 was substituted three times: (i) $\mathbf{x}_1 = \mathbf{s}$, $\mathbf{x}_2 = \cdots = \mathbf{x}_m = \mathbf{0}$; (ii) $\mathbf{x}_1 = \mathbf{z}$, $\mathbf{x}_3 = \mathbf{s} - \mathbf{z}$, $\mathbf{x}_2 = \mathbf{x}_4 = \cdots = \mathbf{x}_m = \mathbf{0}$; (iii) $\mathbf{x}_1 = \mathbf{x}$, $\mathbf{x}_2 = \mathbf{y}$, $\mathbf{x}_3 = \mathbf{s} - \mathbf{x} - \mathbf{y} = \mathbf{0}$. These substitutions led to Equations 3.29, 3.30, and 3.31.

**Function transformation:** This is to replace a function by another one, so that the functional equation is transformed into an easier one.

For illustration, let us consider the following equation:

$$\phi^{-1}\left(\frac{1}{N}\sum_{i=1}^{N}\phi(1/a_i)\right) = 1/\phi^{-1}\left(\frac{1}{N}\sum_{i=1}^{N}\phi(a_i)\right). \tag{3.37}$$

This equation can be transformed, by considering the function $\tilde{\phi}(a) = \phi(1/a)$, into

$$\tilde{\phi}^{-1}\left(\frac{1}{N}\sum_{i=1}^{N}\tilde{\phi}(a_i)\right) = \phi^{-1}\left(\frac{1}{N}\sum_{i=1}^{N}\phi(a_i)\right)$$

As this corresponds to the generalization of the $N$-term Jensen equation (Proposition 3.7), expressions for $\tilde{\phi}$ and $\phi$ are obtained.

**Variable transformation:** In this case, a transformation is applied to a variable to simplify an equation. We illustrate this case with the logarithm equation (Equation 3.15): $\phi(x \cdot y) = \phi(x) + \phi(y)$. We have seen that with the transformations $x = e^u$ and $y = e^v$, we get $\phi(e^u \cdot e^v) = \phi(e^u) + \phi(e^v)$. Then, using the function transformation $\phi(e^u) = \psi(u)$, we get the first Cauchy equation, which is solved, giving $\psi(u) = \alpha u$, and, thus, $\phi(x) = \alpha \log x$.

**Considering a more general equation:** The solution of an equation $A$ is obtained by considering a more general one, $B$. This means that a particular parameterization of $B$ leads to the solution of equation $A$. Accordingly, the solutions of $B$ are solutions of $A$ when the same parameterizations are considered. For example, Proposition 3.6 can be solved with Proposition 3.7, as Equation 3.17 corresponds to Equation 3.20 when $\psi(x) = x$. Therefore, $\phi(x) = \alpha\psi(x) + \beta$ with $\psi(x) = x$ (i.e., $\phi(x) = \alpha x + \beta$) is the solution of Proposition 3.6.

**Variables as constants:** First, the equation is solved taking as a constant one of the variables. Then, in the solution, constants are replaced by functions of the original variables. The proof of Proposition 3.9 includes such a transformation. Note that Equation 3.22 is solved taking $y$ as a constant. This permits us to rewrite $\phi(x, y)$ so that it does not depend on $y$: $\psi(x) = \phi(x, y)$. Such rewriting permits us to reduce Equation 3.22 to the first Cauchy equation, and, thus, $\psi(x) = \alpha x$. However, as the constant $\alpha$ was a function of the *selected* constant $y$, the solution of Equation 3.22 corresponds to $\phi(x, y) = \alpha(y) \cdot x$.

**Separation of variables:** When some variables only appear in one side of an equation, both sides can be rewritten as a function of common variables. Proposition 3.9 also illustrates this case. In the proof of this proposition, Equation 3.24 considers the following equality:

$$\phi(x, y) = \alpha(y) \cdot x = \beta(x) \cdot y.$$

As this equation can be rewritten in a way that both sides do not share a common variable

$$\frac{\phi(x, y)}{x \cdot y} = \frac{\alpha(y)}{y} = \frac{\beta(x)}{x},$$

it means that $\frac{\alpha(y)}{y} = \frac{\beta(x)}{x}$ is equal to a constant $k$. Thus, $\frac{\phi(x,y)}{xy} = k$.

When solving functional equations, these techniques are not used in isolation but combined together. Proposition 3.11 and, to a small extent, Proposition 3.9 illustrate this situation.

## 3.4 Bibliographical Notes

1. **Functional Equations:** Sections 3.1 and 3.2 are based on Aczél's works (in particular, on his books [4] and [6]). Aczel [4] gives extensive references and historical remarks about the development of the field. [4] is an extended and up-to-date (1966) translation of [3]. Formalization of functional equations is given in [4]. Most results (with corresponding proofs) in this chapter are given in both [4] and [6]. [65] is a more recent book on functional equations that also includes most of the examples in this chapter. In particular, the example on functional equations for aggregation (Proposition 3.11) in Section 3.2 is taken from [6] (p. 2; see also [65], p. 157), and the area of the rectangle (from Legendre [223], pp. 293-294) is given in both [4] and [65]. Cauchy's equations were formulated and solved in some restrictive conditions in [66] (pp. 104–113). The Jensen equation is formulated and solved in [204] (p. 176) (see [203]).

   Section 3.3, on the most common methods for solving functional equations, is based on [63] (or [64]).

   In Chapter 4, we review some other results about the use of functional equations for defining aggregation operators. More references on functional equations are given in that chapter.

# 4

# Synthesis of Judgements

*Mesclar ous amb cargols*[1]

---

Catalan saying

In this chapter we study some aggregation operators for numerical information. The description is focused on results based on functional equations. Therefore, not only are the operators given, but also, at least for some of them, their characterization. We refer to these results as *syntheses of judgements*. Although the term could be used for any aggregation operator, we restrict its use to the case of characterizations using functional equations.

To describe the main results, we will assume that there is a set of information sources, denoted by $X = \{x_1, \ldots, x_N\}$, and that each source $x_i$ supplies a numerical value $a_i$. To simplify definitions, we assume that $a_i$ belongs either to the unit interval $I = [0, 1]$ or to the positive real line $\mathbb{R}^+ = [0, \infty)$. For some aggregation operators, we will exclude zero. Note that in some of the results described, other domains are also appropriate (e.g., the whole real line $\mathbb{R}$).

The value $a_i$ supplied by the information source $x_i$ can be expressed by means of a function $f$ that assigns the value $a_i$ to $x_i$. That is, $f(x_i) = a_i$ for all $i \in \{1, \ldots, N\}$.

Using this notation, an aggregation operator is a function $\mathbb{C}(a_1, \ldots, a_N)$, or, equivalently, $\mathbb{C}(f(x_1), \ldots, f(x_N))$, that takes $N$ numerical values and returns another numerical value. Here, as in Section 1.1 we use $\mathbb{C}$ for $\mathbb{C}onsensus$.

## 4.1 Associativity

From a technical point of view, associativity is one of the most important properties when defining operators. This is so because it permits the definition

---

[1] Mix eggs with snails

of an $N$-dimensional operator from a two-dimensional one. For example, when $\mathbb{C}$ is associative, the expression can be computed either using

$$\mathbb{C}(\mathbb{C}(\mathbb{C}(x_1, x_2), x_3), \ldots, x_N)$$

or

$$\mathbb{C}(x_1, \mathbb{C}(x_2, \ldots, \mathbb{C}(x_{N-1}, x_N))).$$

This section is devoted to the study of associativity and the review of some results.

Before going into details on aggregation operators, we review a characterization for associative operators.

**Theorem 4.1.** *Let $I$ be the unit interval; then, $\circ : I \times I \to I$ is a continuous operation in $I$ that satisfies*

$$(a \circ b) \circ c = a \circ (b \circ c) \tag{4.1}$$

*and is cancellative (i.e., $a_1 \circ b = a_2 \circ b$ or $b \circ a_1 = b \circ a_2$ implies $a_1 = a_2$ for any $b \in I$) if, and only if, there exists a continuous and monotone function $\phi : J \to I$ ($J$ has to be open at least from one side) such that*

$$a \circ b = \phi(\phi^{-1}(a) + \phi^{-1}(b)). \tag{4.2}$$

*Proof.* It is obvious that $\circ : I \times I \to I$ is cancellative if and only if $\circ$ is either strictly monotone increasing or strictly monotone decreasing.

We will prove below the case of $\circ$ being strictly monotone increasing. The case of strictly monotone decreasing is similar to the increasing one, and it will not be considered here.

Let $0 < x < 1$ be a positive number; then, since $\circ$ is strictly monotone increasing, we have

$$x \circ x < x. \tag{4.3}$$

Let $0 < c < 1$ be fixed. Then, it follows from associativity that $c \circ c \circ \cdots \circ c$. Using this property, we define the function $f : \mathbb{N} \to (0, 1)$ by

$$f(n) = \underbrace{c \circ c \circ \cdots \circ c}_{n}.$$

where $\mathbb{N}$ is the set of positive integers. It is obvious from the definition of $f$ that $f(n + m) = f(n) \circ f(m)$ for $m, n \in \mathbb{N}$, and that $f$ is strictly monotone decreasing from the fact that $c \circ c < c$.

Next, it follows from the continuity and monotonicity of $\circ$ that there exist $0 < a_2 < 1$ such that $a_2 \circ a_2 = c$. In the same way, there exists $0 < a_n < 1$ such that

$$c = \underbrace{a_n \circ a_n \circ \cdots \circ a_n}_{n}.$$

Let us define the function $f : \mathbb{Q}^+ \to (0, 1)$ by

$$f\left(\frac{m}{n}\right) = \underbrace{a_n \circ a_n \circ \cdots \circ a_n}_{m},$$

where $\mathbb{Q}^+$ is the set of rational numbers.

It follows from the strict monotonicity that $f$ is well defined. In fact, we have $c = a_2 \circ a_2 = a_4 \circ a_4 \circ a_4 \circ a_4$. If $a_2 < a_4 \circ a_4$, then we have $a_2 \circ a_2 < a_4 \circ a_4 \circ a_2 < a_4 \circ a_4 \circ a_4 \circ a_4$. That is a contradiction.

It follows from the definition of $f$ that we have $f(p + q) = f(p) \circ f(q)$ for $p, q \in \mathbb{Q}^+$ and that $f$ is strictly monotone decreasing.

Now, let us consider irrational numbers $x$. In this case, there exists a sequence $x_n$ that tends to $x$. Let us define the function $f : [0, \infty) \to (0, 1]$ by $f(x) = \lim_{n \to \infty} f(x_n)$ and $f(0) = 1$. $f$ is well defined from the strict monotonicity and continuity of $\circ$, and we have $f(x + y) = f(x) \circ f(y)$ for $x, y \geq 0$, and that $f$ is strictly monotone decreasing. Then, if we define a function $\phi : (0, 1] \to [0, \infty)$ by $\phi = f^{-1}$, we have $\phi^{-1}(\phi(x) + \phi(y)) = x \circ y$ for $x, y \in (0, 1]$.

Finally, we can add 0 to the domain of $\phi$ with $\phi(0) = \infty$.

As we have just seen, the inequality given in Equation 4.3 and obtained from the cancellative condition plays an important role in the proof. Equation 4.3 corresponds to the condition for subidempotency of Archimedean t-norms.

We can prove the strictly monotone decreasing case of $\circ$ in a similar way. In such a proof, the inequality $x \circ x > x$ plays the central role. This corresponds to the superidempotency of Archimedean t-conorms. A proof similar to the one of Theorem 4.1 can be given for Theorems 2.46 and 2.49.

The function $\phi$ in Equation 4.2 is unique up to a linear transformation of the variable. That is, $\phi(x)$ might be replaced by $\phi(\alpha x)$ for $\alpha \neq 0$, but no other function is possible. Note that although $\circ$ was not supposed to be symmetric, this follows from Theorem 4.1. Therefore, associativity implies symmetry (i.e., $x \circ y = y \circ x$).

Associative operators have been extensively studied in the literature. Some examples are presented below. The t-norms and t-conorms given in Definitions 2.43 and 2.47 are other examples of such associative operators.

*Example 4.2.* Let $\gamma > 0$, let

$$\phi(a) = 1 - \frac{\gamma}{e^a - 1 + \gamma},$$

and, therefore, let

$$\phi^{-1}(a) = \log\left(\frac{\gamma}{1 - a} + 1 - \gamma\right).$$

Under these conditions, the Hamacher family of associative operators is defined as follows:

$$\mathbb{C}_\gamma(a, b) = \frac{a + b + (\gamma - 2)ab}{1 + (\gamma - 1)ab}.$$

Note that the operators in the Hamacher family do not satisfy unanimity ($\mathbb{C}_1(a, a) \neq a$). In fact, this function is a t-conorm (Definition 2.47) when values are restricted to $[0, 1]$, and, thus, $\max \preceq \mathbb{C}_2$. In the particular case of $\gamma = 1$, it reduces to the algebraic sum

$$\mathbb{C}_1(a, b) = a + b - ab.$$

For $\gamma = 2$, and considering values in $[-1, 1]$, this function corresponds to the rule for combining certainty factors in the expert system PROSPECTOR:

$$\mathbb{C}_2(a, b) = \frac{a + b}{1 + ab}. \tag{4.4}$$

This expression also corresponds to the rule for combining velocities in the theory of relativity.

For $\gamma = \infty$, we have that $\mathbb{C}_\infty$ corresponds to $\perp_{\max}$ (Definition 2.14).

### 4.1.1 Uninorms and Nullnorms

Uninorms and nullnorms are two other well-known families of associative operators. Nevertheless, it has to be said that, in general, these operators are not aggregation operators, in the sense that their outcome is not always between the minimum and the maximum. That is, they do not satisfy Equation 1.1 in Section 1.1:

$$\min(a_1, \ldots, a_N) \leq \mathbb{C}(a_1, \ldots, a_N) \leq \max(a_1, \ldots, a_N). \tag{4.5}$$

#### Uninorms

We now start reviewing some results concerning uninorms. Nullnorms are described in the next section.

**Definition 4.3.** *An operator $UN$ from $[0, 1]^N$ into $[0, 1]$ is a UniNorm if it is associative, symmetric, and has a neutral element $e$ in $(0, 1)$ (i.e., $UN(a, e) = a$).*

Although, as stated, uninorms do not, in general, satisfy Equation 4.5, this equation holds when the neutral element is such that

$$\min(a_1, \ldots, a_N) \leq e \leq \max(a_1, \ldots, a_N).$$

Equivalently, Equation 4.5 holds when there are some $a_i \geq e$ and some $a_i \leq e$.

An important result about uninorms is that two families of uninorms can be constructed in an easy way from t-norms and t-conorms. In short, the aggregation is defined conjunctively (in terms of a t-norm) when all the values to be aggregated are small (less than the neutral element), and disjunctively (in terms of a t-conorm) when all the values are large. In the remaining cases,

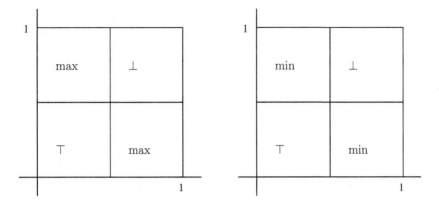

**Fig. 4.1.** Two families of uninorms

which correspond to conflicts between both large and small values, either the minimum or the maximum can be applied. One family of uninorms corresponds to the case of selecting the minimum and the other to the case of selecting the maximum.

Figure 4.1 gives a graphical representation of both families in the case where only two inputs are considered. Recall that, due to associativity, the definition for two inputs is enough for specifying an $N$ input uninorm. Proposition 4.4 gives a formalization of this graphical representation, and gives the proper construction for both families. It shows how to build uninorms from a given pair $(\top, \bot)$, where $\top$ is a t-norm and $\bot$ is a t-conorm.

**Proposition 4.4.** *Given a t-norm $\top$, a t-conorm $\bot$, and $e \in [0,1]$, the following two expressions are uninorms:*

$$UN_{e,\top,\bot}(a_1,\ldots,a_N) = \begin{cases} e \cdot \top(a_1/e,\ldots,a_N/e) & \text{if } \max a_i \le e \\ e + (1-e) \cdot \bot(\frac{a_1-e}{1-e},\ldots,\frac{a_N-e}{1-e}) & \text{if } \min a_i \ge e \\ \min(a_1,\ldots,a_N) & \text{otherwise.} \end{cases}$$
$$(4.6)$$

$$UN^{e,\top,\bot}(a_1,\ldots,a_N) = \begin{cases} e \cdot \top(a_1/e,\ldots,a_N/e) & \text{if } \max a_i \le e \\ e + (1-e) \cdot \bot(\frac{a_1-e}{1-e},\ldots,\frac{a_N-e}{1-e}) & \text{if } \min a_i \ge e \\ \max(a_1,\ldots,a_N) & \text{otherwise.} \end{cases}$$
$$(4.7)$$

Among the uninorms that can be expressed using the proposition above, two special cases are distinguished. They correspond to the case of the pair $(\top, \bot)$ where the t-norm $\top$ is the minimum and where the t-conorm $\bot$ is the maximum. We will denote these uninorms by $UN_e = UN_{e,\min,\max}$ and $UN^e = UN^{e,\min,\max}$.

*Example 4.5.* Let us consider five search engines, each returning a list of relevant URLs together with a relevance degree. Relevance degrees are given in the unit interval. Then, let us consider the merging of such results, taking into account the following:

- All items retrieved (by any of the search engines) will be in the final list.
- A new relevance degree will be computed for each item. This index is defined solely as the combination of the degrees assigned by the five search engines.
- Unavailable degrees will be considered as equal to zero.
- For combining the degrees we consider 0.6 as our central value. For values lower than 0.6, we require negative (towards zero) interaction. For values larger than 0.6, we require positive (towards one) interaction. In case of conflict, to avoid missclassifications, we will assign the largest value to the aggregation.

This situation can be modeled with a uninorm: $UN^{0.6,\top,\bot}$ for a t-norm $\top$ and a t-conorm $\bot$. For example, using Yager's t-norm and t-conorm (Examples 2.45 and 2.48, respectively) we have

$$
UN^{0.6,\top,\bot}(a_1,\dots,a_5) = \begin{cases} 0.6(1 - 1 \wedge (\sum_{i=1}^5 (1 - a_i)^w)^{1/w}) & \text{if } max_i a_i \leq 0.6 \\ 0.6 + 0.4(1 \wedge (\sum a_i^w)^{1/w}) & \text{if } min_i a_i \geq 0.6 \\ a_1 \vee a_2 \vee a_3 \vee a_4 \vee a_5 & \text{otherwise.} \end{cases}
$$
$$(4.8)$$

All uninorms constructed using Proposition 4.4 use a minimum or a maximum in the regions of conflict. That is, the minimum and the maximum are used to combine a pair $(a, b)$, where $a < e$ and $b > e$. Other uninorms can be defined so that such values are combined using an alternative function that is neither the maximum nor the minimum.

The following proposition establishes how to construct uninorms of this form.

**Proposition 4.6.** *Let $e$ be a value in $[0, 1]$, and let $h$ in $[0, 1]$ be a strictly increasing continuous mapping, with $h(0) = -\infty$, $h(e) = 0$, and $h(1) = +\infty$; then, the binary operator $UN$ defined by*

$$UN(a, b) = h^{-1}(h(x) + h(y)), \qquad (4.9)$$

*for all $(a, b)$ in $[0, 1] \times [0, 1] \setminus \{(0, 1), (1, 0)\}$, and for $(a, b)$ in $\{(0, 1), (1, 0)\}$ either $UN(0, 1) = UN(1, 0) = 0$ or $UN(1, 0) = UN(0, 1) = 1$, is a uninorm with neutral element $e$. Uninorms of this form are strictly increasing and continuous on $(0, 1)^2$.*

Therefore, according to this proposition, given any function $h$ satisfying the constraints above, we can construct two different uninorms using Equation 4.9.

One uninorm will have $UN(0, 1) = UN(1, 0) = 0$, and the other, $UN(0, 1) = UN(1, 0) = 1$.

The example below shows such a construction for a particular function $h$.

*Example 4.7.* Let $h_c(x)$ be defined for $x$ in $(0, 1)$ and for $c > 0$ as

$$h_c(x) = \log \left( -\frac{1}{c} \log(1 - x) \right).$$

Then, $h_c^{-1}(x)$ is

$$h_c^{-1}(x) = 1 - e^{-ce^x}.$$

Note that, for $h_c(x)$, the following hold: $h_c(0) = lim_{x \to 0} h_c(x) = -\infty$, $h_c(1) = lim_{x \to 1} h_c(x) = \infty$, and $h_c(1 - e^{-c}) = 0$.

Then, using Proposition 4.6, the following expression is a uninorm with neutral element $e_c = 1 - e^{-c}$:

$$UN_c(x, y) = 1 - e^{-\frac{(\log(1-x))(\log(1-y))}{c}}.$$

Another example of continuous uninorm is the following family of functions.

*Example 4.8.* Let $\lambda > 0$; then, the following expression is a uninorm with neutral element $e_\lambda = 1/(1 + \lambda)$:

$$D_\lambda(x, y) = \frac{\lambda x y}{\lambda x y + (1 - x)(1 - y)}.$$

**Proposition 4.9.** *The uninorm $D_\lambda$ satisfies:*

1. $D_{\lambda_1} \preceq D_{\lambda_2}$, *if* $\lambda_1 \leq \lambda_2$
2. $lim_{\lambda \to 0} D_\lambda = \top_{\min}$
3. $lim_{\lambda \to 1} D_\lambda = \bot_{\max}$
4. *The only self-dual operator in this family is* $D_1$. *That is,* $D_1(x, y) = 1 - D_1(1 - x, 1 - y)$.

*Example 4.10.* Let us reconsider Example 4.5. The use of maximum in the conflicting regions leads to uninorms with discontinuity. In particular, we have discontinuity in $(x, 0.6)$ for $x < 0.6$ (see Equation 4.8).

To avoid such discontinuities, we can use $D_\lambda(x, y)$. As the example requires the neutral element to be 0.6, this means that

$$e_\lambda = 0.6 = 1/(1 + \lambda).$$

So, $\lambda = 1/0.6 - 1 = 2/3$.

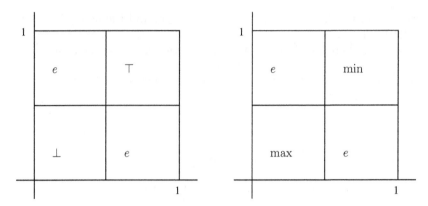

**Fig. 4.2.** A generic family of nullnorms (left) and $NN_{e,\min,\max}$ (right)

### Nullnorms

Nullnorms are operators that are also associative and symmetric; we denote them by $NN$. From the point of view of their properties, instead of having a neutral element $e$ (as uninorms have) they are characterized because there exists an element $e$ such that, for all $x \leq e$, $NN(0,x) = x$, and for all $x \geq e$, $NN(1,x) = x$. This is established in the following definition.

**Definition 4.11.** *An operator $NN$ from $[0,1]^N$ into $[0,1]$ is a NullNorm if it is associative, symmetric, and there is an element $e$ in $(0,1)$ such that*

- *for all $a \leq e$, $NN(0,a) = a$*
- *for all $a \geq e$, $NN(1,a) = a$*

The following result permits us to construct nullnorms from t-norms and t-conorms.

**Proposition 4.12.** *Let $\top$, $\bot$, and $e$ be a t-norm, t-conorm, and a value in $(0,1)$; then, the operator $NN_{e,\top,\bot}$,*

$$NN_{e,\top,\bot}(a_1,\ldots,a_N) = \begin{cases} e \cdot \bot(a_1/e,\ldots,a_N/e) & \text{if } \max a_i \leq e \\ e + (1-e) \cdot \top\left(\frac{a_1-e}{1-e},\ldots,\frac{a_2-e}{1-e}\right) & \text{if } \min a_i \geq e \\ e & \text{otherwise,} \end{cases}$$

(4.10)

*is a nullnorm.*

Figure 4.2 illustrates this proposition as well as the particular nullnorm $NN_{e,\min,\max}$ for the particular case of $N = 2$.

The following proposition shows that all nullnorms can be defined in terms of t-norms, t-conorms, the neutral element, and the median operator.

**Proposition 4.13.** *NN is a nullnorm if and only if there is a t-norm* $\top$, *a t-conorm* $\bot$, *and a neutral element* $e \in [0,1]$ *such that NN can be expressed as*

$$NN(a_1, \ldots, a_N) = M(\top(a_1, \ldots, a_N), \bot(a_1, \ldots, a_N), e)$$

*where M is the median operator.*

Recall that, the median of $(a, b, c)$ is the second largest value in $\{a, b, c\}$. Due to the fact that $\top(a_1, \ldots, a_N) \leq \bot(a_1, \ldots, a_N)$, this proposition is equivalent to considering the following three cases:

- When $a_i \leq e$ for all $a_i$, the output is $\bot(a_1, \ldots, a_N)$
- When $a_i \geq e$ for all $a_i$, the output is $\top(a_1, \ldots, a_N)$
- When some $a_i < e$ and some other $a_i > e$, the output is $e$

Therefore, nullnorms do not satisfy Equation 4.5 (internality). In fact, internality is only satisfied when there are $a_i < e$ and some other $a_i > e$. However, in this case the output is just $e$.

### Uninorms vs. nullnorms

Figure 4.3 illustrates that uninorms and nullnorms have complementary behavior. Note that, for $a, b \leq e$, the uninorm behaves conjunctively (i.e., it yields an output smaller than or equal to $\min(a, b)$) while the nullnorm behaves disjunctively (*i.e.*, it yields a value larger than or equal to $\max(a, b)$). For $a, b \geq e$, the effect is the complement. Therefore, the nullnorm tends to concentrate the outcome around $e$, while the uninorm tends to move away from $e$. This fact is illustrated with arrows, either attracting to $e$ (Figure 4.3, right) or moving away from $e$ (Figure 4.3, left).

This dual nature of uninorms, with conjunctive and disjunctive regions, has been exploited in some applications. For example, the operators for parallel combination of certainty factors in the expert systems MYCIN and PROSPECTOR were uninorms: the certainty factors were defined in $[-1, 1]$, and thus a transformation from $[-1, 1]$ to $[0, 1]$ is needed. In any case, these operators had a conjunctive behavior for values $x, y$ in $[-1, 0]$, and a disjunctive behavior for values $x, y$ in $[0, 1]$. For example, the combination function of MYCIN (in $[-1, 1]$) was defined as

$$\mathbb{C}(x, y) = \begin{cases} x + y - xy & \text{if } 0 \leq \min(x, y) \\ \frac{x+y}{1 - \min(|x|, |y|)} & \text{if } \min(x, y) < 0 < \max(x, y) \\ x + y + xy & \text{if } \max(x, y) \leq 0. \end{cases} \quad (4.11)$$

It can be observed that the expression used for combining positive values $x, y$ is the algebraic t-conorm. So, as stated, this combination function has a disjunctive behavior in this region.

Another example is the combination function in the expert system PROSPECTOR (Equation 4.4) for values in $[-1, 1]$. This function, when data is mapped into $[0, 1]$, corresponds to $D_\lambda$ with $\lambda = 1$.

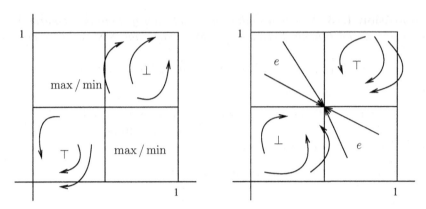

**Fig. 4.3.** Comparison between uninorms (left) and nullnorms (right)

## 4.2 Separability: the Quasi-arithmetic Means

In this section we review some results that rely on associativity. In particular, we study aggregation operators $\mathbb{C}$ that are separable. Informally, $\mathbb{C}$ is separable when the influence of each variable can be made explicit and translated into a new space where no interactions are considered. Then, as there are no interactions in such a space, values can be combined with an associative operator. This is formally stated as follows.

**Definition 4.14.** $\mathbb{C}(a_1, \ldots, a_N)$ *is separable when there exist functions* $g_1, \ldots, g_N$ *and* $\circ$ *such that*

$$\mathbb{C}(a_1, \ldots, a_N) = g_1(a_1) \circ g_2(a_2) \circ \cdots \circ g_N(a_N),$$

*with* $\circ$ *a continuous, associative, and cancellative operator.*

Note that $g_i(a_i)$ only depends on the $i$th source (the source $x_i$ that delivers $a_i$) and on the value $a_i$ supplied by it. As will be seen later on, the weighted mean is an example of a separable function, where the $g_i(a_i)$ is the value $a_i$ supplied by source $x_i$ prorated by a weight associated with $x_i$.

**Proposition 4.15.** *An operator* $\mathbb{C}$ *is separable in terms of monotone increasing functions* $g_1, \ldots, g_N$, *and a continuous, associative, and cancellative* $\circ$, *if and only if it is of the form*

$$\mathbb{C}(a_1, \ldots, a_N) = \phi^{-1}\left(\sum_{i=1}^{N} \phi(g_i(a_i))\right) = \phi^{-1}\left(\sum_{i=1}^{N} \phi_i(a_i)\right),$$

*where* $\phi_i(a) = \phi(g_i(a))$ *for all* $i = 1, \ldots, N$, *and* $\phi^{-1}$ *is the inverse function of* $\phi$.

This proposition is based on Theorem 4.1 (on associativity).

Now, we turn to the where that all influences are measured in the same way. That is, when $g_i = g_j = g$ for all $i, j$. In this case, the following proposition holds.

**Proposition 4.16.** *An operator $\mathbb{C}$ is separable in terms of a unique monotone increasing function $g$ if and only if it is of the form*

$$\mathbb{C}(a_1, \ldots, a_N) = \phi^{-1}(\sum_{i=1}^{N} \phi(g(a_i))). \tag{4.12}$$

Then, in aggregation, we expect that the operator satisfy unanimity, as we expect that, when all information sources supply the same information (i.e., agree on a given value), the outcome is this very value:

$$\mathbb{C}(a, \ldots, a) = a. \tag{4.13}$$

This leads to the following result:

**Proposition 4.17.** *An operator $\mathbb{C}$ is separable in terms of a unique monotone increasing $g$ and satisfies unanimity if and only if is of the form:*

$$\mathbb{C}(a_1, \ldots, a_N) = \phi^{-1}(\frac{1}{N} \sum_{i=1}^{N} \phi(a_i)). \tag{4.14}$$

*Proof.* The proof of this proposition is based on the unanimity condition: $a = \mathbb{C}(a, \ldots, a) = \phi^{-1}(\sum_{i=1}^{N} \phi(g(a)))$. Therefore, $\phi(a) = N\phi(g(a))$ and, thus, $g(a)$ correspond to $\phi^{-1}(\phi(a)/N)$. By replacing $g(a)$ in Equation 4.12 by this expression, the proposition is proved. $\square$

Aggregation operators following Expression 4.14 are the quasi-arithmetic means (also known as generalized $\phi$-means). Note that they have the form of an arithmetic mean $\frac{1}{N}(\sum_{i=1}^{N} b_i)$ once data is mapped by $\phi : I \to J$ into the space $J$ ($b_i = \phi(a_i)$). Then, after aggregation the result is mapped back (by $\phi^{-1}$) into the space $I$. The quasi-arithmetic mean is a family that encompasses several well-known aggregation operators. Different $\phi$ lead to different operators. In particular, $\phi(x) = x$ yields the arithmetic mean, $\phi(x) = \log(x)$ yields the geometric mean, and $\phi(x) = 1/x$ leads to the harmonic mean. Table 4.1 gives the main quasi-arithmetic means. Trigonometric means, which correspond to $\phi$ equal to sin, cos, or tan, are not displayed, although they have also been studied in the literature.

A few properties have been proved for such means. Some of them, corresponding to characterizations of the means, are given in the next sections. A well-known property is that the harmonic ($HM$), the geometric ($GM$), and the arithmetic mean ($AM$) satisfy the following relation:

| Name | Generator function | $\mathbb{C}(a_1, \ldots, a_N)$ |
|------|-------------------|-------------------------------|
| Arithmetic mean | $\phi(x) = x$ | $\frac{\sum_{i=1}^{N} x_i}{N}$ |
| Geometric mean | $\phi(x) = \log x$ | $\sqrt[N]{\frac{\sum_{i=1}^{N} x_i}{N}}$ |
| Harmonic mean | $\phi(x) = 1/x$ | $\frac{N}{\sum_{i=1}^{N} \frac{1}{x_i}}$ |
| Root-mean-square | $\phi(x) = x^2$ | $\sqrt{\frac{\sum_{i=1}^{N} x_i^2}{N}}$ |
| Root-mean-power | $\phi(x) = x^\alpha$ | $\sqrt[\alpha]{\frac{\sum_{i=1}^{N} x_i^\alpha}{N}}$ |
| Exponential mean | $\phi(x) = e^x$ | $\log\left(\frac{\sum_{i=1}^{N} e^{x_i}}{N}\right)$ |
| Radical mean | $\phi(x) = c^{1/x}$ | $\left(\log_c\left(\frac{\sum_{i=1}^{N} c^{1/x_i}}{N}\right)\right)^{-1}$ |
| Basis-exponential mean | $\phi(x) = x^x$ | $m$ s.t. $m^m = \frac{\sum_{i=1}^{N} x_i^{x_i}}{N}$ |
| Basis-radical mean | $\phi(x) = x^{1/x}$ | $m$ s.t. $m^{1/m} = \frac{\sum_{i=1}^{N} x_i^{1/x_i}}{N}$ |

**Table 4.1.** Main quasi-arithmetic means. For basis-exponential and basis-radical mean, the values should satisfy $x_i \geq 1/e$

$$HM \preceq GM \preceq AM.$$

Figure 4.4 illustrates this property for two inputs. Note that, for a pair $a, b$, $AM$ represents $(a + b)/2$, $GM$ represents $\sqrt{ab}$ and $HM$ corresponds to $2ab/(a + b)$. The segment $(a - b)/2$ that is used for the computation of $GM$ is also given in the figure.

Root-mean-powers (also known as the $r$th power mean or generalized mean) are another example of quasi-arithmetic means. They are obtained from Equation 4.14 with a generating function of the form $\phi(x) = x^\alpha$. We will use $RMP_\alpha$ to denote the operators for a particular $\alpha$. So, according to the previous definition:

$$RMP_\alpha(a_1, \ldots, a_N) = \left(\frac{1}{N} \sum_{i=1}^{N} a_i^\alpha\right)^{1/\alpha}.$$

The following proposition can be proved for root-mean-powers:

**Proposition 4.18.** *Let $RMP_\alpha$ be the root-mean-power with parameter $\alpha$. Then, the following holds*

- $RMP_r \preceq RMP_s$ *for $r < s$*

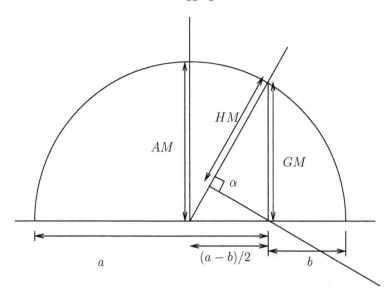

**Fig. 4.4.** Graphical representation that for a pair $(a, b)$ such that, if $a > b$, it holds that $HM(a, b) \leq GM(a, b) \leq AM(a, b)$, where $HM$ stands for the harmonic mean, $GM$ stands for the geometric mean, and $AM$ stands for the arithmetic mean

- $\lim_{\alpha \to 0} RMP_\alpha = GM$
- $\lim_{\alpha \to \infty} RMP_\alpha = \max$
- $\lim_{\alpha \to -\infty} RMP_\alpha = \min$
- *for $\alpha > 1$, we have (Minkowski's inequality)*

$$\Big( \sum_{i=1}^{N} (a_i + b_i)^\alpha \Big)^{1/\alpha} \leq \Big( \sum_{i}^{N} a_i^\alpha \Big)^{1/\alpha} + \Big( \sum_{i}^{N} b_i^\alpha \Big)^{1/\alpha},$$

*and, for $\alpha < 1$, we have*

$$\Big( \sum_{i}^{N} (a_i + b_i)^\alpha \Big)^{1/\alpha} \geq \Big( \sum_{i}^{N} a_i^\alpha \Big)^{1/\alpha} + \Big( \sum_{i}^{N} b_i^\alpha \Big)^{1/\alpha}.$$

*Naturally, equality holds for $\alpha = 1$.*

## 4.3 Aggregation and Measurement Scales

When measurement scales are taken into consideration, the outcome of a function should be *consistent* with the scale, and changes to the scale should lead to *consistent* changes on the outcome. For example, if we are aggregating

| Name | $\alpha$ | $\mathbb{C}(a_1, \ldots, a_N)$ |
|------|----------|-------------------------------|
| Arithmetic mean | $\alpha = 1$ | $\frac{\sum_{i=1}^{N} x_i}{N}$ |
| Root-mean-square | $\alpha = 2$ | $\sqrt{\frac{\sum_{i=1}^{N} x_i^2}{N}}$ |
| Harmonic mean | $\alpha = -1$ | $\frac{N}{\sum_{i=1}^{N} \frac{1}{x_i}}$ |

**Table 4.2.** Some root-mean-powers

data representing the execution time of a particular program on different computers, the outcome of the aggregation should not depend on whether the time is expressed in hours, minutes, or seconds.

Such an idea of consistency is formalized in terms of the permissible transformations in the desired scale. Then, when a function behaves properly with respect to a set of transformations $\Phi$, we say that the function is $\Phi$-invariant or $\Phi$-meaningful. This concept is defined below.

**Definition 4.19.** *Let $\Phi$ be a set of transformations on $E$; then, a function $\mathbb{C} : E^N \to E$ is $\Phi$-invariant (or $\Phi$-meaningful or $\Phi$-ordinally stable) under $\Phi$ if, for any $\phi \in \Phi(E)$ and any $x_i \in E^N$, we have*

$$\mathbb{C}(\phi(x_1), \ldots, \phi(x_N)) = \phi(\mathbb{C}(x_1, \ldots, x_N)).$$

Now, we study aggregation operators in some particular scales. The operators should be invariant to permissible transformations. For example, in the case of ratio scales, aggregation should be invariant to transformations of the form $\phi(x) = rx$ for positive $r$. This property, known as positive homogeneity, is matematically expressed as follows:

$$\mathbb{C}(ra_1, \ldots, ra_N) = r\mathbb{C}(a_1, \ldots, a_N) \tag{4.15}$$

for $r > 0$

We will consider below not only conditions corresponding to permissible transformations, but also others that seem reasonable for aggregation. This is the case for unanimity.

Let us turn again into positive homogeneity. We will give below some results satisfying this property.

**Proposition 4.20.** *An operator $\mathbb{C}$ is separable in terms of a unique monotone increasing $g$ and satisfies unanimity and positive homogeneity if and only if $\mathbb{C}$ is either the root-mean-power with parameter $\alpha \neq 0$ (RMP$_\alpha$) or the geometric mean.*

| Runtime system | 227-mtr | 202-jess | 201-compress | 209-db | 222-mpegaudio | 228-jack | 213-javac | GM |
|---|---|---|---|---|---|---|---|---|
| Sun JDK 1.5.0 Client VM | 325 | 221 | 204 | 43.4 | 251 | 192 | 96.1 | 162.13 |
| Sun JDK 1.4.2 Client VM | 318 | 186 | 199 | 43.6 | 249 | 181 | 90.9 | 154.51 |
| Kaffe | 32 | 21.3 | 191 | 24.8 | 101 | 32.9 | 21.3 | 41.95 |

**Table 4.3.** Performance comparison on a Pentium 4 computer. Execution times of the seven benchmark programs in SPEC JVM98 for three Java runtime systems. Times are given in seconds. The average time using the geometric mean (GM)

Recall that according to Proposition 4.18, the geometric mean corresponds to $\lim_{\alpha \to 0} RMP_\alpha$.

Table 4.2 gives some of the aggregation functions that can be expressed as root-mean-powers for some values of $\alpha$.

*Example 4.21.* Let us consider the problem of assessing the performance of some Java runtime systems. To do so, each program from the SPEC JVM98 is executed on each runtime system. Then, such execution times should be aggregated. Table 4.3 represents the execution times of the seven benchmark programs in SPEC JVM98 for three java runtime systems.

Then, when a separable operator $\mathbb{C}$ is used for computing the mean value, $\mathbb{C}$ should be either the root-mean-power or the geometric mean, as $\mathbb{C}$ should satisfy unanimity and positive homogeneity. This is because, when execution times are equal, the outcome should correspond to such a value, and because the value should be consistent when time is represented in seconds, minutes, or any other unit.

Table 4.3 gives the performance of the runtime systems when the aggregation function is the geometric means.

Another property is reciprocity, which might be required in ratio scales. It is expressed as follows:

$$\mathbb{C}(1/a_1, \ldots, 1/a_N) = 1/\mathbb{C}(a_1, \ldots, a_N). \qquad (4.16)$$

It is illustrated with an example.

*Example 4.22.* Let us consider the case of assessing the relative importance of two criteria (price vs. comfort) for buying a particular product. Then, if $a_1, \ldots, a_5$ are the subjective evaluations from five members of the same family, we can assess their relative importance by $\mathbb{C}(a_1, \ldots, a_5)$.

Reciprocity establishes in this setting that the aggregation is consistent, considering either price vs. comfort (when values $a_i$ are used) or comfort vs. price (when values $1/a_i$ are used).

When reciprocity is added to unanimity and separability, the following result can be proved:

**Proposition 4.23.** *An operator* $\mathbb{C}$ *is separable in terms of a unique monotone increasing g and satisfies unanimity and reciprocity (Equation 4.16) if and only if* $\mathbb{C}$ *is of the form*

$$\mathbb{C}(a_1,\ldots,a_N) = \exp\omega^{-1}(\frac{1}{N}\sum_{i=1}^{N}\omega(\log(a_i))), \qquad (4.17)$$

*with* $\omega$ *an arbitrary odd function. Here* $\exp$ *is the exponential function, and* $\exp a$ *corresponds to* $e^a$.

Now, we consider, at the same time, positive homogeneity and reciprocity.

**Proposition 4.24.** *An operator* $\mathbb{C}$ *is separable in terms of a unique monotone increasing g and satisfies unanimity, positive homogeneity (Equation 4.15), and reciprocity (Equation 4.16) if and only if* $\mathbb{C}$ *is the geometric mean:*

$$\mathbb{C}(a_1,\ldots,a_N) = (\prod_{i=1}^{N} a_i)^{1/N}.$$

In Section 2.1.3, it was said that the permissible transformations in interval scales are of the form $\psi(x) = \alpha x + \beta$. When the data to be aggregated belongs to an interval scale, the following proposition can be applied. Note that here we drop the condition about separability.

**Proposition 4.25.** *For* $\alpha > 0$ *and* $\beta \in \mathbb{R}$, *such that* $\alpha a_i + \beta \in [0,1]$, *an operator* $\mathbb{C}$ *satisfies*

$$\mathbb{C}(\alpha a_1 + \beta,\ldots,\alpha a_N + \beta) = \alpha\mathbb{C}(a_1,\ldots,a_N) + \beta$$

*if and only if there exists an operator* $\mathbb{C}'$ *such that, for all* $(a_1,\ldots,a_N)$ *in* $[0,1]$, *we have*

$$\mathbb{C}(a_1,\ldots,a_N) = \begin{cases} a & \text{if } a_i = a \text{ for all } i \\ (a^* - a_*)\mathbb{C}'\left(\frac{a_1-a_*}{a^*-a_*},\ldots,\frac{a_N-a_*}{a^*-a_*}\right) + a_* & \text{otherwise,} \end{cases}$$
$$(4.18)$$

*where* $a^* = \max a_i$ *and* $a_* = \min a_i$.

**Proposition 4.26.** *For* $\alpha > 0$ *and* $\beta \in \mathbb{R}$, *such that* $\alpha a_i + \beta \in [0,1]$, *an operator* $\mathbb{C}$ *satisfies*

$$\mathbb{C}(\alpha a_1 + \beta,\ldots,\alpha a_N + \beta) = \alpha\mathbb{C}(a_1,\ldots,a_N) + \beta$$

*if and only if there exists an operator* $\mathbb{C}'$ *such that, for all* $(a_1,\ldots,a_N)$ *in* $[0,1]^N$, *we have*

$$\mathbb{C}(a_1,\ldots,a_N) = \begin{cases} a & \text{if } a_i = a \text{ for all } i \\ \bar{a} + \sigma\mathbb{C}'\left(\frac{a_1-\bar{a}}{\sigma},\ldots,\frac{a_N-\bar{a}}{\sigma}\right) & \text{otherwise,} \end{cases} \qquad (4.19)$$

*where* $\bar{a} = \sum_i^N a_i/N$ *and* $\sigma = \sqrt{\sum_{i=1}^{N}((a_i - \bar{a})^2)/N}$.

### 4.3.1 Ordinal Scales

In the previous section, we have considered interval and ratio scales. We consider here ordinal scales. This corresponds to the case in which permissible transformations are functions $\phi$ such that $x > y$ implies $\phi(x) > \phi(y)$.

We start by defining the boolean max-min functions. They are used later in this section.

**Definition 4.27.** *Let* $X = \{x_1, \ldots, x_N\}$ *be the set of information sources (reference set), let* $a_i = f(x_i)$ *be the value supplied by source* $x_i$, *and let* $\mathcal{S} = \{S_i\}_{i=1}^m$ *be a family of subsets of* $X$. *Then, the boolean max-min function of* $\mathcal{S}$ *is defined by*

$$B_{\mathcal{S}}(a_1, \ldots, a_N) = \vee_{i=1}^m \wedge_{x_j \in S_i} f(x_j).$$

Note that $f(x_i)$ is only required to take values in a totally ordered set. So, it is valid for $f$ to be a function from $X$ into $L$, where $L$ is a set of ordered categories.

In the following, however, we will assume that the values belong to a real interval $E$, and then consider $\phi$, the group of all increasing bijections $\phi : E \to E$ (an automorphism of $E$). So, we will consider $\Phi$-invariance (Definition 4.19) under increasing bijections.

*Example 4.28.* Let us consider the problem of evaluating students on the basis of their marks in five different subjects: three science and two humanities. The three science subjects are *Mathematics (ML)*, *Physics (P)*, and *Mathematical Logic (ML)*, and the two humanities subjects are *Literature (L)* and *Greek (G)*.

We will consider good students those who have a good mark in at least two scientific and one of the other subjects. The marks are represented by an application $f$ into $[0, 1]$ (the larger the better).

This situation can be modeled using the boolean max-min function given below. It corresponds to the boolean max-min function of $\mathcal{S}$, with $\mathcal{S}$ defined by $\mathcal{S} = \{S_i\}_{i=1}^6$ and $S_1 = \{M, P, L\}$, $S_2 = \{M, ML, L\}$, $S_3 = \{P, ML, L\}$, $S_4 = \{M, P, G\}$, $S_5 = \{M, ML, G\}$, and $S_6 = \{P, ML, G\}$. Note that the sets $S_i$ are the ones with two scientific subjects and either Literature or Greek:

$$\big(f(M) \wedge f(P) \wedge f(L)\big) \vee \big(f(M) \wedge f(ML) \wedge f(L)\big) \vee \big(f(P) \wedge f(ML) \wedge f(L)\big) \vee$$

$$\big(f(M) \wedge f(P) \wedge f(G)\big) \vee \big(f(M) \wedge f(ML) \wedge f(G)\big) \vee \big(f(P) \wedge f(ML) \wedge f(G)\big)$$

Besides $\Phi$-invariance, the property of $\Phi$-comparison is also of interest.

**Definition 4.29.** *A function* $\mathbb{C} : E^N \to \mathbb{R}$ *is* $\Phi$-*comparison meaningful if, for any* $\phi \in \Phi(E)$ *and any* $x_i, x_i' \in E^N$, *we have that*

$$\mathbb{C}(x_1, \ldots, x_N) = \mathbb{C}(x_1', \ldots, x_N') \text{ implies}$$

$$\mathbb{C}(\phi(x_1), \dots, \phi(x_N)) = \mathbb{C}(\phi(x_1'), \dots, \phi(x_N')).$$

The two properties are related, as shown by the following result.

**Proposition 4.30.** *Let* $\mathbb{C} : E^N \to E$ *be a function satisfying unanimity; then,* $\mathbb{C}$ *is* $\Phi$*-comparison meaningful if and only if it is* $\Phi$*-invariant. Invariance implies meaningfulness, and meaningfulness with unanimity implies invariance.*

Now, we give some results directly concerning aggregation operators. The first one shows that $B_S$ is the right operator in ordinal scales when an operator with unanimity and monotonicity is required. The characterization requires that $E$ open.

**Proposition 4.31.** *For open sets* $E$, *an operator* $\mathbb{C} : E^N \to \mathbb{R}$ *satisfies unanimity, satisfies increasing monotonicity in each argument, and is* $\Phi$*-comparison meaningful if and only if there exists a family* $S$ *of subsets of* $X$ *such that* $\mathbb{C} = B_S$.

An alternative characterization exists that avoids $E$ being open. In this case, $\mathbb{C}$ should be continuous.

**Proposition 4.32.** *A continuous operator* $\mathbb{C} : E^N \to \mathbb{R}$ *satisfies unanimity and is* $\Phi$*-comparison meaningful if and only if there exists a family* $S$ *of subsets of* $X$ *such that* $\mathbb{C} = B_S$.

Now, when symmetry is also required, the permitted operators are restricted to be order statistics. Again, two propositions can be proved, one requiring the symmetric function to be continuous and the other requiring increasing monotonicity in each argument at the expense of $E$ being open. Only the second one is given here.

**Proposition 4.33.** *For open sets* $E$, *an operator* $\mathbb{C} : E^N \to \mathbb{R}$ *satisfies symmetry, increasing monotonicity in each argument, and unanimity, and is* $\Phi$*-comparison meaningful if and only if there exists a* $k \in \{1, \dots, N\}$ *such that* $\mathbb{C} = x_{s(k)}$, *where* $x_{s(k)}$ *corresponds to the kth order statistic.*

It is known that, for an odd $N$, one of the order statistics $(OS_{(N+1)/2})$ is the median. We now give a characterization of the median for odd values of $N$. The characterization uses the invariance of the aggregation with respect to a decreasing bijection. As this decreasing bijection can be understood as a kind of negation or complement, we will denote it by *neg*.

**Definition 4.34.** *Let* $neg : E \to E$ *be a decreasing bijection; then, the function* $\mathbb{C} : E^N \to \mathbb{R}$ *is neg-stable if, for any* $x, x' \in E$, *we have that* $f(x_1, \dots, x_N) = f(x_1', \dots, x_N')$ *implies*

$$f(neg(x_1), \dots, neg(x_N)) = f(neg(x_1'), \dots, neg(x_N')).$$

As in the previous cases, two characterizations of the median can be obtained. We offer the one that considers the monotonicity of the arguments.

**Proposition 4.35.** *Let $N$ be odd, and let neg be a decreasing bijection; then, the operator $\mathbb{C} : E^N \to \mathbb{R}$ fulfills symmetry, increasing monotonicity in each argument, and unanimity, and is $\Phi$-comparison meaningful and neg-stable if and only if $\mathbb{C}$ is the median.*

*Example 4.36.* Let us consider again the problem of evaluating students on the basis of the five subjects in Example 4.28, with marks represented by $f$ into $[0, 1]$. Then, when no relevance is given to any subject and unanimity is considered, any order statistic might be selected for aggregating the marks.

When any symmetry can be found in the range of marks (e.g., an average value of $a$ can be compensated by $1 - a$), let the function *neg* express this symmetry. In this case, the assumption of *neg*-stability implies that the aggregation operator is the median.

Informally, the *neg*-stability corresponds to the fact that, when two students have the same average, that average should be kept equal regardless of whether we are considering their original marks or the *symmetric* ones (i.e., regardless of whether we are considering $f(subject)$ or $1 - f(subject)$).

### 4.3.2 Different Data in Different Scales

Section 4.3 is devoted to the consideration of different scales (nominal, ratio, or ordinal). Nevertheless, we have restricted ourselves to the case where all values are described using the same scale, and, thus, the permissible transformations are the same for all values. There are some results in the literature weakening these constraints. That is, operators that permit different data to be represented in different scales. In such a case, the results should be consistent under the changes of scale of the data sources. We consider below one of the results obtained in this context. We focus on the case of $N$ data from $N$ different ratio scales.

**Proposition 4.37.** *Let us consider $N$ evaluations of a given object using $N$ ratio scales with independent units. Then, an operator $\mathbb{C}$ that satisfies unanimity and symmetry is meaningful to all the scales if and only if $\mathbb{C}$ is the geometric mean:*

$$\mathbb{C}(a_1, \ldots, a_N) = \Big( \prod_{i=1}^{N} a_i \Big)^{1/N}.$$

The following example illustrates this proposition.

*Example 4.38.* Let us consider the cities of Cirat and València and their *proximity* to Barcelona. Let us consider the distance of these two cities to Barcelona with respect to three scales: km, miles, and hours. We will consider them as ratio scales. The values under consideration for these distances are as follows:

- Distance(Barcelona, València) = 349 km, 216.85855 miles, 3.916 h (3h 55')
- Distance(Barcelona, Cirat) = 335 km, 208.16 miles, 4.016 (4h 01')

To determine the nearest city, we will aggregate the values using the result of Proposition 4.37, as although the scales are ratio scales, the concrete scales (i.e., km, miles, and hours) are different. So, the aggregation operators should be the geometric mean. The results of the evaluation for the two cities are as follows:

- València: $(349 \cdot 216.85855 \cdot 3.916)^{1/3} = 66.67273$
- Cirat: $(335 \cdot 208.16 \cdot 4.016)^{1/3} = 65.39263$

So, Cirat is the nearest city.

Now, let us consider some transformations on the last scale. That is, let us reconsider the distance expressed in hours. If the scale units are changed from hours to minutes, we have València at 235 minutes and Cirat at 241 minutes, and if we consider the scale in seconds, then València at 14,100 seconds and Cirat at 14,460 seconds from Barcelona. Using the geometric mean, we get the following

Scale in minutes:
- València: $(349 \cdot 216.85855 \cdot 235)^{1/3} = 261.02972$
- Cirat: $(335 \cdot 208.16 \cdot 241)^{1/3} = 256.1453$

Scale in seconds:
- València: $(349 \cdot 216.85855 \cdot 14100)^{1/3} = 1021.8968$
- Cirat: $(335 \cdot 208.16 \cdot 14460)^{1/3} = 1002.7749$

So, naturally, in both cases, Cirat is still the nearest city.

We now show that the use of another aggregation operator would modify our conclusions when the scale changes. In particular, let us consider the arithmetic mean of the three values.

Scale in hours:
- València: $(349 + 216.85855 + 3.916)/3 = 189.92485$
- Cirat: $(335 + 208.16 + 4.016)/3 = 182.39201$

Scale in minutes:
- València: $(349 + 216.85855 + 235)/3 = 266.95285$
- Cirat: $(335 + 208.16 + 241)/3 = 261.3867$

Scale in seconds:
- València: $(349 + 216.85855 + 14100)/3 = 4888.6196$
- Cirat: $(335 + 208.16 + 14460)/3 = 5001.053$

So, in this case, while the time required for traveling was considered in hours or minutes, the nearest city was Cirat. Nevertheless, this is not the case when the the last scale is expressed in seconds. So, the use of arithmetic mean is not sound under changes of scale. As Proposition 4.37 states, the only sound operation is the geometric mean, as it is the only one that ensures that changes in the scales do not alter the order inferred from the outcome.

## 4.4 Weighted Means

The aggregation operators reviewed so far do not include any kind of weight. In this section, we introduce weight. The typical operator is the weighted mean. In this operator, weights are used to represent importances of the information sources.

Proposition 4.17 characterizes the quasi-arithmetic mean, which corresponds to the case in which all functions $g_i$ in Definition 4.14 are equal ($g_i = g$ for all $i$). When this restriction does not apply, we have weighted aggregation operators. We characterize these operators below. Such characterization requires the sensitivity of $\mathbb{C}$ instead of requiring the cancellativity of $\circ$. Sensitivity is defined as follows.

**Definition 4.39.** *An operator $\mathbb{C}$ is sensitive if, for all $k = 1, \ldots, N$, when $a_k \neq a'_k$,*

$$\mathbb{C}(a_1, \ldots, a_{k-1}, a_k, a_{k+1}, \ldots, a_N) \neq \mathbb{C}(a_1, \ldots, a_{k-1}, a'_k, a_{k+1}, \ldots, a_N).$$

That is, changes in one parameter imply changes in the outcome. This property implies the cancellativity of $\circ$.

**Proposition 4.40.** *A sensitive operator $\mathbb{C}$ is separable in terms of functions $g_i$ and satisfies unanimity if and only if it is of the form*

$$\mathbb{C}(a_1, \ldots, a_N) = \phi^{-1}(\sum_{i=1}^{N} \phi_i(a_i)), \tag{4.20}$$

*with $\phi(x) = \sum_{i=1}^{N} \phi_i(x)$.*

From this proposition we obtain results similar to the ones reported above when $g_1 = \cdots = g_N = g$.

**Proposition 4.41.** *A sensitive operator $\mathbb{C}$ is separable in terms of monotone increasing $g_i$ and satisfies unanimity and positive homogeneity (Equation 4.15) if and only if $\mathbb{C}$ is either the weighted root-mean-power,*

$$\mathbb{C}(a_1, \ldots, a_N) = (\sum_{i=1}^{N} p_i a_i^\alpha)^{1/\alpha},$$

*or the weighted geometric mean,*

$$\mathbb{C}(a_1, \ldots, a_N) = \prod_{i=1}^{N} a_i^{p_i},$$

*with $p_i \neq 0$, $\alpha_i \neq 0$, and $\sum_{i=1}^{N} p_i = 1$, but otherwise arbitrary constants.*

**Proposition 4.42.** *A sensitive operator* $\mathbb{C}$ *is separable in terms of monotone increasing* $g_i$ *and satisfies unanimity and reciprocity (Equation 4.16) if and only if* $\mathbb{C}$ *is of the form*

$$\mathbb{C}(a_1, \ldots, a_N) = \exp \omega^{-1}\left(\sum_{i=1}^{N} \omega_i(\log(a_i)))\right), \tag{4.21}$$

*with* $\omega_i$ *arbitrary, continuous, strictly monotone odd functions, and* $\omega(t) = \sum_{i=1}^{N} \omega_i(t)$ *also strictly monotone.*

**Proposition 4.43.** *A sensitive operator* $\mathbb{C}$ *is separable in terms of monotone increasing* $g_i$ *and satisfies unanimity, positive homogeneity (Equation 4.15), and reciprocity (Equation 4.16) if and only if* $\mathbb{C}$ *is the weighted geometric mean,*

$$\mathbb{C}(a_1, \ldots, a_N) = \prod_{i=1}^{N} a_i^{p_i},$$

*with* $p_i \neq 0$, *and* $\sum_{i=1}^{N} p_i = 1$, *but otherwise arbitrary constants.*

The weighted geometric mean $(WGM)$ and the weighted mean $(WM)$ satisfy $WGM_{\mathbf{p}} \preceq WM_{\mathbf{p}}$ for any weighting vector $\mathbf{p}$.

### 4.4.1 Bajraktarević's Means

Bajraktarević's means are a family of operators that generalize weighted means. Their definition has resemblances with a quasi-arithmetic mean with weights (the quasi-weighted mean or quasi-linear mean). While in the weighted mean, $\sum_i p_i a_i$, the weight $p_i$ is constant for all $a_i$, in a Bajraktarević's mean the weight is a function of the $a_i$. Such a function is expressed by $\pi_i(x_i)$ in the next definition.

**Definition 4.44.** *Given functions* $\pi$ *and* $\phi$ *(where* $\phi$ *is monotone increasing with inverse* $\phi^{-1}$ *and* $\pi$ *nonnegative), the Bajraktarević's mean is defined as follows:*

$$\mathbb{C}(a_1, \ldots, a_N) = \phi^{-1}\left(\frac{\sum_{i=1}^{N} \pi_i(a_i)\phi(a_i)}{\sum_{i=1}^{N} \pi_i(a_i)}\right).$$

The Bajraktarević's mean becomes the quasi-weighted mean when $\pi_i(a_i)$ is a constant that solely depends on $i$ ($\pi_i(a_i) = p_i$), and the quasi-arithmetic mean when $\pi_i(a_i) = k$ for all $i$. Families of quasi-weighted means can be generated applying the functions in Table 4.1.

Another example is when $\pi_i(a) = a^q$ and $\phi(a) = a^{p-q}$ for $p > q$ (and $\phi(a) = \log a$ for $p = q$). In this case, we have that the Bajraktarević's mean reduces to

$$\mathbb{C}(a_1,\ldots,a_N) = \left(\frac{\sum_{i=1}^{N} a_i^p}{\sum_{i=1}^{N} a_i^q}\right)^{1/p-q} \tag{4.22}$$

for $p > q$, or to

$$\mathbb{C}(a_1,\ldots,a_N) = \left(\prod_{i=1}^{N} a_i^{a_i^p}\right)^{\frac{1}{\sum_{i=1}^{N} a_i^p}}$$

for $p = q$.

These means can be further particularized to obtain a root-mean-power and a counter-harmonic mean, which corresponds to Equation 4.22 with $q = p - 1$. That is,

$$\mathbb{C}(a_1,\ldots,a_N) = \frac{\sum_{i=1}^{N} a_i^p}{\sum_{i=1}^{N} a_i^{p-1}}.$$

Note that this mean can be understood as the mean of $a_i$ with weights $a_i^{p-1}$.

## 4.5 Bibliographical Notes

1. **Associativity:** Associativity has been extensively studied. Aczél (in [4]) gives detailed information on the authors who have dealt with associative operators. Aczél in [2] presents one of the characterizations. The book by Alsina, Frank, and Sklar [17] is entirely devoted to operators that satisfy this equation, for example, as t-norms. Alternatively, Sander in [346] reviews several results on associative aggregation operators. Other results can be obtained from papers on specific operators, for example, papers on t-norms, t-conorms, uninorms, nullnorms, and so on. See below for some references on uninorms and nullnorms. Section 2.3.1 was devoted to t-norms and t-conorms. See the corresponding bibliographical section for some references on these operators.

   The Hamacher family is described in several papers, e.g., [73]. The original definition seems to be in [180, 210].

   The expression for combining velocities in the theory of relativity appears in Einstein's work [117].

2. **Combination of certainty factors:** The expert system MYCIN is described in [48] and [360]. MYCIN introduces certainty factors and their parallel and sequential combination functions. The parallel combination is given in Equation 4.11. Tsadiras and Margaritis [419] prove that this function is a uninorm. Later, De Baets and Fodor [89] introduce uninorms in an arbitrary domain $[a, b]$, and give a shorter proof for the result given in [419]. In particular, in the original domain $[-1, 1]$, Equation 4.11 is a uninorm generated by the following generator:

$$h(x) = \begin{cases} ln(1+x) & \text{if } x < 0 \\ -ln(1-x) & \text{if } x \geq 0, \end{cases} \tag{4.23}$$

with inverse

$$h^{-1}(x) = \begin{cases} e^x - 1 & \text{if } x < 0 \\ -e^{-x} + 1 & \text{if } x \geq 0. \end{cases} \tag{4.24}$$

De Baets and Fodor [89] show that the operator used in PROSPEC-TOR [99] is also a uninorm. This operator corresponds to $\mathbb{C}_2$ (Equation 4.4). They prove that when this operator is rescaled into $[0,1]$ (from the standard $[-1,1]$), it corresponds to the uninorm $D_\lambda$ for $\lambda = 1$.

Details about the use of these two combination functions and their relationships in the framework of certainty factors are given in [184] (p. 180). This work by Heckerman gives a sound probabilistic interpretation of these factors, showing some inconsistencies of the original definition in MYCIN. Previously, Hájek in [177] and later Hájek and Valdés in [178] also studied the problem of interpretations of certainty factors.

The definition of certainty factors as values "to represent subjective measures of change in belief" appears in [184]. Description of the combination functions in the first expert systems is discussed in some artificial intelligence related books, e.g., [236].

3. **Uninorms:** Uninorms were introduced by Yager and Rybalov in [457] (which is, probably, the oldest of Yager's published works [445]). Mathematical properties of these operators are presented in [148]. Results by Dombi in [97] and Klement, Mesiar, and Pap in [209] are closely related and relevant to uninorms. In fact, [209] points out that the associative compensatory operators they define (rooted in Dombi's work) are the uninorms that are continuous on $[0,1]^2 \setminus \{(0,1),(1,0)\}$. The links between uninorms and previous research are also highlighted in [148]. Other results on uninorms can be found in [88] (on residual operators of uninorms).

Recent reviews on uninorms, nullnorms, and related subjects can be found in [346] and [55]. The former surveys associative and increasing aggregation operators and the latter reviews (in Sections 6.2 and 6.3) uninorms and nullnorms. A related subject, not considered in this chapter, is the introduction of weights into uninorms. This was proposed by Yager and Rybalov in [457] and further studied in [451].

4. **The mean:** The use of average is already present in ancient texts. Plackett in [322] provides some ancient references on the use of the mean, going back to Hipparchus (ca. 190 BC - ca. 120 BC) and Ptolemy (ca. 90 - ca. 168), for the computation of the number of days in a year. An interesting book on the mathematics behind Ptolemy's Almagest is [316].

Gini, in his book [163], published in 1958, also gives some historical references from ancient times. Relationships between means and proportions are described: *"Un'indagine sull'evoluzione del concetto di media dai tempi più remoti fino ai giorni nostri, deve prendere le mosse dallo studio delle proporzioni, poichè inizialmente no si faceva una rigorosa distinzione fra il concetto di media ed il concetto di proporzioni²* " (p. 1, [163]). Arithmetic, geometric, and harmonic means, as well as other means, are described with respect to particular proportions. Pythagoras and the Pythagoreans are mentioned, as they discovered the relationship between musical notes, numbers, and their proportion.

Another relevant classical author is Pappus from Alexandria (fl. c. 300-c. 350). His *Synagoge* (Book III) describes geometrical constructions of some means. Among them, we can find arithmetic, geometric, and harmonic means. We have consulted a French translation of his work [313]. [297] includes some graphical proofs (proofs without words).

In 1755, Simpson [363] argued in favor of the use of the mean. In his paper, he claims that *"some persons, of considerable note, have been of opinion, and even publicly maintained, that one single observation, taken with due care, was as much to be relied on as the Mean of a great number."* He discusses this fact and studies some situations proving that the mean leads to better results, and he concludes recommending *"the use of the method, not only to astronomers, but to all others concerned in making experiments of any kind."*

More recently, Cauchy (1821) [66] defined the mean of $x_1, \ldots, x_N$ as a value $x$ fulfilling internality (Equation 4.5). Also, in 1821, [20] was published. This paper, which is anonymous, is attributed to Svanberg (see [123], p. 157). The author deals with the problem of finding the best average of a number of observations. It distinguishes between means for values that are different in origin and values that are the same in origin but different due to imperfection of instruments or errors due to *"la maladresse ou la négligence de ceux qui ont mis ces instrumens et ces procédés en us-age³"* (p. 181). The paper presents some means, some of them weighted. It also defines an iterative approach using weighted means where weights are determined using previously estimated means. That is,

$$m^k = \frac{\sum_i a_i \frac{1}{(m^{k-1}-a_i)^2}}{\sum_i \frac{1}{(m^{k-1}-a_i)^2}},$$

where $m^0$ is estimated using the weighted mean.

The paper also expresses the idea of mean as the value that differs as little as possible from the values being aggregated: *"La moyenne cherchée est alors une combinaison de ces divers résultats, de laquelle on puisse présumer qu'elle différe moins du véritable et unique résultat que toute*

---

² An investigation on the evolution of the concept of mean throughout history should start with the study of proportions, since initially there was no strict distinction between the concept of mean and the concept of proportion

³ The clumsiness or carelessness of those that put these instruments and procedures into use

*autre combinaison qu'on pourrait faire des mêmes donnés[4]"* (p. 187). This
is related to the aggregation as the object that is located at the least
distance of the ones to be aggregated (as in Figure 1.5).

Chisini (1929) [76] defined the mean of a function $f(x_1, \ldots, x_N)$ as the
value $\bar{x}$ such that $f(x_1, \ldots, x_N) = f(\bar{x}, \ldots, \bar{x})$. De Finetti (1931) [91] re-
considers Chisini's definition for the mean and offers an alternative one: *"si
definisce media di una grandezza in una data distribuzione (di qualunque
natura essa sia) per rapporto a un'assegnata circostanza quell'unico val-
ore della grandezza che si può sostituire alla distribuzione senza alterarvi
la circostanza in parola[5]"* (p. 375). A similar definition was previously
given by Bemporad (1926) [40]: *"Siano $x_1, x_2, \ldots, x_{n-1}, x_n$ $n$ misure in-
dipendenti di una stessa quantità $x$. Assumeremo i postulati seguenti: I.-
Qualunque sia il numero delle misure eseguite, esiste un numero che è da
assumersi come risultato complessivo di tutte le misure. Il risultato complessivo di un numero qualunque di misure è quantità in tutto paragonabile
al risultato singolo di una sola osservazione, e deve quindi essere consid-
erato alla stessa stregua. Esso è dato da una funzione delle $x_1, x_2, \ldots, x_n$,
finita e continua insieme con le sue derivate parziali prime [6]."* Bemporad
added to this definition three additional constraints (conditions II, III,
and IV in [40]) to further restrict the function.

The work by Kolmogorov [214] and Nagumo [286] (1930) had a strong
influence on the field. They studied means taking into consideration decomposability. An operator is decomposable if when considering a sequence of functions

$$\mathbb{C}^{(1)}(a_1), \mathbb{C}^{(2)}(a_1, a_2), \mathbb{C}^{(3)}(a_1, a_2, a_3), \ldots, \mathbb{C}^{(m)}(a_1, a_2, \ldots, a_m), \ldots$$

we have

$$\mathbb{C}^{(m)}(a_1, \ldots, a_k, a_{k+1}, \ldots, a_m) = \mathbb{C}^{(m)}(a, \ldots, a, a_{k+1}, \ldots, a_m)$$

for $k = 1, \ldots, m$ and $a = \mathbb{C}^{(k)}(a_1, \ldots, a_k)$.

More specifically, Kolmogorov [214] and Nagumo [286] independently
studied decomposability and proved that the quasi-arithmetic mean is

---

[4] The mean we look for is, then, a combination of the different results, which can
be presumed that differs less from the true and unique result than any other
combination that can be obtained from the same data

[5] The mean of a physical quantity for a given distribution (of whichever nature)
with respect to a given setting is defined by the single quantity that can replace
the distribution without modifying the above mentioned setting

[6] Let $x_1, x_2, \ldots, x_{n-1}, x_n$ be $n$ independent measures of the same quantity $x$. We
will assume the following axioms: I.- Whichever the number of measures carried
out, there is a number that must be taken as the overall result of all the measures.
The overall result of an arbitrary number of measures is a quantity completely
comparable to the individual result of a single observation, and should be considered in the same way. It is given by a function of $x_1, x_2, \ldots, x_n$, bounded,
continuous and with continuous first partial derivatives

characterized by continuity, symmetry, strict increasingness, and decomposability.

Another example of a basic property studied in relation to means is bisymmetry. That is, $\mathbb{C}(\mathbb{C}(x, y), \mathbb{C}(u, z)) = \mathbb{C}(\mathbb{C}(x, u), \mathbb{C}(y, z))$. This property has been used for characterizing quasi-arithmetic means. See [4] (p. 281).

Historical notes, references, and more recent results on means can be found in the books by Aczél [4], Aczél and Dhombres [11], Chapter 5 of [146], and the following papers [144] and [246], among others. [144] generalizes the results of Kolmogorov and Nagumo by considering non-strict means (instead of strict ones).

5. **Aggregation on numerical scales:** Sections 4.2 and 4.3 are mainly based on [8] by Aczél and Alsina, and [4] and [5] by Aczél. Some properties also come from the book by Hardy, Littlewoord, and Pólya [122]. This book, devoted to inequalities, contains some of the basic results on aggregation in numerical scales.

The separability of an aggregation operator was first considered in [13]. [13] proves Proposition 4.23, and then characterizes the geometric mean in terms of the quasi-arithmetic mean, reciprocity, and positive homogeneity (Proposition 4.24). Proofs of Proposition 4.20 can be found in Jessen [205] and [91], p. 390-392 (see also [4], pp. 150-153, and [122], pp. 68-69).

Example 4.21 uses the data from performance comparison of Java/.NET runtimes given in [361]. The use of aggregation operators to combine performance and times can be found in several other works. For example, [46] compares average execution times to evaluate strategies for garbage collection (in Java), and [49] compares execution times of benchmarks in Java, Fortran, and C.

Table 4.1 is based on [8] and [50] (mainly p. 218). Gini in [163] includes, most of the means in this chapter (e.g., the root-mean-power on p. 138 and the counter-harmonic mean on p. 139). The root-mean-powers are linked with the expressions for computing aggregation as the object that minimizes a distance when the distance is $d_p(a, b) = |a - b|^p$. Some results for the aggregation operators can be found in the Bibliographical Notes of Chapter 1. Nevertheless, the root-mean-powers with parameter $p - 1$ is usually not equal to the object obtained with $d_p$, except for $p = 1$ and $p = 2$. This result is reported in Taguchi [387] (1974). The root-mean-powers were studied previously by Fechner [134] (1878) as generalized measures of location. In fact, he considered measures of both location and dispersion, and, thus, his work also considered the aggregation operators defined in Chapter 1.

Properties of quasi-arithmetic means are mainly based on [51, 50, 122]. The geometric proof that $HM \preceq GM \preceq AM$ (Figure 4.4) is given in [50] (p. 45) and [434]. The first proof is by Pappus of Alexandria in *Synagoge* (Book III) [313]. Proposition 4.25 is given in [251] and [12] (p. 220, case 5b). Proposition 4.26 corresponds to Theorem 2 in [4] (p. 236).

Other conditions than the ones considered here have also been studied in the literature. For example, equations of the form

$$f(a_1^p, ..., a_N^p) = (f(a_1, ..., a_N))^p$$

are considered for one $p$ with $p \neq 0, 1, -1$. See [7]. This work also solves the case of two different equations with two different $p_1$ and $p_2$ ($\log|p_1|/\log|p_2|$ should be irrational and finite). The consideration of two different aggregation operators, one for measures and the other for ratios, was considered by Aczél and Alsina [8] (see also [9]). This corresponds to the following equation ($\mathbb{C}$ and $\mathbb{Q}$ represent the aggregation operators):

$$\mathbb{Q}\left(\frac{a_1}{b_1}, ..., \frac{a_N}{b_N}\right) = \frac{\mathbb{C}(a_1, ..., a_N)}{\mathbb{C}(b_1, ..., b_N)}.$$

The case of aggregation operators for data represented in different scales is studied in [12]. Different cases corresponding to interval and ratio scales are analyzed and characterized. Proposition 4.37 in this chapter corresponds to Corollary 3.1 in [12].

Chapter 6 describe other operators for numerical scales. The bibliographical notes in it are mainly devoted to such operators.

6. **Aggregation in ordinal scales:** Aggregation in ordinal scales has been studied by several authors. Section 4.3.1 is based on the works by Marichal (specially, [249]) and Ovchinnikov (specially, [305]). In particular, Proposition 4.30 is given in [305].

Some of the results presented here have been proved under different assumptions on the range of the functions, e.g., $\mathbb{R}^N$ or ordered sets. See [247, 252, 304, 307] for details. See also [245] for a review of some results in this area.

The operators considered in this section are not the only ones available for ordinal scales. As will be seen in Chapter 6, there are other operators that can also be applied. In particular, the Sugeno integral (see Section 6.4), and some weighted operators, such as the weighted minimum and the weighted maximum (see Section 6.3), are suitable for ordinal scales. As will be shown, such operators do not satisfy symmetry.

7. **Weighted means:** Separability for the weighted means was considered by Aczél in [5]. The results described there are analogous to the ones of the arithmetic mean. Propositions 4.41, 4.42, and 4.43 are proved in that paper.

The weighted root-mean-powers were studied in [35] (1938). Bajraktarević's mean was introduced in [28], and further studied by Bajraktarević in [29, 30] and by Losonczi in [237]. [29] considers weighting functions of the form $p_i\pi(a_i)$, and on p. 73 it gives an expression analogous to Equation 4.22 but with additional weights $p_i$:

$$\left(\frac{\sum_{i=1}^{N} p_i a_i^p}{\sum_{i=1}^{N} p_i a_i^q}\right)^{1/p-q}.$$

Bajraktarević's means have been recently applied in [106]. For properties of the mean, such as relationships with the root-mean-powers and the counter-harmonic means, see [50].

# 5

# Fuzzy Measures

*Sanbonn no ya*[1]

Japanese saying

Most aggregation operators use some kind of parameterization to express additional information about the objects that take part in the aggregation process. Applying the jargon of artificial intelligence, we can say that the parameters are used to represent the background knowledge. For example, it is well known that in the case of the weighted mean, the weights – i.e., the weighting vector – play this role. In an application, we can use them to express the reliability of the information sources (sensors, experts, and so on). For example, when fusing data from sensors, we can express wich sensor is more likely to give data of better quality and which is more likely to give erroneous data. In a similar way, other aggregation functions use other parameterizations.

Among all the existing types of parameters, the *fuzzy measures* are a rich and important family. They are of interest here because they are used for aggregation purposes in conjunction with several fuzzy integrals (e.g., Choquet and Sugeno integrals). Also, they are general in the sense that, when used with some of the integrals, they can generalize some well-known aggregation operators, such as the weighted mean.

It has to be said that the term "fuzzy measure" is not the only one used in the literature. Other names are also present to refer to the same concept. Some of them are non-additive probabilities and capacities.

Due to the importance of fuzzy measures, we devote this chapter to them. We give their definitions, establish some of the main families, and review some of their properties. Chapter 6 shows their use with fuzzy integrals and provides a justification of their interest in aggregation (see Section 6.2).

---

[1] One arrow is weak, but if three arrows come together they are very strong

## 5.1 Definitions, Interpretations, and Properties

We begin with the definition of fuzzy measures. We assume that the set over which the fuzzy measure is defined is finite, as this is the usual case, with aggregation operators. Nevertheless, some of the results given here also hold for an infinite set.

**Definition 5.1.** *A fuzzy measure $\mu$ on a set $X$ is a set function $\mu : \wp(X) \rightarrow [0,1]$ satisfying the following axioms:*

*(i) $\mu(\emptyset) = 0$, $\mu(X) = 1$ (boundary conditions)*
*(ii) $A \subseteq B$ implies $\mu(A) \leq \mu(B)$ (monotonicity)*

The requirement that the measure for the whole set is 1 ($\mu(X) = 1$) is an arbitrary convention; in general, any other value might be used. Nevertheless, this requirement is specially convenient for aggregation purposes, and it is, in fact, a condition analogous to the one for the weighted means to have weights that add to 1. Therefore, unless otherwise stated, we assume this bound throughout this book.

Among fuzzy measures, we distinguish those where $\mu(A)$ is either 0 or 1. Such measures are known as 0-1 fuzzy measures. We define them below, as they will be used later.

**Definition 5.2.** *$\mu$ is a 0-1 fuzzy measure if $\mu$ is a fuzzy measure and, for all $A \subset X$, it holds that $\mu(A) \in \{0,1\}$.*

As can be observed from their definition, fuzzy measures replace the axiom of additivity satisfied by probability measures (see Section 2.2) by a more general one, monotonicity. This implies that probability measures are particular cases of fuzzy ones. They correspond, in fact, to additive fuzzy measures (measures satisfying $\mu(A \cup B) = \mu(A) + \mu(B)$).

Additive measures have the important property that the whole measure can be defined from the values assigned to the singletons. This corresponds to having a mapping from $X$ into $[0,1]$ (a probability distribution) and defining the measure of a set $A$ (the probability of the set $A$) from such a mapping. Let $p$ be the mapping, then $\mu(A) = \sum_{a \in A} p(a)$. From this point of view, weighting vectors can be seen as probability distributions and, thus, they can be used to infer fuzzy measures.

**Definition 5.3.** *Let $\mathbf{p}$ be a weighting vector defined over the set $X$ (i.e., $p : X \rightarrow [0,1]$ with $\sum_{x_i \in X} p(x_i) = 1$); then, the additive fuzzy measure $\mu$ inferred from $\mathbf{p}$ is defined as*

$$\mu(A) = \sum_{a \in A} p(a)$$

*for all $A \subseteq X$.*

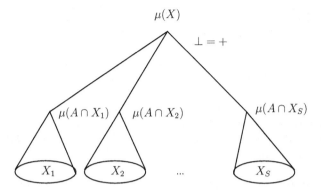

**Fig. 5.1.** $\mu$-inter-additive partition of $X = \{X_1, X_2, \ldots, X_S\}$. $\perp$ is used to represent the operator to combine the measure from the different partition elements (i.e., the addition)

Additivity implies that the measure of a set is the summation of the measures of the elements. Thus, no interaction among the elements with respect to the measure is considered. In fact, we can consider that this approach of the measure of a set is implicitly followed when we use the weighted mean. This consideration is supported by the fact that the Choquet integral (a fuzzy integral described in Section 6.2) with respect to an additive fuzzy measure $\mu$ inferred from a weighting vector $p$ is equivalent to a weighted mean with respect to $p$ (see Theorem 6.24).

In additive measures, addition is used to *combine* the measure of the singletons. A natural generalization to this is to use addition to combine measures of larger sets. In this case, for pairs $x_i, x_j$ we might have $\mu(\{x_i, x_j\}) \neq \mu(\{x_i\}) + \mu(\{x_j\})$. Nevertheless, for larger sets we might have $\mu(X \cup Y) = \mu(X) + \mu(Y)$, with $X \cap Y = \emptyset$. Figure 5.1 gives a graphical representation of this kind of measure, and its formalization is given below. In the figure, $\perp$ is used to represent the combination operator (addition). The next definition formalizes this concept.

**Definition 5.4.** *Let $\mu$ be a fuzzy measure on $X$ and let $\mathcal{P} = \{X_1, \ldots, X_s\}$ be a partition of $X$; then, $\mathcal{P}$ is a $\mu$-inter-additive partition of $X$ if*

$$\mu(A) = \sum_{X_i \in \mathcal{P}} \mu(A \cap X_i)$$

*for every $A \in \wp(X)$. If $\mathcal{P}$ is a $\mu$-inter-additive partition of $X$ with at least two elements, then we will say that $\mu$ is an inter-additive fuzzy measure.*

Naturally, if $\mathcal{P} = \{\{x_1\}, \{x_2\}, \ldots, \{x_N\}\}$ is a $\mu$-inter-additive partition of $X$, then $\mu$ is additive.

We now consider an example of fuzzy measure that is neither additive nor $\mu$-inter-additive.

**Definition 5.5.** *Let $X$ be a reference set; then, the measure $\mu^*$ (the strongest fuzzy measure) is defined by*

$$\mu^*(A) = \begin{cases} 0 & \text{if } A = \emptyset \\ 1 & \text{otherwise} \end{cases}$$

This measure can be used to model total ignorance, because the measure is maximal for all subsets of $X$.

An important concept for fuzzy measures is their duality. This is established in the following definition.

**Definition 5.6.** *Let $\mu_1$ and $\mu_2$ be two fuzzy measures on $X$; then, $\mu_1$ and $\mu_2$ are dual conjugates if and only if they satisfy*

$$\mu_1(A) = 1 - \mu_2(X \setminus A) \tag{5.1}$$

*for all $A \subseteq X$.*

The following two propositions hold for fuzzy measures.

**Proposition 5.7.** *The dual of a fuzzy measure is another fuzzy measure.*

**Proposition 5.8.** *Let $\mu$ be an additive fuzzy measure; then, the dual of $\mu$ is $\mu$.*

*Example 5.9.* The dual of measure $\mu^*$ in Definition 5.5 is

$$\mu_*(A) = \begin{cases} 0 & \text{if } A \neq X \\ 1 & \text{otherwise.} \end{cases}$$

This measure is the weakest fuzzy measure.

Another concept worth presenting here is consistency between fuzzy measures.

**Definition 5.10.** *Given two fuzzy measures $\mu$ and $\mu'$ on $X$, we say that $\mu'$ is consistent with $\mu$ when $\mu(A) > \mu(B)$ implies that $\mu'(A) > \mu'(B)$ for all $A, B \subseteq X$. If $\mu'$ is consistent with $\mu$, and $\mu$ is consistent with $\mu'$, then we say that $\mu$ and $\mu'$ are consistent fuzzy measures.*

*Example 5.11.* Let us consider the fuzzy measures $\mu$ and $\mu'$ defined as in Table 5.1. We have that $\mu'$ is consistent with $\mu$. Nevertheless, $\mu$ is not consistent with $\mu'$ because $\mu'(\{x_2, x_3\}) < \mu'(\{x_1, x_3\})$ but $\mu(\{x_2, x_3\}) = \mu(\{x_1, x_3\})$.

| | $\mu$ | $\mu'$ |
|---|---|---|
| $\{x_1\}$ | 0.1 | 0.2 |
| $\{x_2\}$ | 0.2 | 0.3 |
| $\{x_3\}$ | 0.5 | 0.5 |
| $\{x_1, x_2\}$ | 0.2 | 0.3 |
| $\{x_2, x_3\}$ | 0.5 | 0.5 |
| $\{x_1, x_3\}$ | 0.5 | 0.7 |
| $\{x_1, x_2, x_3\}$ | 1.0 | 1.0 |

**Table 5.1.** Fuzzy measures $\mu$ and $\mu'$ for Example 5.11.

### 5.1.1 Interpretations

The meaning of fuzzy measures has been studied by several authors since their inception. Interpretations are tightly related with particular data uses, and, additionally, some families of fuzzy measures permit some specific interpretations. The most rellevant case corresponds to probabilities. A related case is the distorted probability (described in Section 5.4), which can be seen as a probability distribution that has been distorted by a function. Finally, 0-1 fuzzy measures can be understood in terms of coalitions, e.g., a parliament; that is, given a 0-1 fuzzy measure $\mu$, we have that $\mu(A)$ is 1 if and only if the coalition $A$ is able to pass a bill in the parliament.

This section presents some of the main interpretations. They are described below.

1. **Fuzziness:** A fuzzy measure $\mu$ for a set $A$, i.e., $\mu(A)$, corresponds to the grade that an element in the reference set $X$ belongs to the set $A$. Here, fuzziness is understood as a kind of uncertainty that is different from randomness. Such an approach follows the interpretation of membership in fuzzy sets, their being different from probability distributions.
2. **Importance:** $\mu(A)$ stands for the degree of importance, or weight, of the set A when computing the aggregated value for $X$. This interpretation is specially suited when combining criteria or experts. A related definition is the one that says that $\mu(A)$ measures the power of $A$ to make the decision alone (without the remaining criteria in $X \setminus A$).
3. **Probability:** Several interpretations have been constructed using probability distributions as their cornerstone. They have mainly addressed the belief and plausibility measures (a particular type of fuzzy measure described in Section 5.2). The following are examples of interpretations based on probabilities.
   a) *Belief and plausibility as inner and outer measures.* In a probability space $(X, \mathcal{X}, P)$, the probability $P$ is not necessarily defined on $\wp(X)$ but instead only on $\mathcal{X}$. In this case, belief (denoted by $Bel$) and plausibility ($Pl$) measures can be defined on $\wp(X)$ as extensions of $P$ on $\mathcal{X}$. They are defined as follows:

$$Bel(A) = \sup\ \{P(X)|X \subseteq A \text{ and } X \in \mathcal{X}\}$$

$$Pl(A) = \inf\ \{P(X)|X \supseteq A \text{ and } X \in \mathcal{X}\}$$

Although, for all probability spaces $(X, \mathcal{X}, P)$, the functions $Bel$ and $Pl$ are belief and plausibility measures, the converse is not true. Nevertheless, it is possible to establish such a converse relation in terms of belief functions on a set of formulas rather than on sets.

b) *Probability interval induced by a belief measure.* Each belief measure $Bel$ with its dual plausibility $Pl$ induces a probability interval $P_{Bel}$:

$$P_{Bel} = \{p|Bel(A) \le p(A) \le Pl(A) \text{ for all } A \in \wp(X)\}.$$

Although every belief function is a lower envelope of a probability interval, not every lower envelope is a belief function.

c) *Belief as information loss.* Belief is understood as a probability that has suffered from a process of information loss. This process consists on transferring support from sets $A$ into larger sets $B$ such that $B \ge A$.

d) *Distorted probabilities.* Fuzzy measures on $X$ can be interpreted in terms of a set of probability distributions on disjoint sets of $X$ and a distortion function that, eventually, combines the probabilities and modifies the values.

e) *Mapping between spaces.* In this case, a fuzzy measure $\mu$ in a space $(X, 2^X)$ is represented in terms of an additive measure $\lambda$ on a measurable space $(\Theta, 2^\Theta)$ and a mapping $\nu$ from one space to the other. That is, when $\nu : \wp(X) \to \wp(\Theta)$, we have $\mu(A) = \lambda(\nu(A))$. It has been proved that the Choquet integral is consistent with this interpretation.

When used in aggregation, fuzzy measures are defined to quantify aspects related to the information sources that supply information. According to this, $X$ is equal to the set of information sources. In this case, the interpretations considered above correspond to.

1. **Fuzziness:** $\mu(A)$ is the grade that the correct answer is given by an information source that belongs to $A$. When the grade is bounded and monotonic, conditions in Definition 5.1 hold.
2. **Importance:** $\mu(A)$ stands for the degree of importance of the set $A$. In this setting, condition (i) in Definition 5.1 means that when the set of sources is empty their importance is 0 and that the maximal importance is obtained when all the sources are considered (and the maximal importance is 1); condition (ii) in Definition 5.1 means that the more sources we have, the greater is the importance.
3. **Probability:** When the measure $\mu$ is additive, we have a probability distribution on the information sources. Then, $\mu(A)$ is the probability of the sources in $A$ being correct. In contrast, when the measure is not additive, we have a generalized probability on the set of information sources.

| $\mu(\emptyset) = 0$ | $\mu(\{M, L\}) = 0.9$ |
|---|---|
| $\mu(\{M\}) = 0.45$ | $\mu(\{P, L\}) = 0.9$ |
| $\mu(\{P\}) = 0.45$ | $\mu(\{M, P\}) = 0.5$ |
| $\mu(\{L\}) = 0.3$ | $\mu(\{M, P, L\}) = 1$ |

**Table 5.2.** Fuzzy measure on the set $X = \{M, P, L\}$ following Example 5.12

An alternative way to interpret fuzzy measures is from an *operational* point of view. As said before, fuzzy measures are often used in conjunction with fuzzy integrals. In this case, the fuzzy integral of the characteristic function of a set $A$ corresponds to the fuzzy measure of $A$. This is, given $A \subseteq X$, let $f(x_i) = 1$ if and only if $x_i \in A$; then, the fuzzy integral of $f$ with respect to $\mu$ is $\mu(A)$. Accordingly, and independently of any denotational semantics for $\mu(A)$, we can define $\mu$ on the basis of the desired outcome for the fuzzy integral.

An illustrative example of fuzzy measures frequently used in the literature is the one defined in the following example. It is similar to Example 4.28.

*Example 5.12.* The director of a high school has to evaluate the students according to their level in mathematics $(M)$, physics $(P)$, and literature $(L)$. The evaluation consists of obtaining a final rating as an average of the ratings of the three subjects. For each student, the final rating depends on the importance given to the subjects. To settle these importances, a fuzzy measure is used. Here, $X$ is the set of all subjects (i.e., $X = \{M, P, L\}$), and $\mu(A)$ is the importance of a particular set of subjects $A$. The definition of the measure considers the following elements.

1. Boundary conditions:
   $\mu(\emptyset) = 0$, $\mu(\{M, P, L\}) = 1$
   The importance of the empty set is 0. The set consisting of all objects has maximum importance.
2. Relative importance of scientific versus literary subjects:
   $\mu(\{M\}) = \mu(\{P\}) = 0.45$, $\mu(\{L\}) = 0.3$
   The importance of mathematics and physics is greater than the importance of literature.
3. *Redundancy* between mathematics and physics:
   $\mu(\{M, P\}) = 0.5 < \mu(\{M\}) + \mu(\{P\})$
   Mathematics and physics are similar subjects. The importance of the set containing both should not be larger than their addition.
4. Support between literature and scientific subjects:
   $\mu(\{M, L\}) = \mu(\{P, L\}) = 0.9 > \mu(\{P\}) + \mu(\{L\}) = 0.45 + 0.3 = 0.75$
   $\mu(\{M, L\}) = \mu(\{P, L\}) = 0.9 > \mu(\{M\}) + \mu(\{L\}) = 0.45 + 0.3 = 0.75$
   Mathematics and literature are complementary subjects.

An outline of this fuzzy measure is given in Table 5.2.

Now, we consider another example.

**Definition 5.13.** *Let $X$ be a set, and let $M$ be a set of fuzzy measures on $X$ such that, for all $\mu_1, \mu_2 \in M$, we have $\mu_1(\{x\}) = \mu_2(\{x\})$ for all $x \in X$; then, we define the minimal and the maximal measure in $M$ as follows:*

1. *$\mu$ is the minimal measure of $M$ if, for all $A \in \wp(X)$ and all $\mu' \in M$, we have $\mu(A) \leq \mu'(A)$. Note that the minimal measure of a set can be built from the measure of the singletons in the following way: $\mu(A) = max_{a \in A} \mu(\{a\})$ if $A \neq X$ and $\mu(X) = 1$.*
2. *$\mu$ is the maximal measure of $M$ if, for all $A \in \wp(X)$ and all $\mu' \in M$, we have $\mu(A) \geq \mu'(A)$. Note that maximal measures satisfy $\mu(A) = 1$ for all $|A| > 1$.*

Denoting the measure for the singletons as a mapping $v$ from $X$ into $[0, 1]$, we denote the corresponding minimal measure by $\mu_{MN(v)}$, and the corresponding maximal measure by $\mu_{MX(v)}$.

### 5.1.2 Properties

An important issue when defining fuzzy measures in practical applications is the number of parameters required. It is clear that additive measures require only $|X|$ parameters (the values on the singleton) and that unrestricted fuzzy measures require $2^{|X|}$. Note that the number of parameters is, in fact, reduced by 1 or 2 when boundary conditions are considered: $|X| - 1$ for additive measures (as they add to 1) and $2^{|X|} - 2$ for unconstrained ones (as $\mu(\emptyset) = 0$ and $\mu(X) = 1$). Additionally, when defining a measure, besides supplying a certain number of values, some checking is required to ensure that such values satisfy the monotonicity constraints. Note that, in general, there exist $|X|!$ different monotonic sequences of subsets of $X$.

Unconstrained fuzzy measures are an example of measures that not only need to fix the values, but also to check them. In this case, $2^{|X|} - 2$ values are required, and all sequences have to be checked. In contrast, for additive measures, only $|X| - 1$ values are required and the only checking is whether their sum is equal to 1. Also, for the minimal and the maximal measure of a mapping $v$, we only need the mapping (that is, $|X|$ values), and to check that there is no $x_i$ such that $v(x_i) > 1$.

Fuzzy measures can be rewritten in an alternative way through the Möbius transform. The Möbius transform of a fuzzy measure $\mu$ is a function $m : \wp(X) \to \mathbb{R}$ such that $m(\emptyset) = 0$, $\sum_{A \subseteq X} m(A) = 1$, and, if $A \subset B$, then $\sum_{C \subseteq A} m(C) \leq \sum_{C \subseteq B} m(C)$.

**Definition 5.14.** *Let $\mu$ be a fuzzy measure; then, its Möbius transform $m$ is defined as*

$$m_\mu(A) := \sum_{B \subseteq A} (-1)^{|A| - |B|} \mu(B) \tag{5.2}$$

*for all $A \subset X$.*

| | |
|---|---|
| $m(\emptyset) = 0$ | $m(\{M, L\}) = 0.15$ |
| $m(\{M\}) = 0.45$ | $m(\{P, L\}) = 0.15$ |
| $m(\{P\}) = 0.45$ | $m(\{M, P\}) = -0.4$ |
| $m(\{L\}) = 0.3$ | $m(\{M, P, L\}) = -0.1$ |

**Table 5.3.** Möbius transform of the measure given in Table 5.2.

Note that the function $m$ is not restricted to the $[0, 1]$ interval.

Given a function $m$ that is a Möbius transform, we can reconstruct the original measure as follows:

$$\mu(A) = \sum_{B \subseteq A} m(B)$$

for all $A \subseteq X$.

*Example 5.15.* Let $\mu$ be the fuzzy measure in Example 5.12 (and outlined in Table 5.2); then, the Möbius transform of $\mu$ is given in Table 5.3.

When a measure is additive, the Möbius transform on the singletons corresponds to the probability distribution, and it is zero for non-singletons.

**Proposition 5.16.** *Let $\mu$ be an additive fuzzy measure and let $m$ be its Möbius transform; then, $m(A) = 0$ for all $|A| > 1$. Moreover, let $p(x_i) = m(\{x_i\})$ for all $x_i \in X$; then, $p$ is a probability distribution, or a weighting vector, that infers $\mu$.*

Taking into account the Möbius transform, it is possible to define a family of fuzzy measures on the basis of the largest set $A$ with non-null $m(A)$. This family of fuzzy measures is called $k$-order additive fuzzy measures. $k$ corresponds to the cardinality of such a largest set $A$.

**Definition 5.17.** *Let $\mu$ be a fuzzy measure and let $m$ be its Möbius transform; then, $\mu$ is a $k$-order additive fuzzy measure if $m(S) = 0$ for any $S \subseteq X$ such that $|S| > k$, and there exists at least one $S \subseteq X$ with $|S| = k$ such that $m(S) \neq 0$.*

It is easy to see that any fuzzy measure can be represented as a $k$-order additive fuzzy measure with an appropriate value of $k$. Thus, if $[\mu]_k^X$ is the set of all $k$-order additive fuzzy measures on $X$, $\{[\mu]_k^X\}_{k=1,\dots,|X|}$ is a partition of the set of all fuzzy measures on $X$.

This family of measures can be seen as a generalization of additive ones, as $[\mu]_1^X$ is the set of additive measures. In fact, understanding the Möbius transform as a function that makes explicit the interactions between the information sources, $k$-order additive fuzzy measures stand for measures where the interactions can only be expressed up to dimension $k$, but not for larger dimensions. When $k = 2$, only binary interactions are allowed, while when $k = |X|$, all kinds of interactions are permitted.

It is clear that the value of $k$ corresponds to the complexity of the measure, and, thus, the number of parameters needed for its determination increases when $k$ increases. The next proposition makes this fact concrete.

**Proposition 5.18.** *Let $X$ be a set of cardinality $N$. Then, a $k$-order additive fuzzy measure requires*

$$\sum_{j=1}^{K} \binom{N}{j}$$

*parameters in order to be defined.*

A few other properties of fuzzy measures are also of interest. The next few definitions establish them.

**Definition 5.19.** *We say that a fuzzy measure $\mu$ is*

1. *$k$-order monotone (or $k$-monotone) for $k \geq 2$, if, for all families of $k$ subsets $A_1, \ldots A_k$ in $X$,*

$$\mu\left(\bigcup_{i=1}^{k} A_i\right) \geq \sum_{\emptyset \neq I \subset \{1,\ldots k\}} (-1)^{|I|+1} \mu\left(\bigcap_{i \in I} A_i\right),$$

   *1-monotonicity is defined as monotonicity.*
2. *totally monotone if it is $k$-monotone for any $k \geq 1$.*
3. *$k$-order alternative (or $k$-alternative) for $k \geq 2$, if for all family of $k$ subsets $A_1, \ldots A_k$ in $X$,*

$$\mu\left(\bigcap_{i=1}^{k} A_i\right) \leq \sum_{\emptyset \neq I \subset \{1,\ldots k\}} (-1)^{|I|+1} \mu\left(\bigcup_{i \in I} A_i\right).$$

2-monotonicity is sometimes known as supermodularity or convexity; 2-alternating fuzzy measures are sometimes called submodular measures.

## 5.2 Belief and Plausibility Measures

The mathematical theory of evidence is based on the belief and plausibility measures. They are fuzzy measures that satisfy some additional constraints (see Definitions 5.20 and 5.21 below), and that can be easily defined using the Möbius transform.

**Definition 5.20.** *A fuzzy measure Bel on a set $X$ is a belief measure (also called a belief function) if and only if it satisfies (i) and (ii) in Definition 5.1 and the following equation:*

$$Bel(A_1 \cup ... \cup A_n) \geq \sum_j Bel(A_j) - \sum_{j<k} Bel(A_j \cap A_k) + ... +$$
$$(-1)^{n+1} Bel(A_1 \cap ... \cap A_n). \tag{5.3}$$

**Definition 5.21.** *A fuzzy measure Pl on a set X is a* plausibility measure *if and only if it satisfies (i) and (ii) in Definition 5.1 and the following equation:*

$$Pl(A_1 \cap ... \cap A_n) \leq \sum_j Pl(A_j) - \sum_{j<k} Pl(A_j \cup A_k) + ... +$$
$$(-1)^{n+1} Pl(A_1 \cup ... \cup A_n). \tag{5.4}$$

Belief and plausibility functions are dual in the sense of Definition 5.6. This is, given a belief measure $Bel$, its dual is a plausibility measure, and the dual of this plausibility measure is $Bel$.

There is an alternative and equivalent definition of belief and plausibility measures based on the basic probability assignment function (*bpa*). This function corresponds to a Möbius transform that is positive and restricted to $[0,1]$ for all $A \subseteq X$. Definitions 5.20 and 5.21 establish concepts equivalent to those defined above in terms of basic probability assignments. This is so because for each dual pair of fuzzy measures there is corresponding *bpa*, and for each *bpa* there is a dual pair of belief and plausibility measures.

**Definition 5.22.** *A function* $m : \wp(X) \to [0,1]$ *is a* basic probability assignment *if and only if*

*(i)* $m(\emptyset) = 0$
*(ii)* $\sum_{A \subseteq X} m(A) = 1$

Given such a function, the two fuzzy measures are built in the following way.

**Proposition 5.23.** *Let m be a basic probability assignment defined on the reference set X. Then, the following holds:*

*1. The function* $Bel : \wp(X) \to [0,1]$ *defined by*

$$Bel_m(A) := \sum_{B \subseteq A} m(B) \qquad for\ all\ A \subseteq X \tag{5.5}$$

*is a belief measure (we will call it the belief measure induced from m).*
*2. The function* $Pl : \wp(X) \to [0,1]$ *defined by*

$$Pl_m(A) := \sum_{B \cap A \neq \emptyset} m(B) \qquad for\ all\ A \subseteq X$$

*is a plausibility measure.*

3. *The belief and plausibility measures induced from m are dual (Bel$_m$(A) =*
   *1 − Pl$_m$(X \ A) for all A ⊆ X).*

The *bpa* can be understood as an assignment of an amount of information that one commits specifically to $A$ (and not to any subset of $A$). This
information might refer to fuzziness, importance, or probability.

The reverse construction, i.e., building the basic probability assignment
from a belief function $Bel$, is given by the following expression:

$$m_\mu(A) = \sum_{B \subseteq A} (-1)^{|A|-|B|} Bel(B). \qquad (5.6)$$

Note that this expression is the Möbius transform of $\mu$ (given in Definition 5.14), and, thus, it can be computed for any fuzzy measure. Nevertheless,
when a measure is not a Belief, Expression 5.6 leads to negative values. This
is the case of Example 5.15. Note that for the set $\{M, P\}$ $m$ is negative:
$m(\{M, P\}) = -0.4 = Bel(\{M, P\}) - Bel(\{M\}) - Bel(\{P\})$.

For belief measures, the following result holds.

**Proposition 5.24.** *Let Bel be a belief measure and Pl be a plausibility mea-*
*sure induced from a basic probability assignment m. Then, if Bel(A) = Pl(A)*
*for all A ⊆ X, m focuses only on singletons (i.e., m(A) = 0 for all |A| > 1).*

In this case, $m$ corresponds to the probability distribution. In fact, additive
measures (probability measures) can be seen as both belief and plausibility
ones, as the additivity axiom implies both inequalities 5.3 and 5.4.

Basic probability assignments are of practical interest because they simplify the definition of a measure. When, instead of an arbitrary fuzzy measure,
a *bpa* is considered for defining either $Bel$ or $Pl$, we need only to check whether
$m$ is positive and whether the values add to 1. This has a cost of $2^N$, where
$N$ is the number of elements. In contrast, for an arbitrary fuzzy measure, we
need to check consistency (i.e., whether $\mu(A) \geq \mu(B)$ for all $A > B$). This
corresponds to several checks (between 0 and $N$) for all $2^N$ subsets. Therefore,
it is easier to have fuzzy measures that are $Bel$ and $Pl$ by construction. This
requires positive $m$ and the rescaling of the addition if the sum is greater than
1.

### 5.2.1 Belief Measures from Unconstrained Ones

We consider the definition of belief measures from a Möbius transform. We
show how this is applied to the fuzzy measure in Example 5.12 and to the
Möbius transform in Example 5.15. We will define a basic probability assignment applying a translation (that leads to $m'$) and a normalization (resulting in $m''$) on the function's Möbius transform. The function $m''$ that the
transformation gives is a basic probability assignment. This is based on the
following result which shows that the translation and normalization maintain
monotonicity on two sets.

| | $\mu$ | $m$ | $m'$ | $m''$ | $\mu_{m''}$ |
|---|---|---|---|---|---|
| $\emptyset$ | 0 | 0 | 0 | 0 | 0 |
| $\{x_1\}$ | $a$ | $a$ | $2a$ | $2a/k$ | $2a/k$ |
| $\{x_2\}$ | $a+b$ | $a+b$ | $2a+b$ | $(2a+b)/k$ | $(2a+b)/k$ |
| $\{x_1,x_2\}$ | $a+b$ | $-a$ | 0 | 0 | $(4a+k)/k$ |
| $\{x_3\}$ | 0 | 0 | $a$ | $a/k$ | $a/k$ |

**Table 5.4.** Fuzzy measure on the set $X = \{M, P, L\}$ that is consistent with the measure in Example 5.12

**Proposition 5.25.** *Let $\mu$ be a fuzzy measure with $m(A) < 0$ for some $A \subseteq X$, and let $m$ be its Möbius transform. Then, the measure $m''$ defined as $m''(\emptyset) = 0$ and as*

- $m'(A) = m(A) - \min_{C \subseteq X} m(C)$ *for all $\emptyset \subset A \subseteq X$*
- $m''(A) = m'(A)/\sum_{D \subseteq X} m'(D)$ *for all $\emptyset \subset A \subseteq X$*

*satisfies*

$$\text{if } \mu(A) > \mu(B) \text{ then } \mu_{m''}(A) > \mu_{m''}(B) \text{ for all } A, B. \tag{5.7}$$

*Proof.* Note that defining

$$m''(\emptyset) = \frac{m(0) - \min_{C \subseteq X} m(C)}{\sum_{S \subseteq X} m(D) - \min_{C \subseteq X} m(C)}$$

would lead to $\mu_{m''}(\emptyset) \neq 0$. Nevertheless, Equation 5.7 is also satisfied when $A = \emptyset$ or $B = \emptyset$. This is so because (i) if $A = \emptyset$, there is no $\mu(B)$ such that $\mu(A) = \mu(\emptyset) = 0 > \mu(B)$ and (ii) if $B = \emptyset$, then $\mu(A) > \mu(\emptyset) = 0$ and $\mu_{m''}(A) > \mu_{m''}(B) = 0$, because having $\mu_{m''}(A) \leq 0$ would require that $\mu(A) \leq \min_{C \subseteq X} m(C)$, but, as $\min_{C \subseteq X} m(C)$ is negative, this is impossible.

The proposition above permits us to obtain the following result on consistency (consistency was defined in Definition 5.10).

**Corollary 5.26.** *For each unconstrained fuzzy measure $\mu$, there exists a belief measure consistent with $\mu$.*

Nevertheless, $\mu$ and $\mu_{m''}$ are not consistent, as we might have $\mu_{m''}(A) > \mu_{m''}(D)$ when $\mu(A) \leq \mu(B)$. Table 5.4 illustrates this case. $\mu_{m''}(\{x_1, x_2\}) > \mu_{m''}(\{x_2\})$ but $\mu(\{x_1, x_2\}) = \mu(\{x_2\})$.

With regard to this proposition, we define a fuzzy measure $Bel$ that is a belief function and that is consistent with the measure $\mu$ in Example 5.12 in the sense that $Bel(A) > Bel(B)$ if and only if $\mu(A) > \mu(B)$.

*Example 5.27.* Let $\mu$ be the fuzzy measure in Example 5.12 and let $m_\mu$ be its Möbius transform; let $m''$ be the basic probability assignment defined from $m$ using Proposition 5.25. Then, the measure $\mu_{m''}$ defined from $m''$ is given in Table 5.5, and the basic probability assignment is given in Table 5.6.

| | |
|---|---|
| $\mu(\emptyset) = 0$ | $\mu(\{M, L\}) = 0.5526$ |
| $\mu(\{M\}) = 0.2237$ | $\mu(\{P, L\}) = 0.5526$ |
| $\mu(\{P\}) = 0.2237$ | $\mu(\{M, P\}) = 0.4474$ |
| $\mu(\{L\}) = 0.1842$ | $\mu(\{M, P, L\}) = 1$ |

**Table 5.5.** Fuzzy measure on the set $X = \{M, P, L\}$ that is consistent with the measure in Example 5.12

| | |
|---|---|
| $m(\emptyset) = 0$ | $m(\{M, L\}) = 0.1447$ |
| $m(\{M\}) = 0.2237$ | $m(\{P, L\}) = 0.1447$ |
| $m(\{P\}) = 0.2237$ | $m(\{M, P\}) = 0$ |
| $m(\{L\}) = 0.1842$ | $m(\{M, P, L\}) = 0.0790$ |

**Table 5.6.** Möbius transform of the measure given in Table 5.5

### 5.2.2 Possibility and Necessity Measures

Possibility and necessity measures are particular cases of belief and plausibility. We define them below and give some of their properties.

**Definition 5.28.** *A fuzzy measure Pos on a set $X$ is a possibility measure if it satisfies*

$$Pos(A \cup B) = \max(Pos(A), Pos(B)). \tag{5.8}$$

*A fuzzy measure Nec on a set $X$ is a necessity measure if it satisfies*

$$Nec(A \cap B) = \min(Nec(A), Nec(B)). \tag{5.9}$$

These equations, together with duality (Equation 5.1), establish a tight relation between the two measures. The relation is established in the next proposition.

**Proposition 5.29.** *Let Nec be a necessity measure and let Pos be its dual possibility measure; then, the following implications hold for all $A \in \wp(X)$:*

$$Nec(A) > 0 \quad implies \quad Pos(A) = 1$$

$$Pos(A) < 1 \quad implies \quad Nec(A) = 0$$

To define possibility measures, two alternative approaches can be used:

1. In a way similar to probability measures, that can be determined by probability distributions (or weighting vectors), possibility measures can be determined by possibility distributions. This is based on Equation 5.8, that ressembles to the additivity axiom ($\mu(A \cup B) = \mu(A) + \mu(B)$) replacing addition by maximum.
2. As necessity and possibility measures are, respectively, belief and plausibility measures, we can use basic probability assignments to define them.

We formalize both approaches below. First, we establish the relation between possibility measures and possibility distributions.

**Definition 5.30.** *A* possibility distribution *is a mapping $\pi$ from $X$ to $[0,1]$ such that $\max \pi(x_i) = 1$.*

**Proposition 5.31.** *Possibility measures and possibility distributions can be built one from the other:*

1. *Every possibility measure Pos is uniquely determined by a possibility distribution function $\pi$, defined as follows:*

$$\pi(x_i) := Pos(\{x_i\}) \text{ for all } x_i \in X$$

2. *Let $\pi$ be a possibility distribution defined over the set $X$, then the function $Pos : X \rightarrow [0,1]$ defined as:*

$$Pos(A) := \max_{a \in A} \pi(a) \text{ for all } A \subseteq X$$

*is a possibility measure (the possibility measure inferred from $\pi$).*

**Proposition 5.32.** *Let $\pi$ be a possibility distribution, then $Pos(A) = \max_{a \in A} \pi(a)$ and $Nec(A) = 1 - \max_{a \notin A} \pi(a)$ are dual (in the sense of Definition 5.6).*

This result implies that the definition of *Pos* only requires a value for each element in $X$. Thus, the number of values required to define *Pos* is $|X|$.

Possibility measures correspond to a restricted type of basic probability assignments: the consonant ones. We define now these assignments and, after that, we show how such assignments can be extracted from the measures. The obtention of the measures from the assignments follows Equation 5.5.

**Definition 5.33.** *A basic probability assignment $m$ is* consonant *if and only if non zero values of $m$ belong to a set of nested subsets of $X$. I.e., there is a complete sequence of nested subsets:*

$$A_1 \subset A_2 \subset ... \subset A_n = X$$

*and $m(B) = 0$, for all $B \neq A_1, A_2, ..., A_n$.*

**Proposition 5.34.** *Let Pos be a plausibility measure on $X = \{x_1, ..., x_n\}$, and let $\pi$ be its corresponding possibility distribution (with no loss of generality, we assume that $1 = \pi(x_1) \geq \pi(x_2) \geq ... \geq \pi(x_n)$); then, the corresponding basic probability assignment $m$ is of the form*

$$m(A) = \begin{cases} \pi(x_i) - \pi(x_{i+1}) & \text{if } A = \{x_1, ..., x_i\} \text{ for } i = 1, ..., n-1 \\ \pi(x_n) & \text{if } A = X \\ 0 & \text{otherwise.} \end{cases}$$

Possibility measures do not satisfy additivity, and are the minimal measure that can be built from $\pi$ (in the sense of Definition 5.13).

## 5.3 ⊥-Decomposable Fuzzy Measures

In previous sections, we have studied additive measures and possibility measures. In both cases, the measure of a set is composed from the measure on the singletons. This is, the measure is composed either from a probability or a possibility distribution. The difference is based on whether the composition is achieved using addition or using maximum. In this section, we describe ⊥-decomposable fuzzy measures. In such measures, the composition is in terms of a t-conorm ⊥.

**Definition 5.35.** *A fuzzy measure $\mu$ on a set $X$ is a ⊥-decomposable fuzzy measure if there exists a t-conorm ⊥ such that, for all $A, B \subseteq X$ with $A \cap B = \emptyset$, it holds*

$$\mu(A \cup B) = \mu(A) \perp \mu(B).$$

Note that, in this definition, the t-conorm guarantees monotonicity.

⊥-decomposable fuzzy measures permit their definition on the basis of a t-conorm and a set of values for the singletons. The only requirement is that the combination of all such values using a t-conorm be equal to 1. This is required for the boundary condition $\mu(X) = 1$.

**Proposition 5.36.** *Let $\perp$ be a t-conorm and let $v : X \rightarrow [0, 1]$ be such that:*

$$v(x_1) \perp ... \perp v(x_N) = 1;$$

*then, the fuzzy measure defined by*

$$\mu_v(A) = \begin{cases} v(x_i) & \text{if } A = \{x_i\} \text{ for any } i = 1, ..., |X| \\ \perp_{x_i \in A} v(x_i) & \text{otherwise} \end{cases}$$

*is a ⊥-decomposable measure.*

An alternative expression exists for $\mu(A)$ when the t-conorm is an Archimedean one with a known increasing generator $g$ (recall Theorem 2.49):

$$\mu_v(A) = \begin{cases} v(x_i) & \text{if } A = \{x_i\} \text{ for any } i = 1, ..., |X| \\ g^{(-1)}(\sum_{x_i \in A} g(v(x_i))) & \text{otherwise.} \end{cases}$$

Note that a probability measure can be seen as a ⊥-decomposable fuzzy measure with $\perp(x, y) = \min(x + y, 1)$ (Lukasiewicz t-conorm). However, note that the reversal is not always true, because it could be the case that $\sum_{x_i \in X} \mu(\{x_i\}) > 1$, and in this case the Lukasiewicz t-conorm is 1. An example of this case is given below. When the equality holds, i.e., $\sum_{x_i \in X} \mu(\{x_i\}) = 1$, the application $v : X \rightarrow [0, 1]$ is a weighting vector, and $\mu$ is additive.

*Example 5.37.* Let $X = \{x_1, x_2, x_3, x_4, x_5\}$, let $\bot(x, y) = \min(x + y, 1)$ be the Lukasiewicz t-conorm, and let $v : X \rightarrow [0, 1]$ be defined by $v(x_1) = 0.05$, $v(x_2) = 0.1$, $v(x_3) = 0.2$, $v(x_4) = 0.4$, and $v(x_5) = 0.8$. The corresponding ⊥-decomposable fuzzy measure is not an additive one. Note, for example, that

$$\mu(\{x_2, x_3, x_4, x_5\}) = 1 \neq \mu(\{x_2, x_3\}) + \mu(\{x_4, x_5\}) = 1.3.$$

Another particular case of these measures is when the t-conorm is the maximum. Then, the corresponding measure is a possibility one.

### 5.3.1 Sugeno λ-measures

Sugeno λ-measures are also an example of ⊥-decomposable fuzzy measures.

**Definition 5.38.** *Let $\mu$ be a fuzzy measure; then, $\mu$ is a Sugeno λ-measure if for some fixed $\lambda > -1$ it holds that*

$$\mu(A \cup B) = \mu(A) + \mu(B) + \lambda\mu(A)\mu(B) \tag{5.10}$$

*for all $A \cap B = \emptyset$*

Therefore, the definition of a measure only requires the values for all the singletons and $\lambda$. The following proposition establishes this fact and gives the expression for the general case.

**Proposition 5.39.** *Let $v : X \rightarrow [0, 1]$ and $\lambda > 1$ be such that*

$$(1/\lambda)(\Pi_{x_i \in X}[1 + \lambda v(x_i)] - 1) = 1 \text{ if } \lambda \neq 0;$$
$$\sum_{x_i \in X} v(x_i) = 1 \qquad\qquad \text{ if } \lambda = 0;$$

*then, the fuzzy measure defined by*

$$\mu(A) = \begin{cases} v(x_i) & \text{if } A = \{x_i\} \\ (1/\lambda)(\Pi_{x_i \in A}[1 + \lambda\mu(x_i)] - 1) & \text{if } |A| \neq 1 \text{ and } \lambda \neq 0 \\ \sum_{x_i \in A} \mu(x_i) & \text{if } |A| \neq 1 \text{ and } \lambda = 0 \end{cases}$$

*is a Sugeno λ-measure.*

As stated, Sugeno λ-measures are a particular case of ⊥-decomposable ones. This is implied by Equation 5.10 and the additional requirement that the Sugeno λ-measure is a fuzzy measure (and thus bounded by 1). As the value 1 cannot be exceeded, Equation 5.10 can be considered as if values $\mu(A)$ and $\mu(B)$ were combined using the following t-conorm:

$$\bot(x, y) = min(1, x + y + \lambda xy). \tag{5.11}$$

This is Sugeno's t-conorm, as can be seen in Example 2.48.

For countable sets, Sugeno $\lambda$-measures are a special subclass of belief and plausibility measures. For $\lambda \geq 0$, the measure is a belief one, and for $\lambda \leq 0$, the measure is a plausibility one. Note that $\lambda = 0$ corresponds to both types and, also, to additive fuzzy measures.

We have seen that a Sugeno $\lambda$-measure is determined from the values $\mu(\{x_i\})$ and $\lambda$. In fact, as shown below, an important result proves that the measure on the singletons completely determines $\lambda$. Accordingly, a measure of this family solely requires $|X|$ values in order to be defined.

**Proposition 5.40.** *Let $\mu$ be a Sugeno $\lambda$-measure; then, for a fixed set of $0 < \mu(\{x_i\}) < 1$, there exists a unique $\lambda \in (-1, +\infty)$ and $\lambda \neq 0$ that satisfies $\mu(X) = 1$, that is, satisfies*

$$\lambda + 1 = \Pi_{i=1}^{N}(1 + \lambda\mu(\{x_i\})).$$

This proposition exploits the fact that

$$\mu(X) = (1/\lambda)(\Pi_{x_i \in X}[1 + \lambda v(x_i)] - 1) = 1.$$

The proposition establishes that, given the $|X|$ values for the singletons, solving a $(n-1)$ degree polynomial, we will find the suitable value for $\lambda$.

*Example 5.41.* Let $\mu$ be a Sugeno $\lambda$-measure on $\{M, P, L\}$, with $\mu(\{M\}) = 0.2237$, $\mu(\{P\}) = 0.2237$, and $\mu(\{L\}) = 0.1842$. Then, using Proposition 5.40, we get $\lambda = 2.3860$. This value for $\lambda$ is obtained as follows:

$$\lambda + 1 = (1 + 0.2237\lambda) \cdot (1 + 0.2237\lambda) \cdot (1 + 0.1842\lambda).$$

Thus,

$$0 = -0.3684\lambda + 0.13245277\lambda^2 + 0.0092176795\lambda^3.$$

The solutions of this equations are

$\lambda_1 = 0$
$\lambda_2 = 2.385385$
$\lambda_3 = -16.754812$

The only acceptable value is $\lambda_2$, as the others violate some of the constraints of fuzzy measures. In particular, $\lambda_1$ implies that

$$\mu(X) = 0.2237 + 0.2237 + 0.1842 \neq 1,$$

and $\lambda_3$ (which is invalid, as it is no larger than -1) leads to negative values for $\mu$. For example,

$$\mu(\{M, P\}) = 0.2237 + 0.2237 + 0.2237 \cdot 0.2237 \cdot \lambda_2 = -0.39103916.$$

In Table 5.7 we show the values obtained for $\mu$ using $\lambda_2$.

Note that the values for the singletons correspond to the values given for the same sets in Example 5.27. Nevertheless, the value for the other sets are not the same, as the measure in Example 5.27 is not a $\lambda$-measure.

| | |
|---|---|
| $\mu(\emptyset) = 0$ | $\mu(\{M, L\}) = 0.5061911$ |
| $\mu(\{M\}) = 0.2237$ | $\mu(\{P, L\}) = 0.5061911$ |
| $\mu(\{P\}) = 0.2237$ | $\mu(\{M, P\}) = 0.5667687$ |
| $\mu(\{L\}) = 0.1842$ | $\mu(\{M, P, L\}) = 1$ |

**Table 5.7.** Sugeno $\lambda$-measure on the set $X = \{M, P, L\}$ with the measures in the singletons equal to the one in Example 5.27

Several results are known about these fuzzy measures. A few are given below. First, we review aspects related to the monotonicity of two measures.

**Proposition 5.42.** *Let $\mu_1$ and $\mu_2$ be two Sugeno $\lambda$-measures with parameters $\lambda_1$ and $\lambda_2$, respectively. Then, if $\mu_1(\{x_i\}) \geq \mu_2(\{x_i\})$ for all $x_i \in X$, it holds that $\lambda_1 < \lambda_2$.*

When $\lambda$ tends to $-1$, the measure tends to be such that $\mu(A) = 1$ for all $A \neq \emptyset$ and $\mu(\emptyset) = 0$. Otherwise, when $\lambda$ tends to $+\infty$, the measure tends to be 0 for all $A \neq X$ and 1 for $A = X$.

Now, we review another result on the relationship between fuzzy measures and their dual ones.

**Proposition 5.43.** *Let $\mu_\lambda$ be a Sugeno $\lambda$-measure; then, the following holds:*

1. *The dual of $\mu_\lambda$ is $\mu_{-\lambda/(\lambda+1)}$. That is, if $\mu_\lambda$ is a belief, then $\mu_{-\lambda/(\lambda+1)}$ is a plausibility.*
2. *The value that separates a set $A$ and its complement $X \setminus A$ is*

$$(-1 + \sqrt{1 + \lambda})/\lambda$$

*That is, for a given $\mu_\lambda$, if $\mu_\lambda(A) \geq (-1 + \sqrt{1 + \lambda})/\lambda$, then $\mu_\lambda(X \setminus A) \leq (-1 + \sqrt{1 + \lambda})/\lambda$. Note that, for additive measures (probabilities), this value is 0.5, because if $P(A) \geq 0.5$, then $P(X \setminus A) = 1 - P(A) \leq 0.5$.*

### 5.3.2 Hierarchically ⊥-Decomposable Fuzzy Measures

In ⊥-decomposable fuzzy measures, the measure for a set $A$ is defined as the combination through the t-conorm ⊥ of the measures for the singletons $\{a_i\} \in A$. In some sense, the combination is *homogeneous* for all the elements in $A$, as all elements are combined using the same t-conorm. Hierarchically, ⊥-Decomposable Fuzzy Measures (HDFM) weaken this constraint, allowing different t-conorms in the combination process. To allow for different t-conorms, elements in $X$ are structured into a hierarchical structure (a dendrogram), and then a t-conorm is attached to each node. The measure for a set $A \subseteq X$ is computed using the hierarchy, and, more precisely, using a kind of *projection* of $A$ on the hierarchy. We will formalize this. We start with an example.

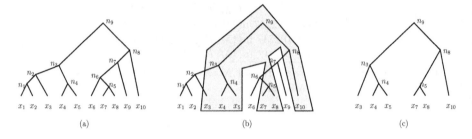

**Fig. 5.2.** A Hierarchically ⊥-Decomposable Fuzzy Measure: (a) hierarchy; (b) projection of a set $A$ on the hierarchy; (c) hierarchy corresponding to the set $A$

*Example 5.44.* Figure 5.2 (a) displays a hierarchical structure to be used in the definition of a hierarchically ⊥-decomposable fuzzy measure. In this example, $X$ consists of the set $X = \{x_1, x_2, x_3, \ldots, x_{10}\}$. The hierarchy consists of the nodes $\{n_1, n_2, \ldots, n_9\}$, where each node $n_i$ has an attached t-conorm $\perp_i$. The definition of the measure for a set $A$ is computed by decomposing $A$ into its components following the outline given by the hierarchical structure for $X$.

Figure 5.2 (b) shows the *projection* of the set $A = \{x_3, x_4, x_5, x_7, x_8, x_9\}$ on the dendrogram of Figure 5.2 (a). Figure 5.2 (c) corresponds to Figure 5.2 (b) once unnecessary elements $x_i$ (elements not in $A$) as well as unnecessary nodes (nodes linking elements in $A$ with elements not in $A$) are removed.

Using the t-conorm $\perp_i$ and the structure of Figure 5.2(c), we can compute the measure for the set $A$. This measure is computed bottom-up. For simplicity, we can consider that we compute a value for each node, and that the value $n_9$ is the measure for the set $A$. That is, the value for node $n_4$ is $\perp_4(\mu(\{x_4\}), \mu(\{x_5\}))$. The value for the node $n_3$ is:

$$\perp_3(\mu(\{x_3\}), value(n_4)) = \perp_3(\mu(\{x_3\}), \perp_4(\mu(\{x_4\}), \mu(\{x_5\}))).$$

The value for node $n_9$, which in this case is $\mu(A)$, corresponds to

$$\perp_9(\perp_3(\mu(\{x_3\}), \perp_4(\mu(\{x_4\}), \mu(\{x_5\})))\perp_8(\perp_5(\mu(\{x_7\}), \mu(\{x_8\}))\mu(\{x_{10}\})))$$

To formalize these measures, we need, first, to formalize the hierarchy of elements.

**Definition 5.45.** *H is a hierarchy of elements X if and only if the following conditions are fulfilled:*

(i) *All the elements in X belong to the hierarchy, and the corresponding nodes are the leaves of the hierarchy:*
   *For all x in X, $\{x\} \in H$.*
(ii) *There is only one root in the hierarchy, and it is denoted by* root. *A node is the root if it is not included in any other node:*
   *if* root $\in H$, *then there is no other node $m \in H$ such that* root $\in m$.

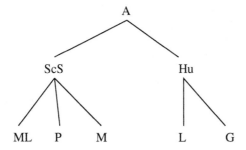

**Fig. 5.3.** Hierarchy of elements corresponding to the subjects Mathematical Logic (*ML*), Physics (*P*), Mathematics (*M*), Literature (*L*) and Greek (*G*). *ScS* stands for Scientific Subjects and *Hu* stands for humanities

*(iii) All nodes belong to one and only one node, except for the root:*
    *if $n \in H$ and $n \neq$ root, then there exists a single $m \in H$ such that $n \in m$.*
*(iv) All nodes that contain only one element are singletons:*
    *if $|h| = 1$, then there exists $x \in X$ such that $h = \{x\}$ for all $h \in H$.*
*(v) All non-singletons are defined in terms of nodes that are in the tree:*
    *if $|h| \neq 1$, then, for all $h_i \in h$, $h_i \in H$.*

Definition 5.45 builds the hierarchy H, defining first a set of nodes, where each node is defined as a set of other nodes. This way, the definition given below for fuzzy measures of this kind is simple. Alternative definitions, where nodes are subsets of the set $X$, are also possible (and simpler), but then the definition of the measure is more complex.

We consider below an example of hierarchy that will be used later to define a hierarchically ⊥-decomposable fuzzy measure. The elements of this hierarchy follow Example 4.28, and are an extension of those in Example 5.12.

*Example 5.46.* The evaluation of students in a high school is based on two sets of subjects: scientific subjects (*ScS*) and Humanities (*Hu*). The former set includes *Mathematics* (*M*), *Physics* (*P*), and *Mathematical Logic* (*ML*). The latter set includes *Literature* (*L*) and *Greek* (*G*).

The hierarchy corresponding to these subjects is as follows (see Figure 5.3).

$$H = \{\{ML\}, \{M\}, \{P\}, \{L\}, \{G\}, ScS, Hu, A\},$$

with

$$ScS = \{\{ML\}, \{M\}, \{P\}\}, Hu = \{\{G\}, \{L\}\}, \text{ and } A = \{ScS, Hu\}.$$

The definition of hierarchically ⊥-decomposable fuzzy measures requires two objects: the extension of a node and a labeled hierarchy. The former is defined over a node in the hierarchy as the set of elements in X that are embedded in that node. The labeled hierarchy assigns to each leaf in the hierarchy a real value in the unit interval, and, for each node that is not a leaf, a t-conorm.

**Definition 5.47.** *Let $H$ be a hierarchy according to Definition 5.45 and let $h$ be a node in $H$; then, the* extension *of $h$ in $H$ is defined as:*

$$EXT(h) := \begin{cases} h & \text{if } |h| = 1 \\ \cup_{h_i \in h} EXT(h_i) & \text{if } |h| \neq 1. \end{cases}$$

**Definition 5.48.** *Let $H$ be a hierarchy according to Definition 5.45; then, a* labeled hierarchy *$L$ for $H$ is a tuple $L =< H, \perp, m >$, where $\perp$ is a function that maps each node $n \in H$ that is not a leaf into a t-conorm, and $m$ is a function that maps each singleton into a value of the unit interval.*

*For simplicity, we will express $\perp(h)$ by $\perp_h$.*

Labeled hierarchies define fuzzy measures. The measure of a set of elements is based on the values that the function $m$ associates with the singletons (the elements of that set), and the t-conorms of the nodes with a nonempty intersection with the set. For a singleton, the value of $m$ is considered the measure of the singleton. For sets, the measure is defined recursively using the nodes in the hierarchy. The following definition describes how the measure is computed from the hierarchy.

**Definition 5.49.** *Let $L =< H, \perp, m >$ be a labeled hierarchy according Definition 5.49; then, the corresponding* Hierarchically $\perp$-Decomposable Fuzzy Measure *(HDFM for short) of a set $B$ is defined as $\mu(B) = \mu_{root}(B)$, where $\mu_A$ for a node $A = \{a_1, ..., a_n\}$ is defined recursively as*

$$\mu_A(B) = \begin{cases} 0 & \text{if } |B| = 0 \\ m(B) & \text{if } |B| = 1 \\ \perp_A(\mu_{a_1}(B_1), ..., \mu_{a_n}(B_n)) & \text{if } |B| > 1. \end{cases}$$

*Here, $B_i = B \cap EXT(a_i)$ for all $a_i$ in $A$.*

**Proposition 5.50.** *When $\mu(X) = 1$, Definition 5.49 leads to a fuzzy measure.*

*Proof.* Note that

(i) the values of the measure belong to the unit interval. This is implied by the fact that the fuzzy measure is built only from the function $m$ (a function into the unit interval) and the t-conorms (which are functions from $[0, 1] \times [0, 1]$ into $[0, 1]$).
(ii) $\mu(\emptyset) = 0$ is implied by the definition.
(iii) Monotonicity is implied by the monotonicity of the t-conorm.

*Example 5.51.* Let us consider the fuzzy measure in Example 5.12. This measure can be represented as a hierarchically decomposable fuzzy measure with $X = \{M, P, L\}$, the hierarchy $H$ as in Figure 5.4, $m$ defined according to Example 5.12 (i.e., $m(M) = m(P) = 0.45$ and $m(L) = 0.3$), and the t-conorms $\perp_{ScS}$ and $\perp_A$ defined as follows:

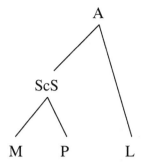

**Fig. 5.4.** Hierarchy of elements for representing the fuzzy measure defined in Example 5.12

- $S_{ScS}(x, y) = (x^w + y^w)^{1/w}$, with $w = (ln2)/(ln0.5 - ln0.45) = 6.5788$
- $S_A(x, y) = f^{(-1)}(f(x) + f(y))$, where $f(x)$ is defined as:

$$f(x) = \begin{cases} 20x & \text{if } x \in [0, 1/2] \\ 3 + 14x & \text{if } x \in [1/2, 3/4] \\ 6 + 10x & \text{if } x \in [3/4, 1]. \end{cases}$$

Note that this hierarchy and the labelling lead to a fuzzy measure that is equivalent to the one in Example 5.12 for all subsets of $X$.

Hierarchically decomposable fuzzy measures compute the measure for a subset B in X in terms of the measure of disjoint subsets of B. Then, the measures are combined using the t-conorm S of the smallest node in the tree that encompasses all the elements in B. The same procedure is applied recursively until we get values for the leaves/singletons (the values in $m$). This is shown in the following example.

*Example 5.52.* Let $X$ and $H$ be as in Example 5.46 (i.e., $X = \{ML, P, M, L, G\}$ and $H$ as in Figure 5.3). Then,

$$\mu(\{P, M, L, G\}) = \perp_A(\mu(\{P, M, L, G\} \cap EXT(ScS)), \mu(\{P, M, L, G\} \cap EXT(L))).$$

This can be further decomposed into

$$\mu(\{P, M, L, G\}) = \perp_A(\perp_{ScS}(m(P), m(M)), \perp_{Hu}(m(L), m(G))).$$

In the definition above we have considered that the nodes of the hiearchy could gather two or more nodes. In fact, it is possible to constrain all nodes to gather only two nodes. This is so because t-conorms are associative. The following example illustrates this situation.

*Example 5.53.* Let $n$ be a node in $H$ defined by subnodes $n_i$, as in: $n = \{n_1, n_2, n_3, \ldots, n_m\}$. Let $\perp_n$ be the t-conorm attached to $n$. Then, this is equivalent to having nodes $n_2^* = \{n_1, n_2\}$, $n_3^* = \{n_2^*, n_3\}$, $\ldots$, $n_m^* = \{n_{n-1}^*, n_m\}$, with t-conorms $\perp_{n_i^*} = \perp_n$.

**Properties**

In the introduction of this section, we said that HDFMs generalize $\perp$-decomposable fuzzy measures. In fact, the latter correspond to one-level HDFMs.

**Definition 5.54.** *Let $\mu$ be a Hierarchically Decomposable Fuzzy Measure on $X$ with labeled hierarchy $L =< H, S, m >$; then, if for each $x \in X$, $\{x\} \in root$, we say that $\mu$ is a one-level HDFM.*

**Proposition 5.55.** *A one level HDFM is a $\perp$-decomposable fuzzy measure with $\perp = S_{root}$.*

**Definition 5.56.** *Let $\mu$ be an HDFM with labeled hierarchy $L =< H, S, m >$; then, $\mu$ is a two-level HDFM if, for each $x \in X$, it holds that there exists an $n \in H$ such that $x \in n$ and $n \in root$.*

**Definition 5.57.** *Let $\mu$ be a two-level fuzzy measure; then, $\mu$ is an additive two-level HDFM if $S_{root}(x, y) = min(1, x + y)$.*

**Proposition 5.58.** *An additive two-level HDFM is an inter-additive fuzzy measure.*

The last result shows that inter-additive fuzzy measures can be seen as a generalization of additive two-level HDFMs. In fact, any HDFM with $\perp_{root} = \hat{+}$ is an inter-additive measure.

## 5.4 Distorted Probabilities

Another family of fuzzy measures is that of distorted probabilities. They are defined in terms of a probability distribution and a function that distorts them. We will review this family below and give some results that link such measures with the ones presented previously. We start by defining when a fuzzy measure is represented by a function and a probability, and, then, restricting the function to be strictly increasing, we reach distorted probabilities.

**Definition 5.59.** *Let $f$ be a real-valued function on $[0, 1]$ and let $P$ be a probability measure on $(X, \wp(X))$. We say that $f$ and $P$ represent a fuzzy measure $\mu$ on $(X, \wp(X))$ if and only if $\mu(A) = f(P(A))$ for all $A \in \wp(X)$.*

**Definition 5.60.** *Let $f$ be a real-valued function on $[0, 1]$. We say that $f$ is strictly increasing with respect to a probability measure $P$ if and only if $P(A) < P(B)$ implies $f(P(A)) < f(P(B))$. We say that $f$ is nondecreasing with respect to a probability measure $P$ if and only if $P(A) < P(B)$ implies $f(P(A)) \leq f(P(B))$.*

Using the two definitions, we define distorted probabilities as follows.

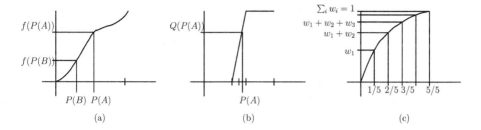

**Fig. 5.5.** Distorted probabilities: (a) computation of $P(A)$ and $P(B)$ for $\mu = f \circ P$; (b) computation of $\mu(A)$ for the measure "at least around 50% of the probability"; (c) discrete representation of the distortion function using a weighting vector **w**

**Definition 5.61.** *Let $\mu$ be a fuzzy measure on $(X, \wp(X))$. We say that $\mu$ is a distorted probability if it is represented by a probability distribution $P$ on $(X, \wp(X))$ and a function $f$ that is nondecreasing with respect to a probability $P$.*

Figure 5.5 (a) illustrates the computation of the measure $\mu$ when $\mu = f \circ P$ for two sets $A$ and $B$ such that $P(A) \leq P(B)$. A few remarks follow with respect to the previous definition.

1. The definition given above uses a nondecreasing function $f$. Nevertheless, alternative definitions with strictly increasing $f$ are also used. If $f$ is strictly increasing, then $\mu = P$ and $\mu' = f \circ P$ are consistent fuzzy measures (according to Definition 5.10). Instead, if $f$ is a nonincreasing function, then $\mu$ is consistent with $\mu'$, but $\mu'$ is not always consistent with $\mu$.

2. For the sake of simplicity, since $X$ is a finite set, a strictly increasing function $f$ with respect to $P$ can be regarded as a strictly increasing function on $[0, 1]$. Points other than $\{P(A)|A \in \wp(X)\}$ are not relevant in our definition, as they are not really used to compute the measure. Note that the function $f$ is only applied to $P(A)$ for all $A \subseteq X$.

3. Distortion functions $f$ can be seen as fuzzy quantifiers. Under this interpretation, $f$ measures to what extent a given probability $P(A)$ satisfies the quantifier. So, in the case of $f$ being a fuzzy quantifier $Q$, the measure $Q \circ P$ stands for "$P(A)$ is $Q$." For example, if we consider the quantifier, "at least around 50%," we have that the fuzzy measure induced by a probability $O$ and the quantifier $Q$ defined by $\mu = Q \circ P$ stands for the measure "at least around 50% of the probability." So, for all sets $A \subseteq X$ with $P(A) \leq 0.4$, we will have $\mu(A) = 0$, and for all $P(A) \geq 0.6$, we will have $\mu(A) = 1$. Besides, for those $A \subseteq X$ with $0.4 \leq P(A) \leq 0.6$, we will have a measure between 0 and 1. Figure 5.5 (b) illustrates this case for a set $A$ such that $0.4 < P(A) < 0.6$.

4. For some applications, it is useful to give a discrete representation of the function in terms of a weighting vector $\mathbf{w} = (w_1, \ldots, w_N)$ (*i.e.*, $w_i \geq 0$

and $\sum_i w_i = 1$). In this case, the function $f$ can be interpolated from the set $\{(i/N, \sum_{j=1,\ldots,i} w_j)\}_{i=\{0,\ldots,N\}}$. This is represented in Figure 5.5 (c) for $N = 5$. This will be studied in more detail in Section 6.1.3.

All fuzzy measures can be represented in terms of a probability distribution and a real-valued function (see Theorem 5.62 below). Nevertheless, not all measures are distorted probabilities. This is so because, for most measures $\mu$, none of the pairs $(f, P)$ that can represent the measure $\mu$ includes a strictly increasing function $f$. In fact, the number of fuzzy measures that are distorted probabilities is rather small in comparison with the total number of fuzzy measures.

**Theorem 5.62.** *For every fuzzy measure $\mu$ on $(X, \wp(X))$, there exist a polynomial $f$ and probability $P$ on $(X, \wp(X))$ such that $\mu = f \circ P$.*

To compare the number of unconstrained fuzzy measures and of distorted probabilities, we classify the measures into sets of consistent fuzzy measures (following Definition 5.10). Then, the following example, illustrates the number of such sets for both types of measures for a set $X$ with 3 elements.

*Example 5.63.* Let $X = \{1, 2, 3\}$, and let $\mu$ such that $\mu(\{1\}) < \mu(\{2\}) < \mu(\{3\})$. Then,

a) when $\mu$ is a distorted probability, either one of the following holds:
$$\mu(\emptyset) < \mu(\{1\}) < \mu(\{2\}) < \mu(\{3\}) < \mu(\{1, 2\}) < \mu(\{1, 3\}) < \mu(\{2, 3\}) < \mu(X)$$
$$\mu(\emptyset) < \mu(\{1\}) < \mu(\{2\}) < \mu(\{1, 2\}) < \mu(\{3\}) < \mu(\{1, 3\}) < \mu(\{2, 3\}) < \mu(X);$$
b) when $\mu$ is an unconstrained fuzzy measure, one of the following holds:
$$\emptyset < \mu(\{1\}) < \mu(\{2\}) < \mu(\{1, 2\}) < \mu(\{3\}) < \mu(\{1, 3\}) < \mu(\{2, 3\}) < \mu(X)$$
$$\emptyset < \mu(\{1\}) < \mu(\{2\}) < \mu(\{3\}) < \mu(\{1, 2\}) < \mu(\{1, 3\}) < \mu(\{2, 3\}) < \mu(X)$$
$$\emptyset < \mu(\{1\}) < \mu(\{2\}) < \mu(\{3\}) < \mu(\{1, 3\}) < \mu(\{1, 2\}) < \mu(\{2, 3\}) < \mu(X)$$
$$\emptyset < \mu(\{1\}) < \mu(\{2\}) < \mu(\{3\}) < \mu(\{1, 3\}) < \mu(\{2, 3\}) < \mu(\{1, 2\}) < \mu(X)$$
$$\emptyset < \mu(\{1\}) < \mu(\{2\}) < \mu(\{1, 2\}) < \mu(\{3\}) < \mu(\{2, 3\}) < \mu(\{1, 3\}) < \mu(X)$$
$$\emptyset < \mu(\{1\}) < \mu(\{2\}) < \mu(\{3\}) < \mu(\{1, 2\}) < \mu(\{2, 3\}) < \mu(\{1, 3\}) < \mu(X)$$
$$\emptyset < \mu(\{1\}) < \mu(\{2\}) < \mu(\{3\}) < \mu(\{2, 3\}) < \mu(\{1, 2\}) < \mu(\{1, 3\}) < \mu(X)$$
$$\emptyset < \mu(\{1\}) < \mu(\{2\}) < \mu(\{3\}) < \mu(\{2, 3\}) < \mu(\{1, 3\}) < \mu(\{1, 2\}) < \mu(X)$$

Thus, among the eight sets obtained for unconstrained fuzzy measures, only two are also distorted probabilities. Therefore, there are six *types*/sets of fuzzy measures that cannot be represented as distorted probabilities.

Table 5.8 compares the sets of consistent measures for different cardinalities of $X$ and for both types of measures. It can be observed that the larger the set $X$, the larger the *gap* between the two sets. Thus, the number of different distorted probabilities (with respect to consistence) is rather small with respect to the number of unconstrained fuzzy measures. $m$-dimensional distorted probabilities have been defined to fill this *gap*.

| $|X|$ | Distorted Probabilities | Unconstrained Fuzzy Measures |
|-------|------------------------|------------------------------|
| 1 | 1 | 1 |
| 2 | 1 | 1 |
| 3 | 2 | 8 |
| 4 | 14 | 70016 |
| 5 | 546 | $\mathcal{O}(10^{12})$ |
| 6 | 215470 | – |

**Table 5.8.** Number of nonempty consistent sets for both distorted probabilities and unconstrained fuzzy measures when $\mu(\{1\}) \leq \mu(\{2\}) \leq \dots$. Here, $\mathcal{O}(10^{12})$ is an estimate

$$
\begin{array}{c|cccc}
\{L\} & 0.3 & 0.9 & 0.9 & 1.0 \\
& & & & \\
\emptyset & 0 & 0.45 & 0.45 & 0.5 \\
\hline
& \emptyset & \{M\} & \{P\} & \{M,P\}
\end{array}
$$

**Fig. 5.6.** Graphical interpretation of the fuzzy measure in Example 5.65 as a two-dimensional distorted probability

### 5.4.1 $m$-Dimensional Distorted Probabilities

In a way similar to $k$-order additive fuzzy measures, those with $k = 1, \dots, |X|$ cover the whole set of measures; $m$-dimensional distorted probabilities also cover this set with different $k$. We define them below.

**Definition 5.64.** Let $\mathcal{P} = \{X_1, X_2, \dots, X_m\}$ be a partition of $X$; then, we say that $\mu$ is an at most $m$-dimensional distorted probability if there exists a function $f$ on $[0,1]^m$ and probabilities $P_i$ on $(X_i, \wp(X_i))$ such that

$$\mu(A) = f(P_1(A \cap X_1), P_2(A \cap X_2), \dots, P_m(A \cap X_m)), \qquad (5.12)$$

where $f$ is strictly increasing with respect to the $i$th axis for all $i = 1, 2, \dots, m$.

We say that an at most $m$-dimensional distorted probability $\mu$ is an $m$-dimensional distorted probability if $\mu$ is not an at most $m - 1$ dimensional one.

As with the case of $k$-order additive fuzzy measures, we have that any fuzzy measure can be represented as a $m$-dimensional distorted probability with an appropriate value of $m$. Therefore, if $[DP]_m^X$ is the set of all $m$-dimensional distorted probabilities on $X$, $\{[DP]_m^X\}_{m \in \{1, \dots, |X|\}}$ is a partition of the set of all fuzzy measures on $X$.

We now reconsider Example 5.12 in the light of distorted probabilities.

*Example 5.65.* The fuzzy measure on $X := \{M, L, P\}$ given in Example 5.12 (and outlined in Table 5.2) is a two-dimensional distorted probability. For

$$
\begin{array}{c|ccc}
1 & 0.3 & 0.9 & 1.0 \\[4pt]
0 & 0 & 0.45 & 0.5 \\[4pt]
\hline
 & 0 & 1/2 & 1
\end{array}
$$

**Fig. 5.7.** The function $f$ in Example 5.65. Note that only relevant values with respect to the probabilities $P_1$ and $P_2$ are displayed

building the distorted probability, we need to consider two sets. One set corresponds to the science subjects $\{M, P\}$ and the other corresponds to the literary subject $\{L\}$. A graphical interpretation of the measure is given in Figure 5.6. In this figure, each axis represents a partition element. Therefore, one axis corresponds to the set $\{L\}$ and the other to the science subjects $\{M, P\}$. The values of the measure are also represented in the figure: for each pair of disjoint sets $(A, B)$, we have the value of the measure for $A \cup B$. For example, the value for $\{L\} \times \{M, P\}$ corresponds to $\mu(\{L\} \cup \{M, P\}) = \mu(X) = 1$. It can be seen that the measure is increasing in both axes. Using the probabilities $P_1$ on the set $\{L\}$ and $P_2$ on the set $\{M, P\}$, defined as $P_1(\{L\}) = 1$ and $P_2(\{M\}) = P_2(\{P\}) = 0.5$, and using the distortion function

$$
f(x, y) := -0.8x^2 + 0.4xy - 0.2yx^2 + 1.3x + 0.3y,
$$

we have $\mu(A) = f(P_1(A \cap \{L\}), P_2(A \cap \{M, P\}))$. Note that the function $f$ is strictly increasing with respect to probabilities $P_1$ and $P_2$, but not strictly increasing in all $[0, 1] \times [0, 1]$. A graphical representation of the function $f$ is given in Figure 5.7. Only the relevant values with respect to the probabilities $P_1$ and $P_2$ are given in the figure.

No relation has been established between $m$-dimensional distorted probabilities and $k$-order additive fuzzy measures. In fact, the space of measures is different, and, thus, in some situations the use of a $k$-order additive fuzzy measure is preferable (as it gives a more compact representation), while in some other situations an $m$-dimensional one is preferable. We show below an example of an $|X|$-order additive fuzzy measure that can be easily represented as a two-dimensional distorted probability.

*Example 5.66.* Let us consider the distorted probability $\mu_{\mathbf{p},\mathbf{w}}$ over $X = \{x_1, x_2, x_3, x_4, x_5\}$ generated by the probability distribution $\mathbf{p} = (0.2, 0.3, 0.1, 0.2, 0.1)$, and a distortion function generated from the weighting vector $\mathbf{w} = (0.1, 0.2, 0.4, 0.2, 0.1)$ (see remark 4 after Definition 5.61). The measure for all subsets of $X$, as well as the Möbius transform of this measure, are given in Table 5.9 (column $\mu_{\mathbf{p},\mathbf{w}}$).

As the Möbius transform is different from 0 for all subsets of $X$, this means that $\mu_{\mathbf{p},\mathbf{w}}$ is a 5-order additive fuzzy measure. That is, there is no $k$-order additive fuzzy measure for $k < 5$ equivalent to $\mu_{\mathbf{p},\mathbf{w}}$.

| $X = \{x_1, x_2, x_3, x_4, x_5\}$ | $\mu_{\mathsf{p,w}}$ | Möbius transform |
|---|---|---|
| { 0 0 0 0 0 } | 0.0 | 0.0 |
| { 0 0 0 0 1 } | 0.04296875 | 0.04296875 |
| { 0 0 0 1 0 } | 0.1375 | 0.1375 |
| { 0 0 0 1 1 } | 0.2375 | 0.05703125 |
| { 0 0 1 0 0 } | 0.06909722 | 0.06909722 |
| { 0 0 1 0 1 } | 0.1375 | 0.02543403 |
| { 0 0 1 1 0 } | 0.3 | 0.09340278 |
| { 0 0 1 1 1 } | 0.5 | 0.07456597 |
| { 0 1 0 0 0 } | 0.18333333 | 0.18333333 |
| { 0 1 0 0 1 } | 0.3 | 0.07369792 |
| { 0 1 0 1 0 } | 0.61666666 | 0.29583333 |
| { 0 1 0 1 1 } | 0.7625 | -0.0278646 |
| { 0 1 1 0 0 } | 0.38333333 | 0.13090278 |
| { 0 1 1 0 1 } | 0.61666666 | 0.09123264 |
| { 0 1 1 1 0 } | 0.81666666 | -0.0934027 |
| { 0 1 1 1 1 } | 0.9 | -0.2537327 |
| { 1 0 0 0 0 } | 0.1 | 0.1 |
| { 1 0 0 0 1 } | 0.18333333 | 0.04036458 |
| { 1 0 0 1 0 } | 0.38333333 | 0.14583333 |
| { 1 0 0 1 1 } | 0.61666666 | 0.09296875 |
| { 1 0 1 0 0 } | 0.2375 | 0.06840278 |
| { 1 0 1 0 1 } | 0.38333333 | 0.03706597 |
| { 1 0 1 1 0 } | 0.70000000 | 0.08576389 |
| { 1 0 1 1 1 } | 0.81666666 | -0.2537326 |
| { 1 1 0 0 0 } | 0.5 | 0.21666667 |
| { 1 1 0 0 1 } | 0.7 | 0.04296875 |
| { 1 1 0 1 0 } | 0.8625 | -0.2166667 |
| { 1 1 0 1 1 } | 0.93090277 | -0.2537326 |
| { 1 1 1 0 0 } | 0.7625 | -0.0059028 |
| { 1 1 1 0 1 } | 0.8625 | -0.2537326 |
| { 1 1 1 1 0 } | 0.95703125 | -0.2537326 |
| { 1 1 1 1 1 } | 1.0 | 0.50746528 |

**Table 5.9.** Fuzzy measure $\mu_{\mathsf{p,w}}$ and its Möbius transform. The first column denotes the subsets of $X = \{x_1, \ldots, x_5\}$ (a 0 in the $i$th column means that $x_i$ is not included, while 1 in the $i$th column means that $x_i$ is included)

## 5.4.2 Properties

A few properties have been proved that establish relationships between this family of measures and some other families. We review them below.

**Proposition 5.67.** *Any fuzzy measure decomposable by means of a continuous Archimedean t-conorm is a distorted probability.*

*Proof.* Let $\mu$ be the decomposable fuzzy measure, and let $\perp$ be the continuous Archimedean t-conorm with generator $g$. Then, $\perp(x, y)$ can be expressed according to Theorem 2.49 as $g^{(-1)}(g(x) + g(y))$ for the strictly increasing function $g$. Now, $\mu$ can be expressed as $f \circ P$, considering (i) the probabilities $p_i = q_i/K$, where $q_i = g(\mu(\{a_i\}))$ and $K = \sum_{a_i \in X} q_i$; and (ii) the distortion function $f(x) = f'(x * K)$, where $f'(x) = g^{(-1)}(x)$.

**Corollary 5.68.** *Any Sugeno $\lambda$-measure is a distorted probability.*

*Proof.* For a Sugeno $\lambda$-measure with $\lambda = 0$, the distorted probability is defined with $f(x) = x$ and $p_i = \mu(\{x_i\})$. In the general case, with $\lambda \neq 0$, we have (i) $f(x) = (e^{x \ln(1+\lambda)} - 1)/\lambda$ and (ii) $p_i = ln(1 + \lambda\mu(\{x_i\}))/ln(1 + \lambda)$

This function $f$ is used in Definition 7.27 to define Sugeno $\lambda$-quantifiers.

**Proposition 5.69.** *Any fuzzy measure that is decomposable by means of the t-conorm $\perp = maximum$ is a distorted probability.*

Additionally, it is easy to show that, in general, distorted probabilities with nonincreasing functions are not decomposable fuzzy measures. The following example illustrates this situation.

*Example 5.70.* Let $\mu$ be a fuzzy measure on $X = \{a, b, c\}$ defined as follows:

$$\mu(\emptyset) = 0, \mu(\{a\}) = 0, \mu(\{b\}) = 0, \mu(\{c\}) = 0$$

$$\mu(\{a, b\}) = 0.2, \mu(\{a, c\}) = 0.4, \mu(\{b, c\}) = 0.4, \mu(\{a, b, c\}) = 1$$

This measure is a distorted probability. Note that, with the probability distribution $p(a) = 0.2$, $p(b) = 0.35$, and $p(c) = 0.45$, and with the function $f$ defined below, we have that $\mu$ can be represented by $f$ and $p$.

$$f(x) = \begin{cases} 0 & \text{if } x < 0.5 \\ 0.2 & \text{if } 0.5 \leq x < 0.6 \\ 0.4 & \text{if } 0.6 \leq x < 0.85 \\ 1.0 & \text{if } 0.85 \leq x \leq 1.0 \end{cases}$$

This function is represented in Figure 5.8. This measure is not a $\perp$-decomposable fuzzy measure because there is no t-conorm such that $\perp(0, 0) \neq 0$. Note that, as $\mu(\{a, b\}) = 0.2$ when $\mu(\{a\}) = 0$ and $\mu(\{b\}) = 0$, we would require $0.2 = \mu(\{a, b\}) = \perp(\mu(\{a\}), \mu(\{b\})) = \perp(0, 0)$.

Distorted probabilities and $m$-dimensional distorted probabilities generalize, respectively, the symmetric and $m$-symmetric fuzzy measures. This is established below. We start by considering the case of symmetric fuzzy measures.

**Definition 5.71.** *Let $\mu$ be a fuzzy measure on $X$. Then, a fuzzy measure is said to be symmetric when the measure of a subset of $X$ depends only on the cardinality of the set, and not on the elements in the set. That is, for all $A, B \subseteq X$, it holds that*

$$if \; |A| = |B| \; then \; \mu(A) = \mu(B).$$

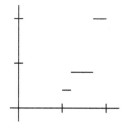

**Fig. 5.8.** Distortion function for Example 5.70.

Note that, if we consider a distorted probability $\mu$ on $X$ generated from $f$, and $P$ such that $p_i = 1/|X|$, then $\mu$ is a symmetric fuzzy measure. Moreover, if the distortion function $f$ is described in terms of a weighting vector (see remark 4 after Definition 5.61), then $\mu(A) = \sum_{j=1}^{|A|} w_k$ is a symmetric fuzzy measure. This is formalized in the next proposition.

**Proposition 5.72.** *Let $\mu$ be a distorted probability on $X$ represented by $f \circ P$ such that $p_i = 1/|X|$; then, $\mu$ is a symmetric fuzzy measure. Let $f$ be represented in terms of a weighting vector $\mathbf{w} = (w_1, \ldots, w_N)$ (i.e., $w_i \geq 0$ and $\sum_i w_i = 1$); then, $\mu$ is a symmetric fuzzy measure.*

*Proof.* Using the approach described in remark 4 after Definition 5.61, we will have the distortion function as defined by interpolation of the points in the set $\{(i/N, \sum_{j=1,\ldots,i} w_j)\}_{i=\{0,\ldots,N\}}$. Therefore, $\mu(A) = \sum_{j=1}^{|A|} w_j$.

Now, we turn into $m$-symmetric fuzzy measures. These measures rely on *sets of indifference*. Roughly, a set $A$ is of indifference if all its elements are indistinguishable with respect to the measure (i.e., if we can replace any element of $A$ with another element of $A$ and the measure does not change). This concept is formalized below.

**Definition 5.73.** *Given a subset $A$ of $X$, we say that $A$ is a set of indifference if and only if*

$$\forall B_1, B_2 \subseteq A, |B_1| = |B_2|$$
$$\forall C \subseteq X \setminus A \quad \mu(B_1 \cup C) = \mu(B_2 \cup C)$$

We now consider $m$-symmetric fuzzy measures for the particular case of $m = 2$; then, we give the general definition.

**Definition 5.74.** *Given a fuzzy measure $\mu$, we say that $\mu$ is an at most 2-symmetric fuzzy measure if and only if there exists a partition of the universal set $\mathcal{P} = \{X_1, X_2\}$, with $X_1, X_2 \neq \emptyset$ such that both $X_1$ and $X_2$ are sets of indifference. An at most 2-symmetric fuzzy measure is a 2-symmetric one if $X$ is not a set of indifference.*

**Definition 5.75.** *Given a fuzzy measure $\mu$, we say that $\mu$ is an at most $m$-symmetric fuzzy measure if and only if there exists a partition of the universal set $\{X_1, \ldots, X_m\}$, with $X_1, \ldots, X_m \neq \emptyset$ such that $X_1, \ldots X_m$ are sets of indifference.*

We then say that $\mu$ is an $m$-symmetric fuzzy measure when it is at most $m$-symmetric but not $(m-1)$-symmetric.

**Proposition 5.76.** *Let $\mu$ be an $m$-symmetric fuzzy measure with respect to a partition $\{x_1, \ldots, x_m\}$. Then, $\mu$ is a $m$-dimensional distorted probability.*

## 5.5 Bibliographical Notes

1. **Fuzzy measures:** The history of measure theory is described in [315]. For a state-of-the-art description of this field, see the *Handbook of Measure Theory* edited by Pap [312] and the collection by Fremlin on *Measure Theory* [137]. See also [118, 216, 331].

   The concept denoted in this chapter by the term "fuzzy measure" is used in several areas with different names. In particular, the names capacities, monotone measures, motone games, and premeasures are common.

   Fuzzy measures were introduced by Sugeno in 1972 [382] in Japanese (in 1974 [384] in English). Fuzzy measures are studied and described in several works. For a general reference books see the monograph by Wang and Klir [427]. For a more specialized book see the edited text by Grabisch, Murofushi, and Sugeno [174]. See also the book by Sugeno and Murofushi [385] (in Japanese). Narukawa in [287] and Radojevic in [330] (see also [329]) proved, independently, that all fuzzy measures can be written as a weighted mean of 0-1 fuzzy measures. $\mu$-inter-additive partitions and inter-additive fuzzy measures were introduced in [283] (see also [282]).

   Capacities were studied by Choquet in [80]. The notion of capacity arose in the problem of electric distribution. Capacities have been studied by several mathematicians. For example, [422] surveyed the notion of capacities before 1937. [62] has studied the capacities of compact sets (finite sets imply compact sets).

   Monotone games were considered by Aumann and Shapley [27]. The term premeasures was used by Šipoš [364]. Some old references dealing with fuzzy measures are listed in [263, 264].

   0-1 fuzzy measures correspond to coalitions or simple games [239, 285]. $m$-Quota games, when bounded by 1, correspond to probabilities. Symmetric games were considered in 1953 (see [239], p. 212). They correspond, when bounded by 1, to symmetric fuzzy measures.

   References for some particular families of measures are described below. See the notes in Chapter 2 on probability measures. Kolmogorov axioms are given in [215].

2. **Interpretations of fuzzy measures:** Several interpretations for fuzzy measures are briefly given in [279]. $\mu(A)$ as a grade is described by Sugeno in [384]. $\mu(A)$ as a degree of importance is common in some papers related to aggregation (see [254], [321]). This meaning was already used for weights in other aggregation operators in [78]. In particular, this work refers to weights (p. 252) as the power of opinion that is a degree of importance (or certainty, or competence) of an opinion. The related definition of $\mu$ as the power of $A$ to make the decision alone was given in [254] (see Section 2.1, p. 626): $\mu(A)$ "can be interpreted as the weight of the degree of importance of the combination A of criteria, or better, its power to make the decision alone (without the remaining criteria)."

   Interpretations based on probabilities have been studied by several authors, mainly in the setting of belief functions [367]. Halpern and Fagin in [120] give a detailed account of such interpretations. Belief functions as inner measures were considered by Dempster in 1967 [94]. The relationship between probability intervals and fuzzy measures has been studied by several authors. Dempster, in [94], considered the class of probabilities compatible with belief functions. [220] gives examples of lower envelopes that are not belief functions. The interpretation of belief as a probability that has suffered from information loss is given in [392]. $N$-dimensional distorted probabilities, defined in [293], permits us to show that all fuzzy measures can be interpreted in terms of a set of probability distributions and a distortion function. The interpretation of fuzzy measures in terms of a mapping between spaces was given by Murofushi and Sugeno in [279].

   Other interpretations not mentioned here include the Transferable Belief Model by Smets [368].

3. **$k$-order additive fuzzy measures:** $k$-order additive fuzzy measures were proposed in [168]. See [169] for additional results. The concept of $k$-order additive was generalized by Mesiar in [261, 262]. The generalization permits us to consider $k$-maxitive fuzzy measures.

4. **Some general aspects:** Example 5.12 is based on [167]. The difficulties for defining fuzzy measures (requiring $2^{|X|}$ values, and checking monotonicity for $|X|!$ different monotonic sequences) was already considered by Sugeno in [384] (p. 13). The problem is solved by defining the Sugeno $\lambda$-measures.

   For the Möbius transform, see [335]. Generalizations of the Möbius transform are given in [261, 262] using operators other than t-norms.

   Definition 5.19 follows [70]. The term convexity was used in [358]

5. **Belief and plausibility measures:** They originated in evidence theory, and were originally proposed by Dempster [94] and developed by Shafer [355].

6. **Possibility and necessity measures:** Shafer [355] and Zadeh [460] introduced them in the context of fuzzy sets.

7. **Decomposable fuzzy measures:** Weber [430] introduced decomposable fuzzy measures in 1984. He uses infinite decomposability. In [431],

references of previous results (with finite and infinite decomposability) are given. For example, Barnard [33] and Dubois and Prade [110] consider the finite case.

In the case where a measure can be determined from a mapping $v$ : $X \to [0,1]$, $v$ is known as the density of the measure.

$\lambda$-measures were introduced by Sugeno in 1973 [383] in Japanese (in 1974 [384] in English). [110] shows that Sugeno $\lambda$-measures are decomposable fuzzy measures. Fung and Ku [158] used, also in 1973, a similar measure, but with $\lambda = -1$. Proposition 5.40 was proved in [226]. See also [388].

Hierarchically S-Decomposable Fuzzy Measures were defined in [398].

8. **Distorted probabilities:** Based on results in experimental psychology around 1948 [326], Edwards defined distorted probabilities (see [115] and [116]) in 1953. Descriptive models using distorted probabilities have been studied in economics. For example, Handa [121] in 1977 and Kahneman and Tversky [206] in 1979 (to develop Prospect theory) used distorted probabilities. In this framework, distortion functions are known as weighting functions (see [325] and [421]). Aumann and Shapley [27] used them in game theory. Distorted probabilities with respect to aggregation have been studied in [189, 190, 293]. [189] and [293] study the proportion of fuzzy measures that are distorted probabilities with respect to the total, the computations leading to Table 5.8.

[293] gives a representation theorem for distorted probabilities when the distortion function is a strictly increasing polynomial. [69] gives necessary and sufficient conditions for the existence of a nondecreasing distortion function. [69] uses the results by Fishburn in [142]. Instead, [293] is based on the results in [351].

Distorted probabilities are equivalent to the $Q$-$p$-decomposable fuzzy measures introduced in [397] to establish that the WOWA operator is a particular case of the Choquet integral.

$m$-dimensional distorted probabilities were introduced in [293]. $m$-symmetric fuzzy measures were defined in [269] and [270]. The proof that $m$-symmetric fuzzy measures are a particular case of $m$-dimensional distorted probabilities is in [291].

9. **Other fuzzy measures:** There exist other families of fuzzy measures, and some generalizations of fuzzy measures. $k$-intolerant fuzzy measures [250] are an example of a family of fuzzy measures. For an example of generalization, see the nonmonotonic fuzzy measures. Introduced by Murofushi, Sugeno, and Machida in [284], they are fuzzy measures where the monotonicity condition has been dropped. A nonmonotonic fuzzy measure can be represented as the substraction of two monotonic fuzzy measures. Formally, a nonmonotonic fuzzy measure is a set function $\mu$ with $\mu(\emptyset) = 0$. For other measures, see the survey in [170].

10. **Aggregation of measures:** An interesting topic not discussed in this book is the aggregation and combination of belief functions. This has been

studied by several authors. Dempster's Rule of Combination is the most widely known combination method. Different methods have been proposed on the basis of different assumptions and different interpretations of belief functions. Halpern and Fagin [120] (see also [119]) differentiate between updating (when belief is understood as generalized probabilities) and combination (when belief is understood as evidence). Dempster's Rule of Combination should be restricted to the latter case according to Halpern and Fagin, and other rules should be applied for generalized probabilities. Chateauneuf [68] has studied the combination of beliefs when they are understood as probability intervals. Chateauneuf also proves that with, the interpretation of beliefs as intervals of probability, the results of Dempster's Rule of Combination are not consistent.

# 6

# From the Weighted Mean to Fuzzy Integrals

Japanese saying

In this chapter we review some aggregation operators for numerical information. While in Chapter 4 description was centered on functional equations, and operators were introduced as a natural consequence of some basic properties (unanimity, positive homogeneity, and so on), here, operators are introduced for greater modeling capabilities and generality. This progression into general aggregation operators leads to a review of operators that are particular cases of Choquet and Sugeno integrals. On the one hand, the Choquet integral generalizes not only arithmetic mean and weighted mean (the most widely used and well-known aggregation operators), but also OWA operators. On the other hand, the Sugeno integral generalizes weighted minimum, weighted maximum, and median operators. In the rest of this chapter we will use *Choquet integral family* to refer to aggregation operators that are generalized by the Choquet integral. In the same way, the *Sugeno integral family* will refer to aggregation operators that the Sugeno integral generalizes.

## 6.1 Weighted Means, OWA, and WOWA Operators

The simplest and most widely used aggregation operators are the arithmetic mean and the weighted mean. Recently (1988), Yager introduced another function, the OWA operator, to model aggregation in intelligent systems. The definition of the three functions is given below. As such definitions require a weighting vector, the section starts by recalling the definition of a weighting vector. All definitions in this chapter, unless stated otherwise, assume the $N$ values $a_1, \ldots, a_N$ to be fused.

[1] Many a little makes a mickle

**Definition 6.1.** *A vector* $v = (v_1...v_N)$ *is a* weighting vector *of dimension* $N$ *if and only if* $v_i \in [0, 1]$ *and* $\sum_i v_i = 1$.

**Definition 6.2.** *A mapping AM:* $\mathbb{R}^N \to \mathbb{R}$ *is an* arithmetic mean *of dimension* $N$ *if* $AM(a_1, ..., a_N) = (1/N) \sum_{i=1}^{N} a_i$.

**Definition 6.3.** *Let* $\mathbf{p}$ *be a weighting vector of dimension* $N$; *then, a mapping WM:* $\mathbb{R}^N \to \mathbb{R}$ *is a* weighted mean *of dimension* $N$ *if* $WM_{\mathbf{p}}(a_1, ..., a_N) = \sum_{i=1}^{N} p_i a_i$.

**Definition 6.4.** *Let* $\mathbf{w}$ *be a weighting vector of dimension* $N$; *then, a mapping OWA:* $\mathbb{R}^N \to \mathbb{R}$ *is an* Ordered Weighting Averaging (OWA) operator *of dimension* $N$ *if*

$$OWA_{\mathbf{w}}(a_1, ..., a_N) = \sum_{i=1}^{N} w_i a_{\sigma(i)},$$

*where* $\{\sigma(1), ..., \sigma(N)\}$ *is a permutation of* $\{1, ..., N\}$ *such that* $a_{\sigma(i-1)} \geq a_{\sigma(i)}$ *for all* $i = \{2, ..., N\}$ *(i.e.,* $a_{\sigma(i)}$ *is the ith largest element in the collection* $a_1, ..., a_N$ *).*

We consider below two situations that can be modeled, respectively, with the weighted mean and the OWA operator.

*Example 6.5.* A university exam on algebra consists on three exercises. Each exercise is evaluated in the $[0, 10]$ interval. The final rating of each student is obtained as a weighted linear combination of his or her three marks. The weights are assigned as follows: 0.5 for the first exercise; 0.25 for the second exercise; and 0.25 for the third exercise. Therefore, a student with 8 points for the first exercise, 6 for the second and 10 for the third will be rated as follows: $8 \cdot 0.5 + 6 \cdot 0.25 + 10 \cdot 0.25 = 8$. This process is modeled using a weighted mean, with weights, $\mathbf{p} = (0.5, 0.25, 0.25)$.

*Example 6.6.* In the Olympic Games, the final rating for a participant in some sports is computed from the rates given by the judges. This final rating is the average of the rates of the judges once the largest and smallest ones are disregarded. This decision making process can be modeled by means of an OWA operator. In the case of five judges, we will use OWA with the weighting vector $\mathbf{w} = (w_1, ..., w_5) = (0, 1/3, 1/3, 1/3, 0)$.

### 6.1.1 Properties

The OWA operator, as well as the weighted mean, give a value that is between the minimum and the maximum of the values to be fused. However, while the OWA can model the minimum and the maximum, the weighted mean cannot. Instead, the weighted mean can be used to model dictatorship (the value of one of the sources is always selected) while OWA cannot. These situations are modeled with the following weighting vectors.

OWA equal to the minimum:

$OWA_{\mathbf{w}}(a_1, \ldots, a_N) = \min(a_1, \ldots, a_N)$ when $\mathbf{w} = (0, 0, \ldots, 0, 1)$

OWA equal to the maximum:

$OWA_{\mathbf{w}}(a_1, \ldots, a_N) = \max(a_1, \ldots, a_N)$ when $\mathbf{w} = (1, 0, \ldots, 0, 0)$

Dictatorship for the $i$th information source:

$WM_{\mathbf{p}}(a_1, \ldots, a_N) = a_i$ when $p_i = 1$ and $p_j = 0$ for all $j \neq i$

An additional difference between OWA and weighted mean is that, in the former, the order of the values $a_i$ is not relevant (order does not affect the result), while in the latter, permutations of the arguments lead to different results. This is because the outcome of the permutation $\sigma$ in the OWA operator is independent of the information sources. Therefore, OWA is a symmetric operator, while weighted mean is not. OWA is also robust, in the sense that it employs all the data minimizing the influence of outliers.

The OWA operator is an *L-estimator* (Definition 2.36). That is, it is a linear combination of order statistics. OWA generalizes all order statistics. Moreover, it is also known that it generalizes, among others, the median (the central value of A), the $k$th minimum, the $k$th maximum, the arithmetic mean, the $\alpha$-trimmed and the $(\alpha, \beta)$-trimmed means, and the $\alpha$-winsorized and $(\alpha, \beta)$-winsorized means.

Let us recall some definitions. The $i$th order statistic (see Definition 2.37), is denoted by $OS_i$. Then, the $k$th maximum is equivalent to $OS_{N-k+1}$, and the $k$th minimum is equivalent to $OS_k$.

**Definition 6.7.** *A mapping* $M: \mathbb{R}^N \to \mathbb{R}$ *is a median of dimension $N$ if*

$$M(a_1, \ldots, a_N) = \begin{cases} \frac{a_{\sigma(N/2)} + a_{\sigma(N/2+1)}}{2} & \text{when } N \text{ is even} \\ a_{\sigma(\frac{N+1}{2})} & \text{when } N \text{ is odd.} \end{cases}$$

*where $\sigma$ is defined as above. Note that when $N$ is odd, $M = OS_{(N+1)/2}$.*

The $(r, s)$-trimmed mean is the mean of values $a_1, \ldots, a_N$ once the $r$ lowest values and the $s$ highest ones are removed. That is, $(a_{\sigma(r+1)} + \cdots + a_{\sigma(N-s)})/(N - r - s)$. The $(r, s)$-winsorized means is the arithmetic mean when the omitted values are replaced by the nearest value to be retained unchanged. That is,

$$\frac{r \cdot a_{\sigma(r+1)} + a_{\sigma(r+1)} + \cdots + a_{\sigma(N-s)} + s \cdot x_{\sigma(N-s)}}{N}.$$

$\alpha$-trimmed and $\alpha$-winsorized corresponds to the previous cases when $r = s$, and when $2\alpha$ is the proportion of the values being omitted. Thus, $\alpha N$ is the number of values to be trimmed at each end.

For $N = 2$, the OWA and the Hurwicz operator ($H(x) = \sigma \max(x) + (1 - \sigma)\min(x)$) are equivalent, while for $N > 2$, the Hurwicz operator is a particular case of OWA.

Order statistics, $k$th maximum, $k$th minimum, trimmed and winsorized means, and median (except for even $N$) are operators based on the ordering among values $a_i$. According to this, they can be used in ordinal scales. For example, in the case of $OS_i$, the element that occupies the $i$th position is selected. In fact, these operators are methods for element selection, and they proceed by considering the input data as a multiset or bag. Therefore, all them satisfy symmetry. Some properties (and characterizations) of these operators were given in Section 4.3.1, devoted to ordinal scales.

### 6.1.2 Interpretation of Weighting Vectors in WM and OWA

From the definitions above, it can be observed that the weighted mean and the OWA operator have similar expressions: both are a linear combination of values with respect to a weighting vector. However, in spite of their similarity, the meaning of the weights is radically different due to the presence of the (ordering) permutation $\sigma$ in the OWA operator.

It is well known that, in the weighted mean, weighting vectors are used to express the reliability of the information sources that have supplied a particular value. That is, $p_i$ corresponds to a measure of the reliability of the $i$th sensor or of the expertise of the $i$th expert. This is not the case with the OWA operator, where weights, due to the ordering $\sigma$, assign importance to elements according only to their position with respect to the others. In this way, a system can reduce the importance of extreme values (or even ignore them, as in Example 6.6), or give greater importance to small values rather than large ones (for example, in the case of a robot that has to avoid collisions). This corresponds to weighting the values rather than weighting the sources.

According to this interpretation, weighting vectors in these two operators are complementary (we will refer to them as $\mathbf{p}$ and $\mathbf{w}$, as in the definitions above), and in some circumstances both are of interest in a single application.

We consider below four scenarios where aggregation operators can be used. These scenarios are later used to illustrate the meaning of the weighting vectors for the weighted mean and OWA operators.

1. *Multicriteria decision making:* Several alternatives are considered, and one of them has to be selected (for example, we want to buy a car, and several brands are considered). Several criteria evaluate each alternative (e.g., comfort, price, security equipment) in the $[0, 1]$ interval (1 being adequate, 0 being inadequate). To select the best alternative, an overall rating is computed for each one. This rating is an average of the criteria.

2. *Fuzzy Constraint Satisfaction Problems:* The optimal solution of a problem has to sastisfy some constraints. For a given possible solution, constraints can be tested, and their evaluation returns a value in the unit interval (0 when a constraint is not satisfied at all, 1 when it is completely satisfied). To have an overall rating of the solution, the evaluations of all constraints are aggregated.

3. *Robot sensing (all data corresponding to the same time instant):* A robot receives the readings of five sensors, each measuring distance to the nearest object. To avoid collisions, the robot estimates the distance of the nearest object by means of a fusion of the readings.
4. *Robot sensing (data obtained at different time instants):* A case similar to the previous one, with the same goals, but with the robot only having a single sensor. In this case, in order to estimate the distance to the nearest object, the robot fuses the last reading with some previous ones.

We consider now the use of weighted mean (WM) and OWA operators for computing the rating or for fusing the sensor readings in the examples above. Then, we interpret the weighting vectors $\mathbf{p}$ and $\mathbf{w}$ in the operators. Recall that we use $\mathbf{p}$ to denote WM weights and $\mathbf{w}$ to denote OWA weights. This difference is semantical, because both weights have the same structure (they follow Definition 6.1).

1. *Multicriteria Decision Making:* The weighting vector $\mathbf{p}$ corresponds to the importance of the criteria (for example, we give more importance to price than to comfort), while $\mathbf{w}$ corresponds to the degree of compensation allowed among criteria. Large compensation (the selection of the largest values) corresponds to evaluating an alternative as *good* (say) when at least one criteria evaluates it as *good*. In contrast, no compensation (the selection of the smallest value) corresponds to assigning a low score to an alternative when at least one criteria is badly rated.
2. *Fuzzy Constraint Satisfaction Problems:* In this case, $\mathbf{p}$ corresponds to the importance of each constraint, while $\mathbf{w}$ corresponds to the degree of compensation between constraints, that is, the degree to which a bad evaluation of a constraint implies a bad evaluation of the solution, or how restrictions have to be addressed so that the solution is considered good.
3. *Robot sensing (all data corresponding to the same time instant):* In this case, $\mathbf{p}$ would be used to express the reliability of each sensor, while $\mathbf{w}$ would be used to determine degree to which small values are important (to avoid collision), independently of the reliability of the sensors. Also, $\mathbf{w}$ can be used to prevent the influence of outliers.
4. *Robot sensing (data obtained at different time instants):* In this case, $\mathbf{p}$ would be used to give more importance to recent data than *old* data, while $\mathbf{w}$ would be used, as in the previous example, to express the importance of small values or to diminish the influence of outliers.

Let us now give an example corresponding to the fuzzy constraint satisfaction problem.

*Example 6.8.* Let us consider two professors $A$ and $B$ who have to teach a course consisting of a tutorial and a training part. A number of fuzzy constraints apply to the number of sessions of the course, the number of sessions given by the professors, and so on. Such constraints are listed below:

- The total number of sessions is six.
- Professor $A$ will give the tutorial, which should consist of about three sessions; three is the optimal number of sessions; a difference in the number of sessions greater than two is unacceptable.
- Professor $B$ will give the training part, consisting of about two sessions.
- Both professors should give more or less the same number of sessions. A difference of one or two is half acceptable; a difference of three is unacceptable.

The constraints of this problem can be described using fuzzy sets. To do so, we need to define the variables, the fuzzy sets (to describe the constraints for the variables), and the constraints. We start defining the variables. Two variables are considered:

- $x_A$: Number of sessions taught by Professor $A$
- $x_B$: Number of sessions taught by Professor $B$

With these variables, the four constraints above are translated into

- $C_1$: $x_A + x_B$ should be about 6
- $C_2$: $x_A$ should be about 3
- $C_3$: $x_B$ should be about 2
- $C_4$: $|x_A - x_B|$ should be about 0

Using fuzzy sets, we can evaluate to what extent any constraint is satisfied. For example, if we have $\mu_6$ to express "about 6," then we can evaluate "$x_A + x_B$ should be about 6" by $\mu_6(x_A + x_B)$. So, given $\mu_6$, $\mu_3$, $\mu_2$, and $\mu_0$, we can compute to what degree a solution pair $(x_A, x_B)$ satisfies all constraints. The corresponding degrees of satisfaction will be:

- $\mu_6(x_A + x_B)$
- $\mu_3(x_A)$
- $\mu_2(x_B)$
- $\mu_0(|x_A - x_B|)$

To completely determine the satisfaction degrees, we need to define the membership functions. We use the following membership function for expressing the fourth constraint:

$$\mu_0(x) = \begin{cases} (2-x)/2 & \text{if } 0 \leq x < 1 \\ 0.5 & \text{if } 1 \leq x < 2 \\ (3-x)/2 & \text{if } 2 \leq x < 3 \\ 0 & \text{if } x \geq 3 \end{cases}$$

For the other three constraints, we use the triangular membership functions represented in Figure 6.1. This corresponds to the following definitions, using Definition 2.41 for the membership functions:

- $\mu_6(x) = \mu_{5,6,7}^t(x)$

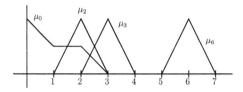

**Fig. 6.1.** Membership functions for Example 6.8

- $\mu_3(x) = \mu_{2,3,4}^t(x)$
- $\mu_2(x) = \mu_{1,2,3}^t(x)$

Let us now consider a few pairs of values $(x_A, x_B)$ and their satisfaction degrees according to the definitions above:

- $(2,2) : (\mu_6(4), \mu_3(2), \mu_2(2), \mu_0'(0)) = (0, 0.5, 1, 1)$
- $(2,3) : (\mu_6(5), \mu_3(2), \mu_2(3), \mu_0'(1)) = (0.5, 0.5, 0.5, 0.5)$
- $(2,4) : (\mu_6(6), \mu_3(2), \mu_2(4), \mu_0'(2)) = (1, 0.5, 0, 0.5)$
- $(3.5, 2.5) : (\mu_6(6), \mu_3(3.5), \mu_2(2.5), \mu_0'(1)) = (1, 0.5, 0.5, 0.5)$
- $(3,2) : (\mu_6(5), \mu_3(3), \mu_2(2), \mu_0'(1)) = (0.5, 1, 1, 0.5)$
- $(3,3) : (\mu_6(6), \mu_3(3), \mu_2(3), \mu_0'(0)) = (1, 1, 0.5, 1)$

In order to rate the set of alternatives with respect to a global satisfaction, we can combine the partial degrees of satisfaction using an aggregation operator. This is, the satisfaction for a pair of solutions $(x_A, x_B)$, denoted by $sat(x_A, x_B)$, will be

$$sat(x_A, x_B) = \mathbb{C}(\mu_6(x_A + x_B), \mu_3(x_A), \mu_2(x_B), \mu_0(|x_A - x_B|)).$$

When no importances are given, we can define $\mathbb{C}$ as the arithmetic mean.

Nevertheless, we might consider that some constraints are more important than others. Let us consider the following situation:

- Professor $A$ is more important than Professor $B$
- The number of sessions equal to six is the most important constraint (but not a *crisp* constraint)
- The difference in the number of sessions taught by the two professors is the least important constraint

We will model this situation by assigning the following weights to the constraints: $\mathbf{p} = (p_1, p_2, p_3, p_4) = (0.5, 0.3, 0.15, 0.05)$. That is, the first constraint, $C_1$, has weight 0.5, the second constraint, $C_2$, has weight 0.3, $C_3$ has weight 0.15, and the $C_4$ has weight 0.05. We can model the aggregation using the weighted mean. Doing so, we obtain the following evaluation for the previous pairs of values:

- $sat(2,2) = WM_{\mathbf{p}}(0, 0.5, 1, 1) = 0.35$
- $sat(2,3) = WM_{\mathbf{p}}(0.5, 0.5, 0.5, 0.5) = 0.5$

- $sat(2, 4) = WM_{\mathbf{p}}(1, 0.5, 0, 0.5) = 0.675$
- $sat(3.5, 2.5) = WM_{\mathbf{p}}(1, 0.5, 0.5, 0.5) = 0.75$
- $sat(3, 2) = WM_{\mathbf{p}}(0.5, 1, 1, 0.5) = 0.725$
- $sat(3, 3) = WM_{\mathbf{p}}(1, 1, 0.5, 1) = 0.925$

So, with this model which considers the importance of each constraint, the best solution is when both professors give the same number of hours, and so the total number of sessions is six and the difference in the number of sessions is zero. The second best solution is when Professor A gives 3.5 sessions and Professor B gives 2.5 sessions, and the third solution is when Professor A gives three sessions and Professor B gives two sessions.

Another matter to be taken into account when considering multiple constraints is compensation. That is, how many values can have a bad evaluation. Let us consider the case where one bad value does not matter. In this case, we can model the aggregation using an OWA operator (if the importance of the constraints are not considered). We show below the results of the OWA operator when a weighting vector $\mathbf{w} = (1/3, 1/3, 1/3, 0)$ is used. This vector stands for the lowest value to be discarded (a weight equal to 1), and all the others having the same weight.

With such a weighting vector, the pairs of solutions are evaluated as follows:

- $sat(2, 2) = OWA_{\mathbf{p}}(0, 0.5, 1, 1) = 0.8333$
- $sat(2, 3) = OWA_{\mathbf{p}}(0.5, 0.5, 0.5, 0.5) = 0.5$
- $sat(2, 4) = OWA_{\mathbf{p}}(1, 0.5, 0, 0.5) = 0.6666$
- $sat(3.5, 2.5) = OWA_{\mathbf{p}}(1, 0.5, 0.5, 0.5) = 0.6666$
- $sat(3, 2) = OWA_{\mathbf{p}}(0.5, 1, 1, 0.5) = 0.8333$
- $sat(3, 3) = OWA_{\mathbf{p}}(1, 1, 0.5, 1) = 1.0$

With this model, the best solution is when both professors give the same number of hours. Since in this case there are three constraints that are completely satisfied, the satisfaction degree is maximum (equal to 1). Two other solutions have a rate of 0.833. They correspond to Professor A giving two sessions and Professor B also giving two sessions, and Professor A giving three sessions and Professor B giving two sessions. Note that, for the pair $(2, 2)$, the first constraint is completely discarded in the evaluation because its satisfaction degree is the lowest (and equal to 0).

### 6.1.3 The WOWA Operator

The Weighted OWA (WOWA) operator was introduced to model situations in which both importance of information sources and importance of values had to be taken into account. The operator aggregates a set of values using two weighting vectors: one corresponding to the vector $\mathbf{p}$ in the weighted mean and the other corresponding to $\mathbf{w}$ in the OWA operator. The definition of the operator is as follows.

**Definition 6.9.** *Let* **p** *and* **w** *be two weighting vectors of dimension* $N$*; then, a mapping* $WOWA: \mathbb{R}^N \to \mathbb{R}$ *is a* Weighted Ordered Weighted Averaging (WOWA) operator *of dimension* $N$ *if*

$$WOWA_{\mathbf{p},\mathbf{w}}(a_1, ..., a_N) = \sum_{i=1}^{N} \omega_i a_{\sigma(i)},$$

*where* $\sigma$ *is defined as in the case of OWA (i.e.,* $a_{\sigma(i)}$ *is the* $i$*th largest element in the collection* $a_1, ..., a_N$*), and the weight* $\omega_i$ *is defined as*

$$\omega_i = w^* \left( \sum_{j \leq i} p_{\sigma(j)} \right) - w^* \left( \sum_{j < i} p_{\sigma(j)} \right),$$

*with* $w^*$ *being a nondecreasing function that interpolates the points*

$$\{(i/N, \sum_{j \leq i} w_j)\}_{i=1,...,N} \cup \{(0,0)\}.$$

*The function* $w^*$ *is required to be a straight line when the points can be inter- polated in this way.*

From now on, we denote by $\omega$ the (weighting) vector $\omega = (\omega_1 \ \ldots \ \omega_N)$. We illustrate this operator by reconsidering Example 6.8.

*Example 6.10.* Let us consider again Professors $A$ and $B$, and the assignment of $x_A$ and $x_B$ sessions to $A$ and $B$, respectively. Now, if we want to evaluate the different alternatives, modeling at the same time the fact that some constraints are more important than others, and that we want some compensation, then we can rate each alternative using the WOWA operator.

Let us consider, as in Example 6.8, that the importance of the constraints is represented by the weights $\mathbf{p} = (p_1, p_2, p_3, p_4) = (0.5, 0.3, 0.15, 0.05)$, and that the compensation is represented by the weighting vector $\mathbf{w} = (1/3, 1/3, 1/3, 0)$. In this case, the WOWA operator permits us to aggregate the satisfaction degrees $\mu_6(x_A + x_B)$, $\mu_3(x_A)$, $\mu_2(x_B)$, and $\mu_0(|x_A - x_B|)$. The results obtained for the pairs in Example 6.8 are as follows:

- $sat(2, 2) = WOWA_{\mathbf{p}}(0, 0.5, 1, 1) = 0.4666$
- $sat(2, 3) = WOWA_{\mathbf{p}}(0.5, 0.5, 0.5, 0.5) = 0.5$
- $sat(2, 4) = WOWA_{\mathbf{p}}(1, 0.5, 0, 0.5) = 0.8333$
- $sat(3.5, 2.5) = WOWA_{\mathbf{p}}(1, 0.5, 0.5, 0.5) = 0.8333$
- $sat(3, 2) = WOWA_{\mathbf{p}}(0.5, 1, 1, 0.5) = 0.8$
- $sat(3, 3) = WOWA_{\mathbf{p}}(1, 1, 0.5, 1) = 1.0$

We can see that the best solution is the pair $(3, 2)$. Thus, this pair is considered better when evaluation uses the weighted mean, the OWA operator or the WOWA operator. This is not always the case: the relative positions of possible solutions might change when using different functions. This is the case with the pair $(2,4)$, which has a fourth position when the combination function is either the weighted mean or the OWA operator, but has a better position when we use the WOWA operator.

This is so because we assign the largest weight to the value $\mu_6(x_A, x_B)$. Thus, when this value is high, as it is in the third and the fourth alternatives, the final outcome is also high. Besides, in relation to compensation, we have that the lowest values have to be discarded. By "lowest values" we mean those that represent one fourth of the total with respect to the weights. That is, the lowest values in $I$, such that $\sum_{i \in I} p_i$ are around 0.25, are discarded. In the case of the satisfaction degree of $(2,4)$, we have that in $(1, 0.5, 0, 0.5)$ we can disregard the 0 (as it is the lowest value); but, as this value has only a weight equal to 0.15, we can also disregard (part of) the value 0.5 (the second lowest value). This improves the final evaluation of the pair $(2,4)$.

### Rationale for WOWA's new weights

Note that the WOWA operator is also a linear combination of the values with respect to a vector (in this case $\omega$). When studied from this point of view, the operator determines a weight $\omega_i$ for each value $a_i$ in terms of the two weighting vectors $\mathbf{p}$ and $\mathbf{w}$, in such a way that the initial weight $p_i$ for $a_i$ is increased (i.e., $\omega_i \geq p_{\sigma(i)}$) if the value $a_i$ is small, and small values have more importance than larger ones (the same holds if the value $a_i$ is large and importance is given to large values). In contrast, when importance is given to large values and $a_i$ is small, we have $\omega_i < p_{\sigma(i)}$ (the same holds if importance is given to small values and $a_i$ is large). Here, small and large should not be understood as an absolute term in the domain, but relative to the other values in $\mathbf{a} = (a_1, ..., a_N)$.

We turn now into the construction of $w^*$. The shape of this function $w^*$ (specially its derivative) shows the relative importance of the elements. An alternative definition for WOWA avoiding such a construction is given in Section 6.1.4. The section also contains an example (see Example 6.16) of the WOWA operator.

### The construction of the interpolation function $w^*$

The WOWA operator defines the weighting vector $\omega$ in terms of differences between pairs of points on the function $w^*$. The points are selected using the weighting vector $\mathbf{p}$, and the function is built using the vector $\mathbf{w}$. The rationale of this construction is described below.

In order to have WOWA operators generalizing the OWA operator, the weights $\omega_i$ are supposed to be equal to $w_i$ when all sources have the same importance. That is, when all weights $p_i$ are equal (i.e., $p_i = 1/N$), $\omega_i = w_i$ for all $i \in [0,1]$. From now on, we denote by $\mathbf{p}^0$ the "same-importance" weighting vector $\mathbf{p}^0 = (p_1^0, \ldots, p_N^0) = (1/N, \ldots, 1/N)$. Note that, when $\mathbf{p} = \mathbf{p}^0$, no interactions between weights are considered, and thus $p_{\sigma(1)}^0 = 1/N$ and $w_1$

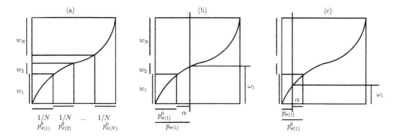

**Fig. 6.2.** Building weights $\omega$ for the WOWA: (a) building $w^*$; (b) extraction of $\omega_1$ when $p_{\sigma(1)} > p^0_{\sigma(1)}$; (c) extraction of $\omega_1$ when $p_{\sigma(1)} < p^0_{\sigma(1)}$

refer, respectively, to the weight for the information source that supplies the largest value in $\{a_1, \ldots, a_N\}$ and the weight for the value itself. Similarly, $p^0_{\sigma(2)} = 1/N$ and $w_2$ refer, respectively, to the weight for the source supplying $a_{\sigma(2)}$ and the weight for this value. In general, $w_i$ and $p_{\sigma(i)}$ are weights referring to the value $a_{\sigma(i)}$ and to the source that supplies this value.

Now, we put the relation between **w** and **p** in a graphical form. To do so, we consider the set of points:

$$\left\{\left(\sum_{j \leq i} p^0_{\sigma(j)}, \sum_{j \leq i} w_j\right)\right\}_{i=1,N} = \left\{\left(\sum_{j \leq i} p^0_{\sigma(j)} = \sum_{j \leq i} 1/N = i/N, \sum_{j \leq i} w_i\right)\right\}_{i=1,N}$$

It is important to underline that differences between $y$-axis coordinates of two consecutive points lead to $w_i$, and that differences between $x$-axis coordinates of two consecutive points lead to the values $p^0_{\sigma(i)}$.

As $\sum_i w_i = 1$ and $\sum_i p^0_i = 1$, these points are in the unit interval, and as $w_i \geq 0$ and $p^0_i \geq 0$ they shape a monotone function. We define $w^*$ as a function that is monotone and interpolate such points. Additionally, as required by the definition of the WOWA operator, the function should be a straight line when the points can be interpolated in this way. Figure 6.2 (a) displays the points as well as the function $w^*$.

Selecting weights $\omega$ from the curve is like strengthening or narrowing the intervals $p^0_i$ on the $x$-axis. A modification of the $p^0_i$ causes a movement of the points over the curve. The values $w_i$ are obtained after such a modification of the values of $p_i$.

Let us consider the case of the largest element $a_{\sigma(1)}$ and the weights $w_1$ and $p_{\sigma(1)}$. In this case, if $p_{\sigma(1)} > p^0_{\sigma(1)} = 1/N$, it is natural that $w_1 > w_1$. Note that, if we increase $p_{\sigma(1)}$ in relation to $p^0_{\sigma(1)}$ (i.e., $p_{\sigma(1)} = p^0_{\sigma(1)} + \alpha$ for $\alpha > 0$), then $w^*(p_{\sigma(1)}) > w^*(p^0_{\sigma(1)})$. Naturally, this value, $w^*(p_{\sigma(1)})$, corresponds to the weight $w_1 = w^*(p_{\sigma(1)}) = w^*(p^0_{\sigma(1)} + \alpha)$. This computation is represented in Figure 6.2 (b). In a similar way, if we decrease $p_{\sigma(1)}$, then $w^*(p_{\sigma(1)}) < w^*(p^0_{\sigma(1)})$. This case is displayed in Figure 6.2 (c).

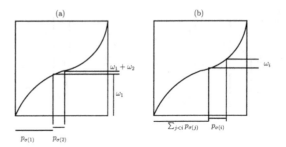

**Fig. 6.3.** Building weights $w$ for the WOWA: (a) extraction of $w_2$; (b) extraction of $w_i$

Let us consider the second largest element $a_{\sigma(2)}$. The corresponding **p** weight is $p_{\sigma(2)}$. In this case, the accumulated value $p^0_{\sigma(1)} + p^0_{\sigma(2)}$ is moved to $p_{\sigma(1)} + p_{\sigma(2)}$. Therefore, we compute $w^*(p_{\sigma(1)} + p_{\sigma(2)})$, which corresponds to $w_1 + w_2$ (see Figure 6.3 (a)), and, thus, $w_2 = w^*(p_{\sigma(1)} + p_{\sigma(2)}) - w^*(p_{\sigma(1)})$. It is important to note that the value $w_2$ will depend not only on $p_{\sigma(2)}$, but also on $p_{\sigma(1)}$. This is natural, as $P_{\sigma(1)}$ might be zero, and in this case, the largest value with no null importance is $a_{\sigma(a)}$.

The computations of all other $w_i$ proceed in a similar way. See Figure 6.3 (b) for the computation of $w_i$. As for $w_2$, the value $w_i$ depends not only on $p_{\sigma(i)}$, but also on all values $p_{\sigma(j)}$ and $w_j$ for $j \leq i$. Note that this construction depends on the permutation $\sigma$ which in turn depends on the values $a_i$ to be aggregated. Therefore, the weights $w_i$ depend on the ordering inferred from the values $a_i$, and, thus, given **p** and **w**, different sets of values lead to different $w_i$.

*Example 6.11.* The weights $\mathbf{w} = (1/3, 1/3, 1/3, 0)$, used in Example 6.10, lead to the following function $w^*$:

$$w^*(x) = \begin{cases} x/0.75 & \text{if } x < 0.75 \\ 1 & \text{if } x \geq 0.75. \end{cases}$$

**The shape of the interpolation function $w^*$**

The construction given above shows that the shape of the function $w^*$, and, if applicable, the shape of its derivative, gives information about the values that are considered more relevant. In particular, the larger the slope of $w^*$, the larger the importance of the corresponding elements. Figures 6.4 (a), (b), and (c) correspond, respectively, to the situation of giving importance to large, medium, or average, and small values. Figure 6.4 (d) corresponds to the situation of equal importance for all values.

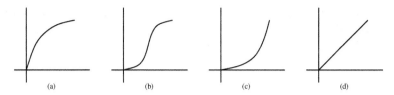

**Fig. 6.4.** Function $w^*$: (a) largest importance to large values; (b) largest importance to medium values; (c) largest importance to medium values; (d) equal importance to all values

## Properties

The WOWA operator generalizes the weighted mean and the OWA operator. In particular, note that, when $\mathbf{p} = (1/N \ \dots \ 1/N)$, it reduces to the OWA operator:

$$WOWA_{\mathbf{p},\mathbf{w}}(a_1, ..., a_N) = OWA_{\mathbf{w}}(a_1, ..., a_N) \text{ for all } \mathbf{w} \text{ and } a_i.$$

Also, when $\mathbf{w} = (1/N \ \dots \ 1/N)$, the operator reduces to the weighted mean:

$$WOWA_{\mathbf{p},\mathbf{w}}(a_1, ..., a_N) = WM_{\mathbf{p}}(a_1, ..., a_N) \text{ for all } \mathbf{p} \text{ and } a_i.$$

This implies that the WOWA operator generalizes all the operators generalized by the weighted mean and the OWA. In particular, when $\mathbf{w} = \mathbf{p} = (1/N \ \dots \ 1/N)$, we get the arithmetic mean:

$$WOWA_{\mathbf{p},\mathbf{w}}(a_1, ..., a_N) = AM(a_1, ..., a_N) \text{ for all } a_i.$$

### 6.1.4 OWA and WOWA Operators and Fuzzy Quantifiers

Alternative definitions for OWA and WOWA operators exist based on fuzzy quantifiers (see Section 2.3.4). Definitions use a kind of fuzzy quantifier to extract the weights. OWA uses a fuzzy quantifier instead of a vector, and WOWA uses a fuzzy quantifier (with interpretation equivalent to that for OWA) and a weighting vector (the $\mathbf{p}$ vector corresponding to that of the weighted mean) The definitions are given below. The one for the fuzzy quantifier and its generator is also included for the sake of completion.

**Definition 6.12.** *A function* $Q : [0, 1] \rightarrow [0, 1]$ *is a* regular nondecreasing fuzzy quantifier *(nondecreasing fuzzy quantifier for short) if (i)* $Q(0) = 0$; *(ii)* $Q(1) = 1$; *and (iii)* $x > y$ *implies* $Q(x) \geq Q(y)$.

Although we will not use it here, the class of *regular increasing monotone (RIM)* quantifiers is well-known. The quantifiers satisfy (i), (ii), and the following (iii) $x > y$ implies $Q(x) > Q(y)$. Sometimes, they also denote nondecreasing quantifiers.

The generating function of a quantifier might be useful for studying its properties. Formally, given a fuzzy quantifier $Q$, a function $q : [0, 1] \to [0, 1]$ is called the generating function of $Q(x)$ if it satisfies

$$Q(x) = \int_0^x q(t)dt,$$

where $q(t) \geq 0$ for all $t \in [0, 1]$, and $\int_0^1 q(t)dt = 1$.

**Definition 6.13.** *Let $Q$ be a regular nondecreasing fuzzy quantifier; then, a mapping $OWA_Q : \mathbb{R}^N \to \mathbb{R}$ is an Ordered Weighting Averaging (OWA) operator of dimension $N$ if*

$$OWA_Q(a_1, ..., a_N) = \sum_{i=1}^N (Q(i/N) - Q((i-1)/N))a_{\sigma(i)},$$

*where $\sigma$ is defined as before.*

From a practical point of view, the use of fuzzy quantifiers in OWA operators is an advantage when the number of elements to be fused is not fixed beforehand. That is, the same quantifier can be used to aggregate values regardless of the number of information sources $N$.

This definition is equivalent to the previous one based on a weighting vector. This is so because, with $w_i$ defined from $Q$ as $w_i = Q(i/N) - Q((i-1)/N)$, we have $OWA_Q(a_1, \dots, a_N)$ equal to $OWA_\mathbf{w}(a_1, \dots, a_N)$. Similarly, $Q$ can be defined as a function that interpolates the points $\{(i/N, Q(i/N))\}$ for $i \in \{0, 1, \dots, N\}$ so that $OWA_Q$ and $OWA_\mathbf{w}$ are equivalent.

Note that the function interpolated from $\{(i/N, Q(i/N))\}$ for $i \in \{0, \dots, N\}$, and, in general, the function $Q$, corresponds to the function $w^*$ in the WOWA definition (Definition 6.9). This makes clear that the function $w^*$ displayed in Figure 6.4 are regular nondecreasing fuzzy quantifiers that can be used with the OWA operator to give, as before, most importance, respectively, to large, medium, or small values, or to assign equal importance to all values. Having said that, it is obvious that the same construction can be applied to the WOWA operator. The corresponding definition for the WOWA operator follows.

**Definition 6.14.** *Let $Q$ be a regular nondecreasing fuzzy quantifier, and let $\mathbf{p}$ be a weighting vector of dimension $N$; then, a mapping $WOWA: \mathbb{R}^N \to \mathbb{R}$ is a Weighted Ordered Weighted Averaging (WOWA) operator of dimension $N$ if*

$$WOWA_{\mathbf{p},Q}(a_1, ..., a_N) = \sum_{i=1}^N \omega_i a_{\sigma(i)},$$

*where $\sigma$ is defined as in the case of the OWA, and the weight $\omega_i$ is defined as*

$$\omega_i = Q(\sum_{j \leq i} p_{\sigma(i)}) - Q(\sum_{j < i} p_{\sigma(i)}).$$

Naturally, this definition is equivalent to the $WOWA_{\mathbf{p},\mathbf{w}}$ in Definition 6.9.

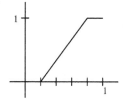

**Fig. 6.5.** Representation of the fuzzy quantifier $Q_2$ in Example 6.15

## Properties

Equivalence between $OWA_w$ and $OWA_Q$ and between $WOWA_{p,w}$ and $WOWA_{p,Q}$ makes clear that the properties obtained for $OWA_Q$ and $WOWA_{p,Q}$ are analogous to the ones obtained for $OWA_w$ and $WOWA_{p,w}$. In particular, WOWA generalizes the weighted mean (when the quantifier is $Q(x) = x$) and the $OWA_Q$ when $\mathbf{p} = (1/N, \ldots, 1/N)$. Also, $OWA_Q$ generalizes the arithmetic mean (with $Q(x) = x$).

OWA with the fuzzy quantifier "for all" (recall from Section 2.3.4 that this is $Q(1) = 1$ and $Q(x) = 0$ for all $x \neq 1$) is equivalent to minimum, and OWA with the fuzzy quantifier "there exists" (recall that this is $Q(0) = 0$ and $Q(x) = 1$ for all $x \neq 0$) is equal to maximum.

The rationale of this is that minimum defines a lower bound in which *all* sources agree, and maximum defines an upper bound in which *only one* source agrees. For other quantifiers, values are between minimum and maximum. The nearer a quantifier is to "there exists" the larger is the output of the OWA operator. When the quantifier is "exactly 50%," the OWA operator is the median (the medium value). This is illustrated in the following example.

*Example 6.15.* Let us consider the quantifiers $Q_1$, $Q_2$, $Q_3$, and $Q_4$ defined as follows (quantifier $Q_2$ is represented in Figure 6.5):

- $Q_1(x) = x^{1/4}$
- $Q_2(x) = \begin{cases} 0 & x \leq 1/5 \\ \frac{5}{3}(x - \frac{1}{5}) & \frac{1}{5} \leq x \leq \frac{4}{5} \\ 1 & x \geq 1/5 \end{cases}$
- $Q_3(x) = x^4$
- $Q_4(x) = x$

Note that $Q_1$ is a quantifier similar to "there exists," with a shape similar to the one in Figure 6.4 (a); $Q_2$ is a quantifier similar to "exactly 50%," with a shape similar to the one in Figure 6.4 (b); $Q_3$ is similar to "for all," with a shape similar to the one in Figure 6.4 (c); and, finally, $Q_4$ corresponds to equal relevance for all inputs (Figure 6.4 (d)).

Let an input vector $\mathbf{a} = \{0.5, 0.25, 0.8, 0.0, 0.75\}$. Then, the OWA operator applied to this vector $\mathbf{a}$ with respect to the quantifiers defined above is as follows:

- $OWA_{Q_1}(0.5, 0.25, 0.8, 0.0, 0.75) = 0.6887$
- $OWA_{Q_2}(0.5, 0.25, 0.8, 0.0, 0.75) = 0.5$
- $OWA_{Q_3}(0.5, 0.25, 0.8, 0.0, 0.75) = 0.1412$
- $OWA_{Q_4}(0.5, 0.25, 0.8, 0.0, 0.75) = 0.4600$

The results show that, as expected, the smallest value is obtained with $Q_3$, and the largest value is obtained with $Q_1$. The value obtained with $Q_2$ is a value in between:

$$OWA_{Q_3}(\mathbf{a}) \leq OWA_{Q_2}(\mathbf{a}) \leq OWA_{Q_1}(\mathbf{a}).$$

Also, $OWA_{Q_4}$ corresponds to the arithmetic mean of the values in $\mathbf{a}$. Note also that $OWA_{Q_4}(\mathbf{a}) \leq OWA_{Q_2}(\mathbf{a})$ because $Q_2$ does not give much importance to extreme values, and, thus, its result corresponds to the average of $(0.25, 0.5, 0.75)$.

Let us consider now the application of the WOWA operator to the same vector $\mathbf{a}$, and compare its results with the OWA and weighted mean.

*Example 6.16.* Let us consider the quantifiers $Q_1(x) = x^{1/4}$ and $Q_3(x) = x^4$, as in the previous example, and let $\mathbf{p}_1 = (0.1, 0.1, 0.6, 0.1, 0.1)$ and $\mathbf{p}_2 = (0.1, 0.1, 0.1, 0.6, 0.1)$ be two weighting vectors. The WOWA operator with $\mathbf{p}_1$ will return a value larger than the one of the OWA operator because $a_3$ is the largest input, and this is the value with the largest weight. In contrast, the WOWA operator with $\mathbf{p}_2$ will return a value smaller than the one of the OWA operator because $a_4$ is the smallest input.

This can be observed in the following computations:

$$WOWA_{\mathbf{p_2}, Q_1} < OWA_{Q_1} < WOWA_{\mathbf{p_1}, Q_1}$$
$$WOWA_{\mathbf{p_2}, Q_2} < OWA_{Q_2} < WOWA_{\mathbf{p_1}, Q_2}$$

because

- $WOWA_{\mathbf{p_1}, Q_1}(0.5, 0.25, 0.8, 0.0, 0.75) = 0.7526$
- $WOWA_{\mathbf{p_1}, Q_2}(0.5, 0.25, 0.8, 0.0, 0.75) = 0.7416$
- $WOWA_{\mathbf{p_2}, Q_1}(0.5, 0.25, 0.8, 0.0, 0.75) = 0.5791$
- $WOWA_{\mathbf{p_2}, Q_2}(0.5, 0.25, 0.8, 0.0, 0.75) = 0.1250$

Below, we give the $\omega$ weighting vectors that are used for each computation of the WOWA operator. It can be observed that the weights $\omega_i$ in the case of computing $WOWA_{\mathbf{p_2}, Q_1}$ are larger for smaller values and smaller for larger values in comparison with the weights used for computing the $OWA_{Q_1}$.

- $\omega$ for $WOWA_{\mathbf{p_1}, Q_1} = (0.880, 0.035, 0.031, 0.028, 0.026)$
- $\omega$ for $WOWA_{\mathbf{p_1}, Q_2} = (0.666, 0.167, 0.167, 0.0, 0.0)$
- $\omega$ for $WOWA_{\mathbf{p_2}, Q_1} = (0.563, 0.106, 0.071, 0.055, 0.205)$
- $\omega$ for $WOWA_{\mathbf{p_2}, Q_2} = (0.0, 0.0, 0.167, 0.167, 0.666)$

When $Q(x) = x$ for all $x$ in $[0, 1]$, the OWA reduces to the weighted mean, because the quantifier states that all elements are equally relevant.

## 6.2 Choquet Integral

Now we consider another tool for aggregation. From a definitional point of view, its main difference with the previous tools is its use of fuzzy measures. Such measures have been studied in detail in Chapter 5, and some interpretations were given in Section 5.1.1.

In the methods seen so far, we have only considered a single weight for each data element (except for WOWA). Besides, we have not (explicitly) considered how to measure the importance of a set of sources. For example, in the case of the multicriteria decision making problem of car selection, we can state that the comfort criterion confort has an importance of 0.3, and that price is more important, and, thus, its weight is 0.5. However, we have not considered the importance of comfort and price when considered together.

In Chapter 5, we have shown that fuzzy measures are functions defined over the parts of a set that satisfy monotonicity (the larger the set, the larger the measure) and boundary conditions (the measure of the whole set is 1). These restrictions permits us to interpret the measure of a set as the measure of its importance: when information sources are added, the importance increases; when all sources are considered, their importance is maximum and equals to 1. Section 5.1.1 considers interpretations of fuzzy measures with respect to aggregation.

Fuzzy measures permit us to incorporate considerations not included in the weights for the weighted means and the OWA operator. In particular, they can be used to express redundancy, complementarity, and interactions among information sources or criteria. Therefore, tools that use fuzzy measures to represent background knowledge permit the consideration of sources that are not independent. The Choquet integral is one of these tools. Sugeno integrals and fuzzy t-conorm integrals (see Sections 6.4 and 6.5) are some other examples.

Below, we give a definition of the Choquet integral together with an equivalent alternative expression. The Choquet integral is defined as the integral of a function $f$ with respect to a fuzzy measure $\mu$. In our case, both the function and the measure are based on the set of information sources $X = \{x_1, \ldots, x_N\}$. The function $f : X \to \mathbb{R}^+$ corresponds to the value that the sources supply (i.e., $f(x_i) = a_i$, using, as before, $a_i$ to denote the $i$th input value) and the fuzzy measure assigns importances to subsets of $X$ (thus, $\mu : \wp(X) \to [0,1]$).

**Definition 6.17.** Let $\mu$ be a fuzzy measure on $X$; then, the Choquet integral of a function $f : X \to \mathbb{R}^+$ with respect to the fuzzy measure $\mu$ is defined by

$$(C) \int f d\mu = \sum_{i=1}^{N} [f(x_{s(i)}) - f(x_{s(i-1)})] \mu(A_{s(i)}), \qquad (6.1)$$

where $f(x_{s(i)})$ indicates that the indices have been permuted so that $0 \leq f(x_{s(1)}) \leq \cdots \leq f(x_{s(N)}) \leq 1$, and where $f(x_{s(0)}) = 0$ and $A_{s(i)} = \{x_{s(i)}, \ldots, x_{s(N)}\}$.

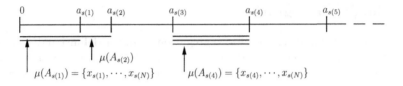

**Fig. 6.6.** Interpreting Choquet integral as a weighting of segments

To denote a Choquet integral, when no confusion exists over the domain $X$, we will use the notation $CI_\mu(a_1, \ldots, a_N) = (C) \int f d\mu$, where, $f(x_i) = a_i$, as before. A Choquet integral can be expressed in an alternative way according to the next proposition.

**Proposition 6.18.** *Let $\mu$ be a fuzzy measure on $X$; then, the Choquet integral of a function $f : X \to \mathbb{R}^+$ with respect to $\mu$ can be expressed as*

$$(C) \int f d\mu = \sum_{i=1}^{N} f(x_{\sigma(i)})[\mu(A_{\sigma(i)}) - \mu(A_{\sigma(i-1)})], \tag{6.2}$$

*or as*

$$(C) \int f d\mu = \sum_{i=1}^{N} f(x_{s(i)})[\mu(A_{s(i)}) - \mu(A_{s(i+1)})], \tag{6.3}$$

*where $\{\sigma(1), \ldots, \sigma(N)\}$ is a permutation of $\{1, \ldots, N\}$ such that $f(x_{\sigma(1)}) \geq f(x_{\sigma(2)}) \geq \cdots \geq f(x_{\sigma(N)})$, where $A_{\sigma(k)} = \{x_{\sigma(j)} | j \leq k\}$ (or, equivalently, $A_{\sigma(k)} = \{x_{\sigma(1)}, \ldots, x_{\sigma(k)}\}$ when $k \geq 1$ and $A_{\sigma(0)} = \emptyset$), and where $s$ and $A_{s(i)}$ are as in Definition 6.17, with $A_{s(N+1)} = \emptyset$.*

Expressions (6.1) and (6.2) outline different aspects of the Choquet integral:

1. Equation 6.1 shows that, for each segment defined by two consecutive values $f(x_{s(i)})$ and $f(x_{s(i-1)})$, the Choquet integral weights the length of the segment $f(x_{s(i)}) - f(x_{s(i-1)})$, according to the measure of all the sources that supply values greater than or equal to $f(x_{s(i)})$. This is so because $A_{s(i)} = \{x_{s(i)}, \ldots, x_{s(N)}\}$. Figure 6.6 illustrates this process. In this figure, the segment $[0, a_{s(1)}]$ is weighted by $\mu(A_{s(1)}) = \mu(\{x_{s(1)}, \ldots, x_{s(N)}\}) = \mu(\{x_{s(j)} | j \geq 1\})$. Similarly, the segment $[a_{s(3)}, a_{s(4)}]$ is weighted by $\mu(A_{s(4)} = \{x_{s(j)} | j \geq 4\})$. This interpretation of the integral is studied in more detail in Section 6.2.1.

2. Equation 6.2 shows that the Choquet integral is a linear combination of values in a way similar to the weighted mean or the OWA operator. Note that each value $f(x_{\sigma(i)})$ is combined with the weight $v_i = (\mu(A_{\sigma(i)}) - \mu(A_{\sigma(i-1)}))$, and the weights define a weighting vector. Note that $\sum_{i=1}^{N} v_i = \sum_{i=1}^{N} (\mu(A_{\sigma(i)}) - \mu(A_{\sigma(i-1)})) = \mu(A_{\sigma(N)}) - \mu(A_{\sigma(0)}) = 1$.

Let $X$ be a reference set, let $(X, \mathcal{A})$ be a measurable space, let $\mu$ be a fuzzy measure on $(X, \mathcal{A})$, and let $f$ be a measurable function $f : X \rightarrow [0, 1]$; then, the Choquet integral of $f$ with respect to $\mu$ is defined by

$$C_\mu(f) := \int_0^\infty \mu_f(r) dr,$$

where $\mu_f(r) := \mu(\{x | f(x) > r\})$.

**Fig. 6.7.** Choquet integral in a continuous domain

The use of fuzzy measures implies that the importance of a set is fixed beforehand, and, therefore, it is not influenced by the values supplied by the sources. That is, $\mu(A)$ does not depend on the actual values of $f(x)$ for $x \in A$. However, the Choquet integral is influenced by the values, and the weight $v_i$ computed for a certain value $f(\{a_{\sigma(i)}\})$ depends on the source $x_{\sigma(i)}$ and on the sources $x_{\sigma(1)}, \ldots, x_{\sigma(i-1)}$ because, as shown in the second point above, $v_i = \mu(A_{\sigma(i)}) - \mu(A_{\sigma(i-1)})$.

Another equivalent expression for the Choquet integral can be given in terms of the simple functions.

**Definition 6.19.** *A function $f : X \rightarrow [0, 1]$ is a simple function if its image $f(X)$ is finite. Let the image of $f(X)$ be $\{a_1, \ldots, a_N\}$, where $0 = a_0 \leq a_1 \leq a_2 \leq \cdots \leq a_N = 1$. Then, there exists a family of sets $X = A_0 \supseteq A_1 \supset A_2 \supset \cdots \supset A_N$ with characteristic functions $\chi_{A_i}$ such that*

$$f(x) = \max_i a_i \chi_{A_i}(x).$$

*This expression is equivalent to*

$$f(x) = \sum_{i=1,N} (a_i - a_{i-1}) \chi_{A_i}(x).$$

For simple functions $f$, the Choquet integral can be expressed by

$$(C) \int f d\mu = \sum_{i=1}^N a_i \mu(A_i).$$

Finally, for the sake of completeness, we give in Figure 6.7 an expression for the Choquet integral when the set $X$ is not finite.

### 6.2.1 Construction of Choquet Integral

We have just given the definition of Choquet integrals where the aggregated value is computed as the integral of a function with respect to a measure. In fact, the weighted mean can also be seen from this point of view. A weighted mean with a weighting vector $\mathbf{p} = (p_1, \ldots, p_n)$ can be interpreted as an integral with respect to an additive measure defined on the singletons by $\mu(\{x_i\}) = p_i$.

**Definition 6.20.** *Let $\mu$ be an additive fuzzy measure; then, the integral of a function $f : X \to \mathbb{R}^+$ (with $a_i = f(x_i)$) with respect to $\mu$ is*

$$WM_\mathbf{p}(a_1, \ldots, a_N) = \int f d\mu = \sum_{x \in X} f(x)\mu(\{x\}). \tag{6.4}$$

*Here, $\mathbf{p} = (\mu(\{x_1\}), \ldots, \mu(\{x_N\}))$.*

Note that this is the expected value of the (multi)set $f(x)$.

For additive fuzzy measures, Equation 6.4 can be rewritten into several equivalent expressions:

$$\int f d\mu = \sum_{x \in X} f(x)\mu(\{x\}) \tag{6.5}$$

$$= \sum_{i=1}^{R} b_i \mu(\{x | f(x) = b_i\}) \tag{6.6}$$

$$= \sum_{i=1}^{N} (a_i - a_{i-1})\mu(\{x | f(x) \geq a_i\}) \tag{6.7}$$

$$= \sum_{i=1}^{N} (a_i - a_{i-1})(1 - \mu(\{x | f(x) \leq a_{i-1}\})), \tag{6.8}$$

where $a_i = f(x_i)$ and where $b_i$ corresponds to the $i$th value in increasing order in the set $\{f(x) | x \in X\}$ (that is, the set of values in the range of $f(x)$) and $R$ is the cardinality of this set. Naturally, $0 < b_1 < b_2 < \cdots < b_R$. Also, note that $b_i \neq b_j$ for $i \neq j$, although it is possible that $f(x_i) = f(x_j)$ for $i \neq j$.

Figure 6.8 illustrates this case. In the figure, $N = 6$ and $X = \{x_1, \ldots x_N\}$. Then, we have $a_1 < a_2 < a_3 = a_4 < a_5 < a_6$, and, thus, $b_1 = a_1 < b_2 = a_2 < b_3 = a_3 = a_4 < b_4 = a_5 < b_5 = a_6$. Thus, $R = 5$.

When the integral is seen as a way to compute the area under the function $f$, each expression corresponds to a different way of computing it. Figures 6.8 (a), (b), and (c) give a graphical interpretation of Expressions 6.5, 6.6, and 6.7. Note that this is not an *area*, because we are considering measures of the elements in $X$.

In Figure 6.8 (a), the area (recall that this is not an area) is decomposed into blocks, each with one element of $X$. Let $x$ be one of the elements in $X$ and let $b_i = f(x)$. The area of each block is $f(x) \cdot \mu(\{x\}) = b_i \cdot \mu(\{x\})$. In Figure 6.8 (b), the integral is decomposed into blocks of elements $x$ with the

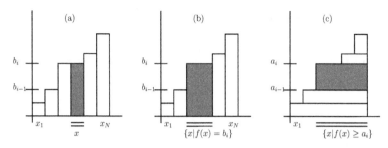

**Fig. 6.8.** Graphical interpretation of several integrals: (a) according to Expression 6.5; (b) according to Expression 6.6; (c) according to Expression 6.7

same $f(x)$. That is, the area of all $x$ such that $f(x) = b_i$ is computed together and is defined as $b_i \cdot \mu(\{x|f(x) = b_i\})$. In Figure 6.8 (c), blocks are defined according to the values $f(x)$. For each $a_i$, elements with a value greater than or equal to $a_i$ are selected. In this way, the area is decomposed into blocks according to the pairs of consecutive values $a_{i-1}$, $a_i$ in the range of $f(x)$. For example, in the case of the consecutive values $a_{i-1}$ and $a_i$, the area of the block is $(a_i - a_{i-1}) \cdot \mu(\{x|f(x) \geq a_i\})$.

When measures are additive, the three expressions are equivalent. However, for non-additive fuzzy measures, they are usually different. From the point of view of aggregation, only Expression 6.7 is meaningful because (6.5) and (6.6) do not always satisfy the constraint that the integral is a value greater than or equal to the minimum and less than or equal to the maximum. That is, among Expressions (6.5), (6.6), and (6.7), only Expression 6.7 satisfies

$$\min(f(x_1), \ldots, f(x_N)) \leq \int f d\mu \leq \max(f(x_1), \ldots, f(x_N)) \qquad (6.9)$$

for all fuzzy measures $\mu$.

Moreover, Expression 6.5 does not use interactions of sources, as the measure is only applied to the singletons. Expression 6.6 exhibits a similar behavior. This is so because in practical applications most values $f(x)$ will be different (the cardinalities of the sets $\{f(x)|x \in X\}$ and $X$ are similar) and, therefore, the measure will be also applied only to singletons.

### 6.2.2 Properties

The Choquet integral satisfies the requirements for aggregation operators. That is, it satisfies unanimity and monotonicity. Therefore, it also satisfies internality. Additionally, it satisfies positive homogeneity ($\mathbb{C}(af) = a\mathbb{C}(f)$ for positive $a$) and monotonicity on the measure (i.e., if $\mu(A) \leq \mu'(A)$ for all $A \subseteq X$, then $CI_\mu(f) \leq CI_\mu(f)$ for all functions $f$). Another property that the Choquet integral satisfies is horizontal additivity. This property is defined as follows.

**Definition 6.21.** *Let $f : X \to [0,1]$; then, for $c \in [0,1]$, let us define $f_c^+$ as follows:*

$$f_c^+ = \begin{cases} 0 & \text{if } f(x) \leq c \\ f(x) - c & \text{if } f(x) > c. \end{cases}$$

*Then, $f = (f \wedge c) + f_c^+$ is a horizontal additive decomposition of $f$.*

The Choquet integral is horizontal additive because, for any decomposition of this type,

$$CI_\mu(f) = CI_\mu(f \wedge c) + CI_\mu(f_c^+).$$

Due to this property, the Choquet integral is also called horizontal integral.

Now, we present a representation theorem for the Choquet integral. This result is similar in spirit to the propositions given in Chapter 4. It establishes that when some conditions are satisfied, the appropriate operator is the Choquet integral. The properties considered are based on comonotonicity.

**Definition 6.22.** *Let $X$ be a reference set, and let $f, g$ functions $f, g : X \to [0,1]$. Then,*

- *$f < g$ when, for all $x_i$,*

$$f(x_i) < g(x_i)$$

- *$f$ and $g$ are comonotonic if, for all $x_i, x_j \in X$,*

$$f(x_i) < f(x_j) \text{ imply that } g(x_i) \leq g(x_j)$$

- *$\mathbb{C}$ is comonotonic monotone if and only if, for comonotonic $f$ and $g$,*

$$f \leq g \text{ imply that } \mathbb{C}(f) \leq \mathbb{C}(g)$$

- *$\mathbb{C}$ is comonotonic additive if and only if, for comonotonic $f$ and $g$,*

$$\mathbb{C}(f + g) = \mathbb{C}(f) + \mathbb{C}(g)$$

Taking into account these properties, the following theorem can be proved.

**Theorem 6.23.** *Let $\mathbb{C}$ be an aggregation operator with the following properties:*

- *$\mathbb{C}$ is comonotonic monotone*
- *$\mathbb{C}$ is comonotonic additive*
- *$\mathbb{C}(1, \ldots, 1) = 1$*

*Then, there exists a fuzzy measure $\mu$ such that $\mathbb{C}(f)$ is the Choquet integral of $f$ with respect to $\mu$.*

It is also true that a Choquet integral satisfies the conditions above. Now, we review some results that relate this integral to WOWA, OWA, and weighted mean.

**Proposition 6.24.** *For every weighting vector* **p**, *we have* $WM_{\mathbf{p}} = CI_\mu$, *with* $\mu$ *defined by*

$$\mu_{\mathbf{p}}(B) = \sum_{x_i \in B} p_i \qquad \qquad \text{for all } B \subseteq X.$$

This proposition shows that the Choquet integral is a proper generalization of weighted mean for non-additive measures. That is, when a measure is additive, the Choquet integral reduces to a weighted mean.

**Proposition 6.25.** *For every weighting vector* **w**, *we have* $OWA_{\mathbf{w}} = CI_\mu$, *with* $\mu$ *defined by*

$$\mu_{\mathbf{w}}(B) = \sum_{i=1}^{|B|} w_i \qquad \qquad \text{for all } B \subseteq X.$$

**Proposition 6.26.** *For every weighting vector* **p** *and every regular non-decreasing fuzzy quantifier* $Q$, *we have* $WOWA_{\mathbf{p},Q} = CI_\mu$, *with* $\mu$ *defined by*

$$\mu_{\mathbf{p},Q}(B) = Q\left(\sum_{x_i \in B} p_i\right) \qquad \qquad \text{for all } B \subseteq X.$$

Note that the last measure corresponds to a distorted probability (Definition 5.61), and that, when $p_i = 1/N$, we have a symmetric measure (Proposition 5.72), and that with $Q(x) = x$ we have a probability distribution.

These propositions show that the weighted mean, OWA, and WOWA are particular cases of Choquet integrals. There are some results that establish the reversal conditions; that is, when a Choquet integral can be reduced to any of the former operators. The first result is about the symmetric Choquet integral, i.e., the Choquet integral that satisfies the symmetry condition.

**Proposition 6.27.** *Any symmetric Choquet integral* $CI_\mu$ *is an OWA operator.*

**Proposition 6.28.** *Any Choquet integral with a distorted probability is a WOWA operator.*

In fact, the quantifier $Q$ (or the function $w^*$) in the WOWA operator corresponds to the distortion function in distorted probability, and the weighting vector **p** corresponds to the probability distribution in the distorted probability.

The following example gives the distorted probability that can be built from the weighting vector **p** and a function $w^*$. This fuzzy measure corresponds to the one used in Example 6.10, where WOWA was considered,

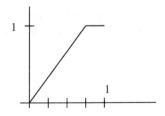

**Fig. 6.9.** Fuzzy quantifier constructed from the OWA weights $\mathbf{w} = (1/3, 1/3, 1/3, 0)$ in Example 6.10

as the function $w^*$ was built in Example 6.11 from the weighting vector $\mathbf{w} = (1/3, 1/3, 1/3, 0)$. For the sake of completeness, we also give the fuzzy measures that make the Choquet integral equal to the OWA operator and the weighted mean.

*Example 6.29.* Let $\mathbf{p} = (0.5, 0.3, 0.15, 0.05)$ be a weighting vector and let $w^*$ be the function shown in Figure 6.9 and defined as follows

$$w^*(x) = \begin{cases} x/0.75 & \text{if } x < 0.75 \\ 1 & \text{if } x \geq 0.75 \end{cases}$$

Then, the distorted fuzzy measure $\mu_{WOWA}$ defined from $\mathbf{p}$ and $w^*$ is given in Table 6.1. The construction of $\mu_{WOWA}$ follows Proposition 6.26. This table also displays the measures $\mu_{WM}$ and $\mu_{OWA}$ constructed according to Propositions 6.24 and 6.25.

Note that the Choquet integral with respect to $\mu_{WM}$ is equivalent to the weighted mean, the Choquet integral with respect to $\mu_{OWA}$ is equivalent to the OWA operator, and the Choquet integral with respect to $\mu_{WOWA}$ is equivalent to the WOWA operator.

Now, we consider the Choquet integral of a crisp set.

**Proposition 6.30.** *Let $A$ be a crisp subset of $X$; then, the Choquet integral of $A$ with respect to $\mu$ is $\mu(A)$. Here, the integral of $A$ corresponds to the integral of its characteristic function, or, in other words, to the integral of the function $f_A$ defined as $f_A(x) = 1$ if and only if $x \in A$.*

The proposition gives some hints about the definition of fuzzy measures.

Let $f_B$ the membership function of a fuzzy set $B \subseteq A$. Then, due to the fact that the Choquet integral is monotone, for all $\mu$, we have $CI_\mu(f_B) \leq CI_\mu(f_A)$. According to this, the Choquet integral can be interpreted as a measure of the fuzzy set $B$.

| set | $\mu_{WM}$ | $\mu_{OWA}$ | $\mu_{WOWA}$ |
|---|---|---|---|
| $\mu(\emptyset)$ | 0.0 | 0.0 | 0.0 |
| $\mu(\{C_4\})$ | 0.05 | 0.3333 | 0.0666 |
| $\mu(\{C_3\})$ | 0.15 | 0.3333 | 0.2 |
| $\mu(\{C_3, C_4\})$ | 0.2 | 0.6666 | 0.2666 |
| $\mu(\{C_2\})$ | 0.3 | 0.3333 | 0.4 |
| $\mu(\{C_2, C_4\})$ | 0.35 | 0.6666 | 0.4666 |
| $\mu(\{C_2, C_3\})$ | 0.45 | 0.6666 | 0.6 |
| $\mu(\{C_2, C_3, C_4\})$ | 0.5 | 1.0 | 0.6666 |
| $\mu(\{C_1\})$ | 0.5 | 0.3333 | 0.6666 |
| $\mu(\{C_1, C_4\})$ | 0.55 | 0.6666 | 0.7333 |
| $\mu(\{C_1, C_3\})$ | 0.65 | 0.6666 | 0.8666 |
| $\mu(\{C_1, C_3, C_4\})$ | 0.7 | 1.0 | 0.9333 |
| $\mu(\{C_1, C_2\})$ | 0.8 | 0.6666 | 1.0 |
| $\mu(\{C_1, C_2, C_4\})$ | 0.85 | 1.0 | 1.0 |
| $\mu(\{C_1, C_2, C_3\})$ | 0.95 | 1.0 | 1.0 |
| $\mu(\{C_1, C_2, C_3, C_4\})$ | 1.0 | 1.0 | 1.0 |

**Table 6.1.** Fuzzy measures for Example 6.29

## 6.3 Weighted Minimum and Weighted Maximum

The operators described in previous sections are based on the existence of a numerical scale where addition and multiplication are applicable (and meaningful). Alternative operators exist that do not rely on these. They only use maximum and minimum (and thus they can operate on ordinal scales). This is the case with weighted minimum, weighted maximum, and Sugeno integral. The last is in some sense the counterpart of the Choquet integral in the ordinal setting. We begin in this section with the weighted minimum and maximum. Section 6.4 is devoted to Sugeno integrals.

In the definition given below, the weighted minimum also requires the existence of a negation function (see Section 2.3.1) over the domain. For example, the negation $neg(x) = 1 - x$ can be used when the values $x$ are in the unit interval. The negation function is used so that the weights can be interpreted as importance (as with the weighted mean). An alternative definition without negation might be given but the meaning of the weights would change (as they would then correspond to the negation of importance).

It has been said that aggregation operators $\mathbb{C}$ are defined so that they satisfy

$$\min(a_1, \ldots, a_N) \leq \mathbb{C}(a_1, \ldots, a_N) \leq \max(a_1, \ldots, a_N).$$

Therefore, minimum and maximum are extreme cases of aggregation operators.

Minimum can be seen as the determination of the most conservative or most pessimistic value; for example, the minimum consensus among information sources or the minimum value for which all information sources agree.

This interpretation is consistent with the use of the quantifier "for all" in the OWA operator that, as shown in Section 6.1.4, is equivalent to minimum. Moreover, minimum can be seen as intersection or conjunction when applied to certainty degrees. This was shown in Section 2.3.4.

In contrast, the maximum can be seen as the selection of the most progressive, innovative, or optimistic value: it corresponds to an extreme opinion, as only one information source needs to supply this value. This interpretation corresponds to the consideration of the fuzzy quantifier "there exists" in the OWA operator. In this case, the OWA operator behaves like a union or disjunction of the certainty degrees.

Weighted minimum and weighted maximum permit us to include weights in the aggregation process. In this case, weights are represented by means of a weighting vector, but now, instead of having weights that are positive and add to one, they are positive but at least one is maximal. That is, one of the weights corresponds to the maximum value in the ordinal scale. In the case where the unit interval is used, at least one weight should be equal to 1. From now on, we will call the weighting vectors used in the weighted mean and OWA the *probabilistic weighting vectors* (they can be understood as probability distributions); and we will call the ones used in weighted minimum and weighted maximum the *possibilistic weighting vectors* (they can be understood as possibility distributions).

In this way, while the weighted mean is the aggregation of data with respect to a probabilistic weighting vector (a probability distribution), the weighted minimum and the weighted maximum are the aggregation of data with respect to a possibilistic weighting vector (a possibility distribution).

**Definition 6.31.** *A vector* $\mathbf{v} = (v_1...v_N)$ *is a* possibilistic weighting vector *of dimension N if and only if* $v_i \in [0,1]$ *and* $\max_i v_i = 1$.

**Definition 6.32.** *Let* $\mathbf{u}$ *be a weighting vector of dimension N; then, a mapping WMin:* $[0,1]^N \rightarrow [0,1]$ *is a* weighted minimum *of dimension N if* $WMin_{\mathbf{u}}(a_1, ..., a_N) = \min_i \max(neg(u_i), a_i)$.

**Definition 6.33.** *Let* $\mathbf{u}$ *be a weighting vector of dimension N; then, a mapping WMax:* $[0,1]^N \rightarrow [0,1]$ *is a* weighted maximum *of dimension N if* $WMax_{\mathbf{u}}(a_1, ..., a_N) = \max_i \min(u_i, a_i)$.

In these definitions, the values in the possibility distribution can be understood as certainty degrees. Alternative definitions can be given with a weighting vector $\mathbf{v} = (v_1, \ldots, v_N)$ where $v_i = neg(u_i)$. Note that having such a vector $\mathbf{v}$, negation is not required for the weighted minimum.

The problem of combining constraints in a fuzzy CSP problem can also be solved using weighted minimum or weighted maximum. This is shown in the next example.

*Example 6.34.* Let us consider again Example 6.8, with the same formulation but with the importance of the constraints given by a possibilistic weighting vector. Let this vector be $\mathbf{u} = (1, 0.5, 0.3, 0.1)$. Note that the weighting vector in Example 6.8 is not suitable here because the possibilistic weighting vector should be such that at least one of the weights is equal to 1 ($\max_i u_i = 1$). The evaluations for the possible solutions with such a vector $\mathbf{u}$ using the WMin are as follows:

- $sat(2, 2) = WMin_{\mathbf{u}}(0, 0.5, 1, 1) = 0$
- $sat(2, 3) = WMin_{\mathbf{u}}(0.5, 0.5, 0.5, 0.5) = 0.5$
- $sat(2, 4) = WMin_{\mathbf{u}}(1, 0.5, 0, 0.5) = 0.5$
- $sat(3.5, 2.5) = WMin_{\mathbf{u}}(1, 0.5, 0.5, 0.5) = 0.5$
- $sat(3, 2) = WMin_{\mathbf{u}}(0.5, 1, 1, 0.5) = 0.5$
- $sat(3, 3) = WMin_{\mathbf{u}}(1, 1, 0.5, 1) = 0.7$

Note that the computation of these evaluations need $neg(\mathbf{u})$. We have used $neg(x) = 1 - x$, which corresponds to $neg(\mathbf{u}) = (0, 0.5, 0.7, 0.9)$.

In contrast, in the case of the WMax, the evaluations are as follows:

- $sat(2, 2) = WMax_{\mathbf{u}}(0, 0.5, 1, 1) = 0.5$
- $sat(2, 3) = WMax_{\mathbf{u}}(0.5, 0.5, 0.5, 0.5) = 0.5$
- $sat(2, 4) = WMax_{\mathbf{u}}(1, 0.5, 0, 0.5) = 1$
- $sat(3.5, 2.5) = WMax_{\mathbf{u}}(1, 0.5, 0.5, 0.5) = 1$
- $sat(3, 2) = WMax_{\mathbf{u}}(0.5, 1, 1, 0.5) = 0.5$
- $sat(3, 3) = WMax_{\mathbf{u}}(1, 1, 0.5, 1) = 1$

These results show that when using the weighted minimum the pair $(3, 3)$ is the best pair, but that when using the weighted maximum it is indistinguishable from the pairs $(2, 4)$ and $(3.5, 2.5)$.

We now give another example to illustrate the application of weighted minimum and weighted maximum. In this case, we show its use in fuzzy systems.

*Example 6.35.* Let us consider fuzzy inference systems where the rules follow the structure given in Section 2.3.5. That is, we have a set of rules $\{R_i\}_{i=1,\ldots,N}$ of the form

$$R_i: \textbf{IF} \quad x \text{ is } A_i \textbf{ THEN } y \text{ is } B_i.$$

According to what has been described in Section 2.3.5, when rules are disjunctive, the output of the fuzzy system, given $x = x_0$, is

$$\tilde{B} = \vee_{i=1}^{N}(B_i \wedge A_i(x_0)), \tag{6.10}$$

which, for a particular $y_0$ in the domain of $Y$, leads to the following membership degree:

$$\tilde{B}(y_0) = \vee_{i=1}^{N}(B_i(y_0) \wedge A_i(x_0)).$$

This expression corresponds to the weighted maximum (Definition 6.33) of the values $B_i(y_0)$ with respect to the weights $A_i(x_0)$. That is,

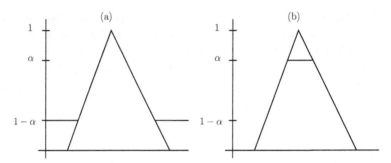

**Fig. 6.10.** Considering uncertainty $\alpha$ in a fuzzy set $B_i$ with a triangular membership function: (a) conjunctive rules; (b) disjunctive rules

$$\tilde{B}(y_0) = WMax_{\mathbf{u}}(B_1(y_0), \ldots, B_N(y_0)), \tag{6.11}$$

where the weighting vector $\mathbf{u}$ is defined as $\mathbf{u} = (A_1(x_0), \ldots, A_N(x_0))$.

Note that the weighting vector is independent of the value $y_0$. Thus, for a given $x_0$, the same aggregation operator with the same weights is applied to all $y_0$ in $Y$. Similarly, when rules are conjunctive, we have

$$\tilde{B}(y_0) = \wedge_{i=1}^{N}\big(\mathcal{I}(A_i(x_0), B_i(y_0))\big). \tag{6.12}$$

In this case, with an appropriate selection of the implication $\mathcal{I}$, we can rewrite the expression in terms of a weighted minimum (Definition 6.32). In particular, this is possible using the Kleene-Dienes implication. This implication (Section 2.3.2, Equation 2.17) is defined as $\mathcal{I}(x, y) = \max(1 - x, y)$. Thus, the following equation holds for such implications:

$$\tilde{B}(y_0) = \wedge_{i=1}^{N}\big(\mathcal{I}(A_i(x_0), B_i(y_0))\big) = \wedge_{i=1}^{N} \max(1 - A_i(x_0), B_i(y_0)). \tag{6.13}$$

The weighting vector $\mathbf{u}$ defined as $\mathbf{u} = (A_1(x_0), \ldots, A_N(x_0))$ permits us to rewrite $\tilde{B}(y_0)$ as follows:

$$\tilde{B}(y_0) = WMin_{\mathbf{u}}(B_1(y_0), \ldots, B_N(y_0)). \tag{6.14}$$

This example shows that, in fuzzy inference systems, $WMax$ and $WMin$ are used for combining the certainty degrees of the output fuzzy sets $B_i(y_0)$ and the degrees of satisfaction of the rules $A_i(x_0)$. When the rules are conjunctive, the combination is a weighted minimum, and the fuzzy sets $B_i$ are transformed into $B_i \vee (1 - A_i(x_0))$. Figure 6.10 (a) illustrates this fact when $\alpha = A_i(x_0)$ and when $B_i$ is represented as a triangular fuzzy set.

When the rules are disjunctive, the combination is a weighted maximum, and the fuzzy sets $B_i$ are transformed into $B_i \wedge A_i(x_0)$. Figure 6.10 (b) illustrates this fact when $\alpha = A_i(x_0)$.

We have reviewed two operators that somehow correspond to a translation of the weighted mean (which combines values with respect to a probability distribution) into operators that combine values with respect to possibility distributions. This translation also exists for the OWA operator. The resulting operators are OWMin and OWMax. Their definitions are similar to Definitions 6.32 and 6.33, but include an ordering permutation, as in the case of the OWA (see Definition 6.4).

### 6.3.1 Properties of Weighted Minimum and Maximum

The aggregation operators described in this section are not directly related to the ones described in previous sections. Nevertheless, for some particular weights, equalities can be given. For example, when $\mathbf{u} = (1, 1, \ldots, 1)$, we have that $WMIN_{\mathbf{u}} = \min$ and $WMAX_{\mathbf{u}} = \max$.

### 6.3.2 Dealing with Symbolic Domains

Note that, although definitions are given in the unit interval, the operators can be applied to symbolic ordinal domains. In this case, both weights and values have to be in the same domain, say $L$, and negation, when needed, is an involutive function that reverses the ordering in $L$. Because (according to the following proposition) conditions on negations for symbolic domains completely determine them, once $L$ and the weights are known, the weighted minimum and the weighted maximum are completely determined. There is no need to define the negation function.

**Proposition 6.36.** *Let* $L = \{l_0, \ldots, l_r\}$ *with* $l_0 <_L l_1 <_L \cdots <_L l_r$ *be a finite ordinal scale; then, there exists only one negation function,* $neg : L \to L$, *satisfies*

*(N1) if* $x <_L x'$ *then* $neg(x) >_L neg(x')$ *for all* $x, x'$ *in* $L$.
*(N2)* $neg(neg(x)) = x$ *for all* $x$ *in* $L$.

*This negation function is defined by*

$$neg(x_i) = x_{r-i} \text{ for all } x_i \text{ in } L. \tag{6.15}$$

## 6.4 Sugeno Integrals

In the same way that the Choquet integral generalizes the OWA and the weighted mean, the Sugeno integral generalizes the weighted minimum (and the weighted maximum). While the $WMin$ (and $WMax$) express importance or reliability through weighting vectors, the Sugeno integral uses fuzzy measures. Recall that this shift from weighting vectors to fuzzy measures occurs when moving from the weighted mean to the Choquet integral.

Sugeno integrals, due to the combination of weights and values through the minimum, is defined for functions into $[0, 1]$ and normalized fuzzy measures.

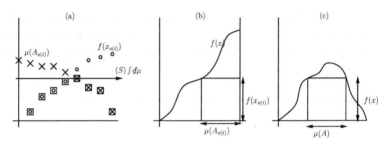

**Fig. 6.11.** Graphical interpretation of Sugeno integrals

**Definition 6.37.** *Let $\mu$ be a fuzzy measure on $X$; then, the Sugeno integral of a function $f : X \rightarrow [0, 1]$ with respect to $\mu$ is defined by*

$$(S) \int f d\mu = \max_{i=1,N} \min(f(x_{s(i)}), \mu(A_{s(i)})), \qquad (6.16)$$

*where $f(x_{s(i)})$ indicates that the indices have been permuted so that $0 \leq f(x_{s(1)}) \leq \dots \leq f(x_{s(N)}) \leq 1$ and $A_{s(i)} = \{x_{s(i)}, \dots, x_{s(N)}\}$.*

**Proposition 6.38.** *The Sugeno integral of a function $f : X \rightarrow [0, 1]$ with respect to a fuzzy measure $\mu$ can be equivalently expressed by*

$$\max_{i} \min(f(x_{\sigma(i)}), \mu(A_{\sigma(i)})),$$

*where $A_{\sigma(k)} = \{x_{\sigma(j)} | j \leq k\}$ (or, equivalently, $A_{\sigma(k)} = \{x_{\sigma(1)}, \dots, x_{\sigma(k)}\}$ when $k \geq 1$ and $A_{\sigma(0)} = \emptyset$), and where $\sigma$ is a permutation such that $f(x_{\sigma(i)}) \geq f(x_{\sigma(i+1)})$ for $i \geq 1$.*

To denote a Sugeno integral of a function $f$ over the domain $X$, we will use, when no confusion arises, the notation $SI_\mu(a_1, \dots, a_N)$ for $(S) \int f d\mu$. Here, $a_i = f(x_i)$ as before.

The Sugeno integral can be interpreted in a way similar to the weighted maximum (note that expressions are similar, but use the measure instead of the possibility distribution). The difference is that now each value $f(x_{s(x)})$ is weighted according to the weight (the measure) of all the sources that support a value (i.e., that supply a value larger or equal than $f(x_{s(i)})$); that is, according to $\mu(A)$.

Figure 6.11 (a) gives a graphical interpretation of the operation of Sugeno integrals. The figure displays the values $f(x_{s(i)})$, the measures for $\mu(A_{s(i)})$, and the combination $min(f(x_{s(i)}), \mu(A_{s(i)}))$ for all $x_i \in X$. The largest of these values is the result of the Sugeno integral. In this figure, the values are denoted by dots, the measures are denoted by crosses, and the values $min(f(x_{s(i)}), \mu(A_{s(i)}))$ are denoted by squares.

The graphical interpretation of the Sugeno integral shows that the outcome is obtained by "saturation." We can see that the integral selects the

Let $X$ be a reference set, let $(X, \mathcal{A})$ be a measurable space, let $\mu$ be a fuzzy measure on $(X, \mathcal{A})$, and let $f$ be a measurable function $f : X \rightarrow [0, 1]$; then, the Sugeno integral of $f$ with respect to $\mu$ is defined by

$$S_\mu(f) := \sup_{r \in [0,1]} [r \wedge \mu_f(r)],$$

where $\mu_f(r) := \mu(\{x | f(x) > r\})$.

**Fig. 6.12.** Sugeno integral in a continuous domain

| set | {Tokyo} | {Kyoto} | {Nagano} | {Tokyo, Kyoto} | {Kyoto, Nagano} | {Tokyo, Nagano} | X |
|-----|---------|---------|----------|----------------|-----------------|-----------------|---|
| $\mu$ | 0.7 | 0.5 | 0.2 | 0.9 | 0.6 | 0.8 | 1 |

**Table 6.2.** Fuzzy measure for the traveler example: Satisfaction degree for visiting Tokyo, Kyoto, and Nagano

importance that overcomes (or saturates) certain thresholds. Due to the ordering, the threshold is decreasing as long as inputs are increasing. In this way, the integral finds a trade-off (or compromise) between the importance or reliability of the sets (i.e., $\mu(A)$) and the values that the members of the sets have assigned (i.e., the value $\max_{x \in A} f(x)$).

For a continuous monotone or convex function, the Sugeno integral computes the length of the square with maximum area, as shown in Figure 6.11 (b) and (c). The definition of the Sugeno integral for continuous functions is given in Figure 6.12.

The following example illustrates the Sugeno integral.

*Example 6.39.* Let us consider a decision making problem. There is a traveler who intends to visit three cities in Japan, and considers different locations for staying. The three cities, represented by $X = \{x_1, x_2, x_3\}$, correspond to Tokyo, Kyoto, and Nagano.

Let us consider the degree of satisfaction of the traveler when visiting such cities. Such a degree is expressed by means of a fuzzy measure. Note that the measure is monotone, as the greater the number of cities visited, the greater the satisfaction. Boundary conditions mean that no cities visited correspond to the lowest satisfaction (equal to zero), and that all cities being visited correspond to maximum satisfaction (equal to one). A fuzzy measure for the satisfaction of the visitor is given in Table 6.2.

Then, for finally deciding where to stay, the degree of satisfaction $\mu$ should be combined with the accessibility of the town from the visitor's location. To do so, we should express the accessibilities. In other words, for a given visitor's

| set | $Tokyo$ | $Kyoto$ | $Nagano$ |
|-----|---------|---------|----------|
| $f$ | 0.8 | 0.4 | 0.5 |
| $\mu_f$ | 0.7 | 1 | 0.8 |

**Table 6.3.** $f$: Accessibility degrees to $x_i$ from Tsukuba; $\mu_f$: Satisfaction degree for visiting $x_i$ and all those city that are at least as accessible as $x_i$

location, the degree of accessibility is a function of the town to be visited. That is, if the traveler is located in Tsukuba, then the most accessible town is Tokyo, followed by Nagano and finally Kyoto. This situation is modeled by a function $f : X \rightarrow D$ such that $f(Tokyo) > f(Nagano) > f(Kyoto)$. Table 6.3 gives measures for such accessibility from Tsukuba. The values for the measure are expressed using the same terms as the degree of satisfaction. This is, $f(x)$ is comparable with $\mu(A)$.

Once we have defined $\mu$ and $f$, we can consider the degree of satisfaction of visiting a particular city $x_i$ and all the cities at least as accessible as $x_i$. With regard to accessibility, we might consider that if we visit $x_i$, it is meaningful to visit also cities that are *easier* to visit than $x_i$. For example, if we stay in Tsukuba and we visit Nagano, then it is reasonable to visit also Tokyo, which is nearer. For a given $x_i$, the set of such cities is $\{x|f(x) \geq f(x_i)\}$. In the case of visiting Nagano, $\{x|f(x) \geq f(Nagano)\} = \{Tokyo, Nagano\}$. Then, we can compute the degree of satisfaction by applying the measure $\mu$ to such a set. So, the degree of satisfaction for visiting $x_i$ and nearer cities is defined by

$$\mu_f(x_i) = \mu(\{x|f(x) \geq f(x_i)\}).$$

Accordingly, $\mu_f(Nagano) = \mu(\{Tokyo, Nagano\}) = 0.8$. Similarly, we compute the values for $Tokyo$ and $Kyoto$. They are given in Table 6.3.

Now, we consider an overall degree of being in the town of Tsukuba. To do so, we consider, for all cities $x_i$, the degrees $f(x_i)$ and $\mu_f(x_i)$. That is, the accessibility of $x_i$ and the satisfaction of visiting $x_i$ as well as cities more easily accessible. To combine the two values, as in some sense we want both to be satisfied, we apply a conjunctive approach. That is, we combine them with the minimum (which, as seen in Section 2.3.1, can be used to model conjunction). The rationale of the approach is given below by considering two cases.

Case $f(x_i) \geq \mu_f(x_i)$: In this case, it is easier to access $x_i$ than the traveler's satisfaction. Accordingly, the overall degree for $x_i$ cannot be larger than $\mu_f(x_i)$.

Case $\mu_f(x_i) \geq f(x_i)$: In this case, accessibility is easier than the corresponding satisfaction. Nevertheless, the overall degree of the city cannot exceed that of the cities visited.

So, in both cases we consider the minimum of the two values. Note that this operation is valid because both degrees are expressed in the same terms (i.e., they are comparable). Thus, using $\wedge$ to denote the minimum, we use $f(x_i) \wedge \mu_f(x_i)$ to denote the evaluation of visiting $x_i$.

Taking everything into account, the place $x_i$ with the largest evaluation $f(x_i) \wedge \mu_f(x_i)$ stands for the evaluation of staying in Tsukuba. This largest evaluation corresponds to the Sugeno integral of $f$ with respect to $\mu$, which in this example is equal to

$$SI_\mu(f) = \max_{x_i} f(x_i) \wedge \mu_f(x_i) = 0.7.$$

Note that this overall satisfaction equal to 0.7 is obtained for $x_i = Tokyo$ (i.e., $\arg\max_{x_i} f(x_i) \wedge \mu_f(x_i) = Tokyo$), and the cities to be visited will be $\{Tokyo, Nagano\}$. So, using the Sugeno integral, we visit the cities where accessibility equals satisfaction; cities are selected so that satisfaction compensates accessibility. In a more general case, with a larger set $X$, we add more and more cities, increasing satisfaction and decreasing accessibility until both degrees are equal.

Note that in this example we have used satisfaction degrees in $[0, 1]$. Nevertheless, the same would apply if we consider other ordinal scales. In particular, the same applies if we consider an ordered set $L = \{l_0, \ldots, l_r\}$ with $l_0 <_L l_1 <_L \cdots <_L l_r$.

*Example 6.40.* In Example 6.35, we have seen that the WMin and WMax can be used to combine the certainty degrees of the output fuzzy sets. In the case where there are interactions among the rules, Sugeno integrals might be used for defining appropriate fuzzy measures.

*Example 6.41.* Let us consider again Example 6.10 (Section 6.1.2). In this case, we will use the Sugeno integral, considering three different fuzzy measures. In particular, the measures are defined in Table 6.1. They are reproduced in Table 6.4.

- $sat(2, 2) = SI_{\mu WM}(0, 0.5, 1, 1) = 0.5$
- $sat(2, 3) = SI_{\mu WM}(0.5, 0.5, 0.5, 0.5) = 0.5$
- $sat(2, 4) = SI_{\mu WM}(1, 0.5, 0, 0.5) = 0.5$
- $sat(3.5, 2.5) = SI_{\mu WM}(1, 0.5, 0.5, 0.5) = 0.5$
- $sat(3, 2) = SI_{\mu WM}(0.5, 1, 1, 0.5) = 0.5$
- $sat(3, 3) = SI_{\mu WM}(1, 1, 0.5, 1) = 0.85$

- $sat(2, 2) = SI_{\mu OWA}(0, 0.5, 1, 1) = 0.66$
- $sat(2, 3) = SI_{\mu OWA}(0.5, 0.5, 0.5, 0.5) = 0.5$
- $sat(2, 4) = SI_{\mu OWA}(1, 0.5, 0, 0.5) = 0.5$
- $sat(3.5, 2.5) = SI_{\mu OWA}(1, 0.5, 0.5, 0.5) = 0.5$
- $sat(3, 2) = SI_{\mu OWA}(0.5, 1, 1, 0.5) = 0.666$
- $sat(3, 3) = SI_{\mu OWA}(1, 1, 0.5, 1) = 1.0$

- $sat(2, 2) = SI_{\mu WOWA}(0, 0.5, 1, 1) = 0.5$

| set | $\mu_{WM}$ | $\mu_{OWA}$ | $\mu_{WOWA}$ |
|---|---|---|---|
| $\mu(\emptyset)$ | 0.0 | 0.0 | 0.0 |
| $\mu(\{C_4\})$ | 0.05 | 0.3333 | 0.0666 |
| $\mu(\{C_3\})$ | 0.15 | 0.3333 | 0.2 |
| $\mu(\{C_3, C_4\})$ | 0.2 | 0.6666 | 0.2666 |
| $\mu(\{C_2\})$ | 0.3 | 0.3333 | 0.4 |
| $\mu(\{C_2, C_4\})$ | 0.35 | 0.6666 | 0.4666 |
| $\mu(\{C_2, C_3\})$ | 0.45 | 0.6666 | 0.6 |
| $\mu(\{C_2, C_3, C_4\})$ | 0.5 | 1.0 | 0.6666 |
| $\mu(\{C_1\})$ | 0.5 | 0.3333 | 0.6666 |
| $\mu(\{C_1, C_4\})$ | 0.55 | 0.6666 | 0.7333 |
| $\mu(\{C_1, C_3\})$ | 0.65 | 0.6666 | 0.8666 |
| $\mu(\{C_1, C_3, C_4\})$ | 0.7 | 1.0 | 0.9333 |
| $\mu(\{C_1, C_2\})$ | 0.8 | 0.6666 | 1.0 |
| $\mu(\{C_1, C_2, C_4\})$ | 0.85 | 1.0 | 1.0 |
| $\mu(\{C_1, C_2, C_3\})$ | 0.95 | 1.0 | 1.0 |
| $\mu(\{C_1, C_2, C_3, C_4\})$ | 1.0 | 1.0 | 1.0 |

**Table 6.4.** Fuzzy measures for Example 6.29

- $sat(2,3) = SI_{\mu_{WOWA}}(0.5, 0.5, 0.5, 0.5) = 0.5$
- $sat(2,4) = SI_{\mu_{WOWA}}(1, 0.5, 0, 0.5) = 0.666$
- $sat(3.5, 2.5) = SI_{\mu_{WOWA}}(1, 0.5, 0.5, 0.5) = 0.666$
- $sat(3,2) = SI_{\mu_{WOWA}}(0.5, 1, 1, 0.5) = 0.6$
- $sat(3,3) = SI_{\mu_{WOWA}}(1, 1, 0.5, 1) = 1.0$

### 6.4.1 Properties

In this section, we review a few results related to Sugeno integrals. We start by establishing that the weighted minimum and the weighted maximum are particular cases of the Sugeno integral. We show the measures that permit us to establish this relation.

**Proposition 6.42.** *The Sugeno integral generalizes both weighted minimum and weighted maximum.*

1. *A weighted maximum with a possibilistic weighting vector* **u** *is equivalent to a Sugeno integral with the fuzzy measure*

$$\mu_{\mathbf{u}}^{wmax}(A) = \max_{a_i \in A} u_i.$$

2. *A weighted minimum with a possibilistic weighting vector* **u** *is equivalent to a Sugeno integral with the fuzzy measure*

$$\mu_{\mathbf{u}}^{wmin}(A) = 1 - \max_{a_i \notin A} u_i.$$

Given a possibilistic weighting vector $\mathbf{u}$, the two fuzzy measures generated above are dual, as shown in Proposition 5.32. That is

$$\mu_{\mathbf{u}}^{wmin}(A) = 1 - \mu_{\mathbf{u}}^{wmax}(X \setminus A),$$

and when $\mathbf{u}$ is a possibility distribution, $\mu_{\mathbf{u}}^{wmin}$ defines a necessity distribution.

**Proposition 6.43.** *The following hold for all functions $f$, $g$, and for any fuzzy measure $\mu$:*

1. *if $f \leq g$, then $SI_{\mu}(f) \leq SI_{\mu}(g)$*
2. *$SI_{\mu}(a, \ldots, a) = a$ for any constant $a \in [0,1]$*
3. *$SI_{\mu}(f + a) \leq SI_{\mu}(f) + SI_{\mu}(a)$ for any constant $a \in [0,1]$*
4. *$SI_{\mu}(f \vee g) \geq (SI_{\mu}(f) \vee SI_{\mu}(g))$*
5. *$SI_{\mu}(f \wedge g) \leq (SI_{\mu}(f) \wedge SI_{\mu}(g))$*

The next proposition is analogous to Proposition 6.30 for the Choquet integral.

**Proposition 6.44.** *Let $A$ be a crisp subset of $X$; then, the Sugeno integral of $A$ with respect to $\mu$ is $\mu(A)$.*

Now, we consider two representation theorems for the Sugeno integral. We will use the comonotonic monotone property established in Definition 6.22 and the definitions given below.

**Definition 6.45.** *Let $X$ be a reference set, let $a$ be a value in $[0,1]$, and let $f, g$ functions $f, g : X \to [0,1]$. Then,*

- *$\mathbb{C}$ is minimum homogeneous if and only if, for comonotonic $f$ and $g$,*

$$\mathbb{C}(a \wedge f) = a \wedge \mathbb{C}(f)$$

- *$\mathbb{C}$ is comonotonic maxitive if and only if, for comonotonic $f$ and $g$,*

$$\mathbb{C}(f \vee g) = \mathbb{C}(f) \vee \mathbb{C}(g)$$

The following results can be proved.

**Proposition 6.46.** *Let $\mathbb{C}$ be an aggregation operator with the following properties:*

- *$\mathbb{C}$ is comonotonic monotone*
- *$\mathbb{C}$ is comonotonic maxitive*
- *$\mathbb{C}$ is minimum homogeneous*
- *$\mathbb{C}(1, \ldots, 1) = 1$*

*Then, there exists a fuzzy measure $\mu$ such that $\mathbb{C}(f)$ is the Sugeno integral of $f$ with respect to $\mu$.*

**Proposition 6.47.** *Let $\mathbb{C}$ be an aggregation operator with the following properties:*

- $\mathbb{C}(f \vee g) = \mathbb{C}(f) \vee \mathbb{C}(g)$
- $\mathbb{C}(a \wedge f) = a \wedge \mathbb{C}(f)$
- $\mathbb{C}(1, \ldots, 1) = 1$

*Then, there exists a possibility measure Pos such that $\mathbb{C}(f)$ is the Sugeno integral of $f$ with respect to Pos.*

Both Choquet and Sugeno integrals integrate functions with respect to fuzzy measures. Nevertheless, for most of the functions, their results are different. The following proposition establishes a measure of the difference between the two results.

**Proposition 6.48.** *Let $f : X \to [0, 1]$; then,*

$$|SI_\mu(f) - CI_\mu(f)| \leq 1/4.$$

## 6.5 Fuzzy Integrals

In this section we review two different approaches that have been defined to encompass in a unified framework the two families of integrals reviewed in this chapter. First, we will present the fuzzy t-conorm integral, and then, the twofold integral. The interest of the t-conorm integral is mainly conceptual. In the case of the twofold integral, we give an example.

### 6.5.1 The Fuzzy t-Conorm Integral

The Fuzzy Integral was defined to put Choquet integral and Sugeno integral into a unified framework. That is, a more general integral that encompasses both was introduced. The new integral uses pseudo addition and multiplication. The operators can be replaced by addition and multiplication in the Choquet integral, or maximum and minimum in the Sugeno integral.

Formally, the fuzzy integral is defined over a tuple called a t-conorm system for integration, and an operator $-_\Delta$ built on one of the elements of the tuple. Then, three different spaces are considered, each with an associated t-conorm (t-conorms are used here as pseudo-additions). The following spaces are considered:

1. *The space of values of integrands ($F$):* This domain is denoted by $D = [0, 1]$, and thus the function to integrate is such that $f : X \to D$. The corresponding t-conorm is denoted by $\Delta$. So, $F = (D, \Delta)$.
2. *The space of values of measures ($M$):* Denoting the domain by $T = [0, 1]$, we have $\mu : \wp(X) \to T$. The corresponding t-conorm is $\bot$. Therefore, $M = (T, \bot)$.
3. *The space of values of integrals ($I$):* In this case, the domain is denoted by $\bar{T} = [0, 1]$, and the corresponding t-conorm is $\underline{\bot}$. Thus, $I = (\bar{T}, \underline{\bot})$.

$$\underbrace{\underbrace{\sum_{i=1,N} \Big(a_{s(i)} - a_{s(i-1)}\Big)}_{F} \underbrace{\mu(A_{s(i)})}_{M}}_{I}$$

$$\underbrace{\sum_{i=1,N} \underbrace{\Big(a_{s(i)} - a_{s(i-1)}\Big)}_{F} \underbrace{\mu(A_{s(i)})}_{M}}_{\underbrace{\phantom{xxxxxxxx}}_{I}}{}^{\textstyle I^*}$$

**Fig. 6.13.** The spaces in the fuzzy integrals

In some situations, it is possible to consider that two of the spaces can collapse into a single one:

1. When the integral is understood as the measure of a fuzzy set $A$, and this measure is defined to be an integral of the membership function of the fuzzy set, then $I = ([0,1], \bot)$ should be equal to $M = ([0,1], \bot)$. This is so because the integrals, which are valued in $I$ with respect to a measure $\mu : \wp(X) \to M$, are an extension of the measure $\mu$.
2. When the integral is understood as some kind of expected value of the function to integrate $f$, then the space of the integral $([0,1], \bot)$ and the integrand $([0,1], \Delta)$ should be the same.

The spaces are illustrated in Figure 6.13 (left), using the expression of the Choquet integral. The examples given in this chapter fall into the second category: the integral is a kind of expected value.

The decomposition presented here is not the only one possible. Other authors include an additional space. This situation is represented in Figure 6.13 (right), and the new space, denoted by $I^*$, is where the products of $F$ by $M$ are operated. In this case, $I^*$ is an internal space for operation. Together with these three spaces, a product-like operation $\otimes : D \times T \to \bar{T}$ is considered. This is the pseudo multiplication mentioned above.

The t-conorms presented here and the operation $\otimes$ define, when some conditions are fulfilled, a t-conorm system. This is formally defined as follows.

**Definition 6.49.** $\mathcal{F} = (\Delta, \bot, \underline{\bot}, \otimes)$ *is a* t-conorm system for integration *if and only if*

1. $\Delta, \bot,$ *and* $\underline{\bot},$ *are continuous t-conorms that are the maximum or Archimedean.*
2. $\otimes : ([0,1], \Delta) \times ([0,1], \bot) \to ([0,1], \underline{\bot})$ *is a product-like operation fulfilling*
   *a)* $\otimes$ *is continuous on* $(0,1]^2$
   *b)* $a \otimes x = 0$ *if and only if* $a = 0$ *or* $x = 0$
   *c) when* $x \bot y < 1$, $a \otimes (x \bot y) = (a \otimes x) \underline{\bot} (a \otimes y)$ *for all* $a \in [0,1]$
   *d) when* $a \Delta b < 1$, $(a \Delta b) \otimes x = (a \otimes x) \underline{\bot} (b \otimes x)$ *for all* $x \in [0,1]$.

The definition does only consider t-conorms that are either the maximum or Archimedean. In the case where $\Delta, \bot,$ and $\underline{\bot}$ are continuous Archimendean t-conorms, we will denote their generators by $k$, $g$, and $h$.

According to the type of t-conorms $\Delta$, $\bot$, and $\underline{\bot}$, four types of t-systems can be distinguished:

Type (i): $\Delta, \perp$, and $\underline{\perp}$ are Archimedean t-conorms.
Type (ii): $\Delta, \perp$, and $\underline{\perp}$ are the maximum.
Type (iii): $\underline{\perp}$ is an Archimedean t-conorm, and at least one of the others is the maximum.
Type (iv): $\underline{\perp}$ is the maximum, and at least one of the others is Archimedean.

Theoretical results show that only types (i) and (ii) are meaningful. In the other cases, the product-like operation becomes constant, or the product is a function that only depends on one of the arguments (the other arguments do not affect the result). So, in such cases, the integral is not much useful for aggregation.

We review now some results relating to the four types of t-systems, and we show that only (i) and (ii) are of interest.

**Proposition 6.50.** *Let $\mathcal{F} = (\Delta, \perp, \underline{\perp}, \otimes)$ be a t-system of type (i). Then,*

$$a \otimes x = 1 \text{ for all } a > 0 \text{ and } x > 0$$

*or*

$$a \otimes x = h^{(-1)}(k(a) \cdot g(x)) \text{ for all } a \text{ in } [0,1] \text{ and for all } x \text{ in } [0,1], \quad (6.17)$$

*where $k$, $g$, and $h$ are generators of $\Delta, \perp$, and $\underline{\perp}$ and where $h^{(-1)}(x)$ stands for the quasi-inverse of $h$.*

This case corresponds to Type (i) above. When a t-system is defined with three Archimedean t-conorms, and the product-like operator $\otimes$ is defined according to Equation 6.17, we will say that we have an Archimedean t-system. Note that, in this case, the t-system is completely determined by the generators $(k, g, h)$.

**Proposition 6.51.** *Let $\mathcal{F} = (\Delta, \perp, \underline{\perp}, \otimes)$ be a t-system of type (ii). Then, when $\otimes$ is a nondecreasing operator (i.e., if $x > y$, then $z \otimes x \geq z \otimes y$ for all $x$, $y$, and $z$ in $[0,1]$ and in relation to the first argument of $\otimes$), equations (c) and (d) in Definition 6.49 hold.*

This situation corresponds to Type (ii) above. In this case, a non-decreasing operator $\otimes$ satisfying (a) and (b) in Definition 6.49 defines a t-conorm system for integration, with $\Delta = \perp = \underline{\perp} = maximum$. This is a *maximum type t-system* (a *maximum t-system* for short).

**Proposition 6.52.** *Let $\mathcal{F} = (\Delta, \perp, \underline{\perp}, \otimes)$ be a t-system of type (iii). Then,*

$$a \otimes x = 1 \text{ for all } a > 0 \text{ and } x > 0.$$

**Proposition 6.53.** *Let $\mathcal{F} = (\Delta, \perp, \underline{\perp}, \otimes)$ be a t-system of type (iv). If $\perp$ is an Archimedean t-conorm, then*

$$a \otimes x = a \otimes 1 \text{ for all } a \text{ in } [0,1] \text{ and } x > 0.$$

Let $X$ be a reference set, let $(X, \mathcal{A})$ be a measurable space, let $\mu$ be a fuzzy measure on $(X, \mathcal{A})$, let $f$ be a measurable function $f : X \to [0, 1]$, and let $\mathcal{F} = (\Delta, \perp, \underline{\perp}, \otimes)$ be a t-system for integration; then, the fuzzy t-conorm integral (or fuzzy t-integral) of $f$ based on $(\Delta, \perp, \underline{\perp}, \otimes)$ with respect to $\mu$ is defined as

$$(\mathcal{F}) \int f \otimes d\mu := lim_{n \to \infty} (\underline{\perp}) \int f_n \otimes d\mu,$$

where $\{f_n\}$ is a nondecreasing sequence of simple functions which pointwise converge to $f$.

---

**Fig. 6.14.** Fuzzy t-conorm integral in a continuous domain

*If $\Delta$ is an Archimedean t-conorm, then*

$$a \otimes x = 1 \otimes x \ for \ all \ x \ in \ [0, 1] \ and \ a > 0.$$

*Furthermore, if $\Delta$ and $\perp$ are both Archimedean t-conorms, then*

$$a \otimes x = 1 \otimes 1 \ for \ all \ a > 0 \ and \ x > 0.$$

These results show that only for types (i) and (ii) the product-like operator is really a function of the two arguments. Therefore, we only consider types (i) and (ii) below. The definition is based on the substraction operator $-_\Delta$ constructed from the t-conorm $\Delta$ (recall Definition 2.51).

**Definition 6.54.** *Let $\mu$ be a fuzzy measure on $X$, and let $\mathcal{F} = (\Delta, \perp, \underline{\perp}, \otimes)$ be a t-system for integration. Then, the fuzzy t-conorm integral (or fuzzy t-integral) of a function $f : X \to [0, 1]$ based on $(\Delta, \perp, \underline{\perp}, \otimes)$ with respect to $\mu$ is defined by:*

$$(\mathcal{F}) \int f \otimes d\mu = \underline{\perp}_{i=1}^{N}(a_i -_\Delta a_{i-1}) \otimes \mu(A_{s(i)}),$$

*where $a_i = f(x_{s(i)})$ with $f(x_{s(i)}) \le f(x_{s(i+1)})$ and $a_0 = f(x_{s(0)}) = 0$, and $A_{s(i)} = \{x_{s(i)}, \ldots, x_{s(N)}\}$.*

This definition can be interpreted in a way similar to the one for the Choquet integral displayed in Figure 6.8 (c). See Figure 6.15. In this case, $(a_i -_\Delta a_{i-1})$ corresponds to a way to measure the height of the block, and $\mu(A_{s(i)})$, as before, corresponds to the measure of the elements comprising the block. The operator $\otimes$ is a way to combine the two values to evaluate the block.

For the Choquet integral, the evaluation roughly corresponds to the area, while for the Sugeno integral, it corresponds to the shortest length between $a_i$ (note that $a_i -_\Delta a_{i-1} = a_i$) and $\mu(A_{s(i)})$. Figure 6.14 corresponds to the definition of this integral for continuous functions.

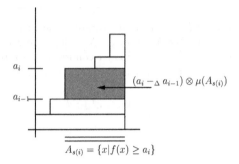

$$A_{s(i)} = \{x | f(x) \ge a_i\}$$

**Fig. 6.15.** Interpretation of the fuzzy integral

The expression of the Fuzzy t-integral can be rewritten for the two types of t-conorms considered above: Archimedean and maximum.

**Proposition 6.55.** *Let $\mu$ be a fuzzy measure on $X$, and let $\mathcal{F} = (\Delta, \perp, \underline{\perp}, \otimes)$ be a t-system for integration. Then, the following holds for the fuzzy t-integral of a function $f : X \to [0, 1]$ based on $\mathcal{F} = (\Delta, \perp, \underline{\perp}, \otimes)$ with respect to $\mu$:*

*Type (i): When the system is Archimedean, and $k$, $g$, and $h$ are the generators of $\Delta$, $\perp$, and $\underline{\perp}$, the following equality holds:*

$$(\mathcal{F}) \int f \otimes d\mu = h^{-1}\left(min(h(1), (C) \int k \circ f \; d(g \circ \mu))\right),$$

*where $\circ$ is function composition.*

*Type (ii): When the system is a maximum t-system and $\otimes$ is a t-norm, the following equality holds:*

$$(\mathcal{F}) \int f \otimes d\mu = \bigvee_{i=1,N} a_i \otimes \mu(A_{s(i)}).$$

The first expression reduces to the Choquet integral when $k = g = h = id$, and the second expression reduces to the Sugeno integral when $\otimes$ is the minimum. With $\otimes$ being the product, the Fuzzy t-integral computes the area of the square with maximum area.

Some particular cases of the expression above are known in the literature as Choquet-like and Sugeno-like fuzzy integrals. They are formally defined as follows:

1. Choquet-like is a fuzzy t-integral with type (i) t-system, and such that $a \otimes x = 1$ if and only if $a = 1$ and $x = 1$. This is equivalent to $h(1) = k(1) \cdot g(1)$.

   Therefore, when the generators of the Archimedean t-conorms $\Delta$, $\perp$, and $\underline{\perp}$ are, respectively $k, g$, and $h$, we have that the Choquet-like integrals have the following expression:

$$h^{-1}\big(h(1) \wedge (C) \int k \circ fd(g \circ \mu)\big). \tag{6.18}$$

For example, for $h(x) = k(x) = x^\alpha$ and $g(x) = x$, we have

$$\big(1 \wedge (C) \int f^\alpha d\mu\big)^{1/\alpha} \tag{6.19}$$

or, using Equation 6.2,

$$\big(\sum_{i=1}^{N}(f(x_{\sigma(i)}))^\alpha[\mu(A_{\sigma(i)}) - \mu(A_{\sigma(i-1)})]\big)^{1/\alpha}. \tag{6.20}$$

2. Sugeno-like is a fuzzy t-integral with type (ii) t-system, with $\otimes$ a t-norm.

### 6.5.2 Twofold Integral

The twofold integral is an alternative generalization for both Choquet and Sugeno integrals. Informally, the t-conorm integral builds its generalization by assuming that the two fuzzy measures in the Choquet and Sugeno integrals can collapse into a single measure in the generalized integral. In contrast, in the twofold integrals, the two fuzzy measures are kept as they are.

The rationale of this approach is that the semantics of both measures are different. In particular, the Choquet integral is seen as a "probabilistic flavor" measure, and the Sugeno integral is seen as a "fuzzy flavor" measure. So, the definition keeps both measures, including both fuzzy and probabilistic flavors. We will use $\mu_C$ to denote the measure that corresponds to the one in the Choquet integral (the one with the probabilistic flavor), and $\mu_S$ for the one in the Sugeno integral (fuzzy flavor).

**Definition 6.56.** *Let $\mu_C$ and $\mu_S$ be two fuzzy measures on $X$; then, the twofold integral of a function $f : X \to [0, 1]$ with respect to the fuzzy measures $\mu_S$ and $\mu_C$ is defined by*

$$TI_{\mu_S,\mu_C}(f) = \sum_{i=1}^{n}\big((\bigvee_{j=1}^{i} f(x_{s(j)}) \wedge \mu_S(A_{s(j)}))(\mu_C(A_{s(i)}) - \mu_C(A_{s(i+1)}))\big),$$

*where $f(x_{s(i)})$ indicates that the indices have been permuted so that $0 \le f(x_{s(1)}) \le \cdots \le f(x_{s(n)}) \le 1$, and where $A_{s(i)} = \{x_{s(i)}, \cdots, x_{s(n)}\}$ and $A_{s(n+1)} = \emptyset$.*

Now, we turn to the properties of this integral. We start by considering the relation between the twofold integral and the Choquet and Sugeno integrals. We show that this integral is a proper generalization of the Sugeno and Choquet integrals, as, for a particular measure, the twofold integral reduces to one of the others. In particular, generalization is obtained using the measure $\mu^*$ given in Definition 5.5, corresponds to ignorance ($\mu^*(A) = 1$ when $A \neq \emptyset$, and $\mu(\emptyset) = 0$).

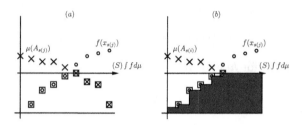

**Fig. 6.16.** Graphical representation of fuzzy integrals: (a) Sugeno integral; (b) twofold integral

**Proposition 6.57.** *The twofold integral satisfies the following properties.*

*When $\mu_C = \mu^*$, the twofold integral reduces to the Sugeno integral:*

$$TI_{\mu_S,\mu_C}(a_1,\ldots,a_n) = SI_{\mu_S}(a_1,\ldots,a_n).$$

*When $\mu_S = \mu^*$, the twofold integral reduces to the Choquet integral:*

$$TI_{\mu_S,\mu_C}(a_1,\ldots,a_n) = CI_{\mu_C}(a_1,\ldots,a_n).$$

*When $\mu_C = \mu_S = \mu^*$, the twofold integral reduces to the maximum:*

$$TI_{\mu_S,\mu_C}(a_1,\ldots,a_n) = \bigvee(a_1,\ldots,a_n).$$

Additionally, the twofold integral satisfies the basic properties of aggregation operators. That is, it is monotonic, satisfies unanimity, and, therefore, yields a value between the minimum and the maximum.

**Proposition 6.58.** *Let $X$ be a finite set and let $\mu_C$ and $\mu_S$ be two fuzzy measures on $X$; then,*

*(i) for all functions $f$ and $g$ over $X$ such that $g(x) \geq f(x)$ for all $x \in X$,*

$$TI_{\mu_S,\mu_C}(g) \geq TI_{\mu_S,\mu_C}(f);$$

*(ii) for all $\mathbf{a} = (a,\ldots,a)$,*

$$TI_{\mu_S,\mu_C}(\mathbf{a}) = a;$$

*(iii) for all functions $f$ on $X$,*

$$\min_{x \in X} f(x) \leq TI_{\mu_S,\mu_C}(f) \leq \max_{x \in X} f(x).$$

The next few propositions establish some additional properties of the twofold integral.

**Proposition 6.59.** *Let $A$ be a subset of $X$ and let $f_A$ be its characteristic function; then, the twofold integral of $f_A$ with respect to the two measures $\mu_S$ and $\mu_C$ is equal to*

$$TI_{\mu_S,\mu_C}(f_A) = \mu_S(A) \cdot \mu_C(A).$$

Let $X$ be a reference set, let $(X, \mathcal{A})$ be a measurable space, and let $\mu_C$ and $\mu_S$ be fuzzy measures on $(X, \mathcal{A})$. Then, for a measurable function $f : X \to [0,1]$, let us define $\phi_f : [0,1] \to [0,1]$ by

$$\phi_f(x) := \bigvee_{0 \leq r \leq x} (r \wedge \mu_S(\{f > r\})).$$

Note that $\phi_f(1) = S_{\mu_S}(f)$ and $\phi_f$ is nondecreasing, so the cardinality of noncontinuous points of $\phi_f$ is at most countable. $\phi_f$ permits us to define a Lebesgue-Stjeltjes measure $\nu_{\phi_f}$ on the real line by

$$\nu_{\phi_f}([a,b]) := \phi_f(b+0) - \phi_f(a-0).$$

Then, the *twofold integral* of a measurable function $f : X \to [0,1]$ with respect to fuzzy measures $\mu_S$ and $\mu_C$ is defined by

$$TI_{\mu_S, \mu_C}(f) = \int_0^1 \mu_C(f > a) d\nu_{\phi_f}(a).$$

**Fig. 6.17.** Twofold integral in a continuous domain

**Proposition 6.60.** *For all $f$, the following inequalities hold for the Choquet and Sugeno integrals:*

$$TI_{\mu_S, \mu_C}(f) \leq CI_{\mu_C}(f)$$
$$TI_{\mu_S, \mu_C}(f) \leq SI_{\mu_S}(f)$$

Additionally, it can be proved that the following relation holds:

**Proposition 6.61.** *For all $f$, $\mu_C$, and $\mu_S$,*

$$TI_{\mu_S, \mu_C}(f) = CI_{\mu_C}\big(f \wedge SI_{\mu_S}(f)\big).$$

Figure 6.16 gives a graphical representation of the twofold integral. The figure includes (left) the representation of the Sugeno integral (already given in Figure 6.11) and (right) the representation of the twofold integral (filled part). Figure 6.17 gives the definition of the twofold integral in a continuous domain.

## 6.6 Hierarchical Models for Aggregation

Aggregation operators can be combined to obtain hierarchical models. This corresponds to computing partial aggregations, and then combining them into an overall score.

Hierarchical models are appropriate to decompose a complex problem into simpler ones and to improve modularity. Additionally, when the number of

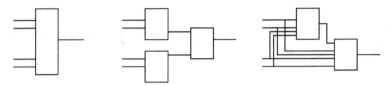

**Fig. 6.18.** Hierarchical models for aggregation with 4 inputs: (a) non-hierarchical case, (b) hierarchical case, (c) hierarchical case with duplicated inputs (overlapping hierarchical model)

information sources is large, the number of parameters to tune the system might become very large. This is the case with fuzzy integrals, which require fuzzy measures. As we have seen in Chapter 5, fuzzy measures are power set functions, and therefore, for $N$ sources, $2^N$ parameters have to be fixed. The use of hierarchical models defined in terms of fuzzy integrals permits us to reduce the number of parameters.

Figure 6.18 illustrates this case for four input variables. The figure includes the case of one-step (one aggregation operator) and two-step (two partial aggregation operators and a general combination) models. In this case, when fuzzy integrals are used, the definition of the parameters (fuzzy measures) requires $2^4 - 2$ (considering boundary conditions) values for the nonhierarchical model, but only six values for the parameters of the hierarchical model.

More complex models can also be considered. This is the case in Figure 6.18 (c), where the inputs are duplicated. In this case, one aggregation combines four inputs, and the other aggregation combines the original four inputs and the result of the previous aggregation. So, the number of parameters is even larger than in the nonhierarchical model (i.e., $(2^4 - 2) + (2^5 - 2)$). In case of dupplication in the inputs, we call the models overlapping hierarchical models. The term separated hierarchical model will be used for the nonoverlapping case.

The twofold integral studied in the previous section can be considered in the light of hierarchical models. Proposition 6.61, which linked the twofold integral with the Choquet and Sugeno integrals, permits us to consider a two-step hierarchical model with the Sugeno integral in the first step and the Choquet integral in the second step. This hierarchical model corresponds to the one in Figure 6.18 (c).

Some results have been obtained about the modeling capabilities of such hierarchical models. We consider multistep hierarchical models in which the aggregation operator is the Choquet integral.

**Definition 6.62.** Let $f : X \rightarrow \mathbb{R}^+$ be a function that represents the data to be aggregated. Then,

- If $\mathbb{C}(f(x_1), \ldots, f(x_N)) = f(x_i)$ for some $x_i$, then $\mathbb{C}$ is a 0-step Choquet integral.

- If $\mathbb{C}_j$ for $j \in M = \{1, \ldots, m\}$ are $k_j$-step Choquet integrals, then, given a fuzzy measure $\mu$ on $M$, the function

$$\mathbb{C}(f(x_1), \ldots, f(x_N)) =$$
$$CI_\mu(\mathbb{C}_1(f(x_1), \ldots, f(x_N)), \ldots, \mathbb{C}_m(f(x_1), \ldots, f(x_N)))$$

is a $k$-step Choquet integral for $k = 1 + \max\{k_j | j \in M\}$.

An important result about hierarchical models is that when we have an $m$-step overlapping model, it can always be reduced to an equivalent two-step model. Moreover, some constraints can be given to the corresponding fuzzy measures.

**Theorem 6.63.** *The following conditions hold:*

*(i) Every multistep Choquet integral is a monotone increasing, positively homogeneous, piecewise linear function.*

*(ii) Every monotone increasing, positively homogeneous, piecewise linear function on a full-dimensional convex set in $\mathbb{R}^N$ is representable as a two-step Choquet integral such that the fuzzy measures of the first step are additive and the fuzzy measure of the second step is a 0-1 fuzzy measure.*

So, in principle, in the case of Choquet integrals, no model more complex than a two-step one is needed, as all other models can be reduced to it.

# 6.7 Bibliographical Notes

1. **General references on aggregation operators:** General references for aggregation operators include the books by Grabisch, Nguyen, and Walker [172], Bullen, Mitrinovic, and Vasic [50], Bullen [51], and Calvo, Mayor and Mesiar [56]. See also [111], [55], and [52], which give reviews of aggregation operators. The books on fuzzy measures and integrals, [174], [427], and [385], are also adequate. The chapter by Benvenuti, Mesiar in Vivona [41] in the *Handbook on Measure Theory* by Pap [312] includes definitions for fuzzy integrals and some of their properties. General books on measure and integration are also appropriate. Some of them include some integrals (e.g., König [216] describes the Choquet integral).

   Applications of aggregation operators can be found in several papers. [389, 414] consider the use of fuzzy integrals to combine the results of fuzzy-rule-based systems. The former considers both Choquet and Sugeno integrals, and the latter focuses on the Sugeno integral. [374] gives an exhaustive comparison of operators (67 parameterized operators) with respect to their performance in image retrieval; not only aggregation operators, but also other operators, such as t-norms and t-conorms, are considered. Example 6.8 is based on [366] and [396].

   For other applications, see the corresponding bibliographical notes in Chapter 1.

2. **Median:** Definition 6.7 is the standard definition for the median. Nevertheless, other definitions are possible when $N$ is even. This matter is briefly discussed in the bibliographical notes of Chapter 1. The results by Jackson (1921) [201] are recalled there, as they are of interest for alternative definitions for even $N$.

3. **OWA operators:** The first definition of the OWA operator is given in [442]. See also [444]. Example 6.6, on the use of OWA to model the score system in the Olympic Games, is due to Yager. [145] presents a characterization of the operators. Some of their mathematical properties, as well as some applications, are described in [456].

   The use of fuzzy quantifiers to define weights appears in [442] and [444]. Liu in [228] proposed the study of the generating functions of fuzzy quantifiers. Liu studied, among other properties, their orness.

   Several generalizations exist for the OWA operators. For example, the quasi-OWA operator (introduced by Fodor, Marichal, and Roubens in [145]):

   $$\phi^{-1}\left(\sum_i w_i\phi(a_{\sigma(i)})\right)$$

   Ralescu and Ralescu [328] (Example 2, p. 326) introduced in 1997 the Geometric OWA (GOWA). This operator, named in [74] for its similarity with the geometric mean, a geometric mean of order statistics, corresponds to

   $$\Pi_i a_{\sigma(i)}^{w_i}$$

   where $w_i$ define a weighting vector ($\sum_i w_i = 1$ and $w_i \geq 0$). This operator has been further studied in [185, 231, 440]

   Other generalizations of the OWA operator include, the Nonmonotonic OWA [449], the BADD-OWA [454], the Generalized OWA, the Induced OWA (IOWA) [271, 455], and the Induced GOWA (IGOWA) [75]. The BADD-OWA (BAsic Defuzzification Distribution OWA) corresponds to the counter-harmonic mean. The Generalized OWA, introduced by Yager in [452], corresponds to the root-mean-power (generalized mean) of order statistics. For a given $\alpha$, it is defined as

   $$\left(\sum_i w_i a_{\sigma(i)}^{\alpha}\right)^{1/\alpha}$$

   The IOWA is an OWA operator where the ordering is defined in terms of a priority vector $\mathbf{b} = (b_1 \ldots b_N)$, which corresponds to the priority of $x_i$. The IOWA of $\mathbf{a}$ with respect to $\mathbf{w}$ and $\mathbf{b}$ is $\sum_i w_i a_{\sigma(i)}$ where $\{\sigma(1), ..., \sigma(N)\}$ is a permutation of $\{1, ..., N\}$ such that $b_{\sigma(i-1)} \geq b_{\sigma(i)}$ for all $i = \{2, ..., N\}$. This operator, named by Yager and Filev in [455], was introduced by Mitchell and Estrakh [271] (see also [347] and [272]) in 1997. The Induced-Choquet integral (I-COA) was introduced in [450].

4. **On the weights for the OWA operator and weighted means:** Considering data with extreme values as outliers and removing it before

applying a mean has been used for a long time. For example, the use of the arithmetic mean without the two extreme values – which corresponds to applying the OWA operators with weights $(0, 1/(N-2), \ldots, 1/(N-2), 0)$ – is rather old. Maire and Boscovich used this approach in [243] (1755) for measuring the degrees of a meridian. Svanberg (?) [20], in 1821, reports an example of applying this approach for computing the mean returns from some real estate properties in French provinces.

In 1722, Cotes [85] considered the case of weighting observations according to the reciprocal of their errors. He gives an example (p. 22): *"Sit p locus Objecti alicujus ex Observatione prima definitus, q, r, s ejusdem Objecti loca ex Observationibus subsequentibus; fint insuper P, Q, R, S pondera reciproce proportionalia spatiis Evagationum, per quae fe diffundere possiut Errores ex Observationibus fingulis prodeuntes, quaeque dantur ex datis Errorum Limitibus; & ad puncta p, q, r, s posita intelligantur pondera P, Q, R, S, & inveniatur eorum gravitatis centrum Z: dico punctum Z fore Locum Objecti maxime probabilem, qui pro vero ejus loco tutissime haberi potest[2]."*

Newcomb [299], in 1912, argues that rejecting observations with large residuals results in discontinuity. Then, he proposes using a weight $w = e_0/(e_0 + \Delta)$, where $\Delta$ is the excess of the error above a certain limit $\epsilon_0$. Later, in 1926, Bemporad [40] explicitly establishes the property that equal credibility on measures should imply symmetry in the function. On p. 88 he states: *"Se i risultati delle singole misure presentano ugual grado di attendibilità, il risultato complessivo deve essere funzione simmetrica di essi[3]."*

5. **WOWA operator:** Torra introduced the WOWA operator in [394] and [395]. An interpolation method to build $W^*$ from the weighting vector **w** was considered in [402]. This method adapted the one by Chen and Otto in [71]. As Beliakov [37] pointed out, this method by Chen and Otto is equivalent to [256], but the latter is more efficient. Nevertheless, the adaptation in [402] makes the results of both methods different in the boundaries of $W^*$ (see [404]); that is, near the points $(0,0)$ and $(1,1)$.

---

[2] Let $p$ be the precise position of a particular object according to the first observation, and $q$, $r$, $s$ be the positions of this same object according to subsequent observations; besides let $P$, $Q$, $R$, and $S$ be weights inversely proportional to the spaces of the evagations, through which the errors that progress from each of the observations can be spread, and which are given from the boundaries of the errors given; and let weights $P$, $Q$, $R$, and $S$ be understood as located at points $p$, $q$, $r$, and $s$, and let $Z$ be at the gravity center of these points. I assert that point $Z$ should be the most probable place of the object, which can be said, with complete certainty, as the real place of the object.

[3] If the results of individual measures show the same degree of reliability, the overall result must be a symmetric function of such results

The definition of WOWA using quantifiers is equivalent to the OWA, with importances defined in [447]. In [397], it was proved that the WOWA operator is a particular case of the Choquet integral.

Liu studies some properties of the WOWA in [229], focusing on continuous WOWA. Continuous WOWA was introduced in [410]. $m$-dimensional WOWA was introduced in [291].

6. **The Choquet integral:** The first definition of the Choquet integral was due to Vitali in 1925 [424] for additive measures. Choquet defined it independently in 1953 [80]. The term horizontal integration is used in [216]. The generalized Choquet integral defined by Yager in [452] corresponds to the Choquet-like integral defined in Equation 6.20. Properties of the Choquet integral can be found in general references on fuzzy measures and integrals.

   Comonotonic additivity for Choquet integral was introduced by Dellacherie [93], and, later, Schmeidler [348] proved the representation theorem for a comonotonically additive functional. The representation theorems for functionals on restricted domains are shown by Greco in [175] and by Narukawa, Murofushi, and Sugeno [289].

   An extension of the domain of the Choquet integral is proposed by Šipoš [365]. The Choquet integral with respect to a nonmonotonic fuzzy measure is proposed by Murofushi, Sugeno, and Machida [284].

7. **Weighted minimum and weighted maximum:** Weighted minimum was introduced by Yager in 1981 [441]. Previously, in 1976, Negoita and Flondor [295] had given a generalization of WMin. The introduction of weighted maximum was due to Dubois in [109] (see also [111] and [112]). [112] gives expressions of both WMin and WMax in terms of the median, and the expressions show that they are Sugeno integrals. A characterization of these operators can be found in [147]. OWMax and OWMin were introduced in [113]. Yager [443] defined later the Ordinal OWA. This corresponds to a particular OWMax: when data is ordered in an increasing way, weights are decreasing. Ordinal OWA corresponds to the Sugeno integral with respect to a symmetric fuzzy measure. So, it is analogous to the OWA operator that corresponds to the Choquet integral with respect to a symmetric fuzzy measure. [308] studies characterizations of these operators.

   Proposition 6.36 is proved in [14].

8. **The Sugeno integral:** The Sugeno integral was introduced by Sugeno in 1974 [384] (see [382] (1972) for a previous definition in Japanese). The graphical interpretation in Figure 6.11 (a) was given in [458], and the interpretation in Figure 6.11 (c) was given in [427]. Properties of Sugeno integrals are studied in detail in [427], and detailed references are given there. Propositions 6.46 and 6.47 are proved in [288]. Proposition 6.48 is proved in [279]. Example 6.39 is based on [411]. [207] proves that the Sugeno integral can be expressed in terms of medians. This result is not reported in this book.

9. **Fuzzy integrals:** [164] is one of the papers to consider four different spaces in integration, as given in Figure 6.13, $I$, $I^*$, $F$, and $M$. The t-conorm system for integration and the fuzzy t-conorm integral (or fuzzy t-integral) were introduced in [280]. Naturally, Choquet-like and Sugeno-like integrals also appear in that paper. The terms also appear later in [173, 172].

The twofold integral was introduced in [407]. Its properties were studied in [292]. Its graphical representation (Figure 6.16), and its generalization to the continuous case (Figure 6.17) is given in [294]. Other fuzzy integrals and other generalizations of Choquet and Sugeno integrals also exist. The general fuzzy integral is one of such generalizations [41].

10. **Hierarchical models:** Aggregation operators that combine partial results previously obtained by other aggregation operators have been studied for some time. For example, the *symmetrical mean* is a hierarchical model. This operator, defined as an arithmetic mean of all permutations of a root-mean-powers with weights $(\alpha_1, \dots, \alpha_N)$ (with $\sum \alpha_i = 1$), was studied by Muiheard in 1903 [275]. Its definition is

$$\frac{1}{N!} \sum_{\pi} a_{\pi(1)}^{\alpha_1} \cdots a_{\pi(N)}^{\alpha_N}.$$

The operator reduces to the arithmetic mean with weights $(1, 0, \dots, 0)$, and to the geometric mean with weights $(1/N, \dots, 1/N)$ (see [122], p. 45).

Bullen, Mitrinovic and Vasic review in [50] (p. 191) the *mixed means* (a combination of root-mean-powers). The oldest related result is the Carlson function [61] (1970).

The multistep Choquet integral, as defined in this book, corresponds to the definition in [276], which is an extension of the two-step integral in [265]. [265] proves that the weighted mean of Choquet integrals is a Choquet integral. Theorem 6.63 is given in [276] and [277]. Conditions on when a Choquet integral is decomposable in a hierarchical model are given by Fujimoto, Murofushi, and Sugeno in [155]. See also [151] for additional details. The first work in this direction was [281].

[292] studies multistep representations of Sugeno and twofold integrals in terms of the Choquet integral with constant. The Choquet integral with constant $b$ of a function $f$ with respect to a fuzzy measure $\mu$ is defined by $CI_\mu(f) + b$. The Choquet integral with constant is defined in [277].

Other hierarchical models are also present in the literature. For example, Calvo, Mesiarová, and Valášková [58] present a hierarchical model that generalizes the twofold integral. This generalization permits them to construct a dual of the twofold integral. [413] proposes another hierarchical model called the meta-knowledge model. In this definition, partial aggregations can be used to modify the fuzzy measures embedded in some Choquet integrals. The term meta-knowledge comes from hierarchical fuzzy systems. In such settings, partial inferences permit us to modify the fuzzy rule base of another fuzzy system. Magdalena in [242] proposes such meta-knowledge in a fuzzy system.

11. **Other aggregation operators:** For references on the median, arithmetic mean, weighted mean, and quasi-arithmetic mean, see Chapter 4. The Hurwicz operator is described in [132].

12. **Operators for linguistic labels:** Although in this chapter we have focused on operators for numerical information, some of the operators such as the weighted minimum/maximum and the Sugeno integral can also be applied to ordinal data. In Chapter 4, a few other aggregation operators were defined taking into account ordinal scales. Other operators exist for ordinal data.

    Linguistic aggregation operators encompass operators used for the aggregation of linguistic labels. Linguistic labels can be considered as enriched ordinal scales, because each category in the ordinal scale might have associated with it some additional information. For example, in fuzzy-rule-based systems, each label has associated with it a fuzzy set on a given domain.

    [403] classifies such aggregation operators into three categories according to the underlying scale for the labels: (i) explicit quantitative or fuzzy scales, (ii) implicit numerical scale, (iii) no additional scale, with operators only considering the qualitative scale. The operator in [403] (which aggregates the labels, taking into account a semantics based on antonyms [95] and extended negation functions [393]) and operators that aggregate the fuzzy numbers belong to (i). Linguistic OWA [186], Linguistic WOWA [395], and Induced Linguistic operators [439] belong to (ii). Aggregation operators for ordinal scales such as weighted minimum or the Sugeno integral belong to (iii). Other operators that rely on operations on ordinal scales, e.g., t-norms and t-conorms, are also in (iii). This is the case for the ordinal weighted mean defined in [164] on t-norms and t-conorms [255].

    The Linguistic OWA operator was based on the convex combination of linguistic terms proposed in [92]. It was further studied in [187].

# 7

# Indices and Evaluation Methods

*(...) som aixi, i fins i tot en els moments
que és millor estar-se quiet,
tenim el desfici de prendre decisions.*[1]

P. Calders, [54] (p. 94)

This chapter reviews some of the existing tools for evaluating aggregation methods and their parameters. We focus on some indices for fuzzy measures (Shapley and Banzhaf), an interaction index, and the degree of disjunction. Other methods exist. The influence function and other tools such as gross-error sensitivity and local-shift sensitivity developed in robust statistics (see Section 2.2.6) are of interest here. The tools permit us to have some knowledge on how a particular estimator might behave when embedded in a real system. In particular, we have seen that the influence function of the arithmetic mean is unbounded while that of the median is bounded.

Graphical representations are another example of such tools. For example, we can represent graphically the outcome of a binary operator in the $[0, 1] \times [0, 1]$ region. This corresponds to a 3D representation. Alternatively, we can consider the representation of a subset of inputs. In the case of functions not satisfying unanimity (such as t-norms, t-conorms, or uninorms), we can consider the diagonal (i.e., $\mathsf{T}(x, x)$ for $x \in [0, 1]$). In the case of a binary aggregation operator $\mathbb{C}$, we might consider $\mathbb{C}(x, neg(x))$ for some negation $neg$. This function permits us to visualize the compensation between $x$ and $neg(x)$. We will denote this function by $\mathcal{C}$. That is, $\mathcal{C}_{\mathbb{C}}(x) = \mathbb{C}(x, neg(x))$:

*Example 7.1.* Let us consider the function $\mathcal{C}$ for some aggregation functions:

- $\mathcal{C}_{AM}(x) = (x + 1 - x)/2 = 1/2$
- $\mathcal{C}_{GM}(x) = \sqrt{x(1 - x)}$

---

[1] (...) that's the way we are, and even in the case when it is better to be still, we are eager to make decisions.

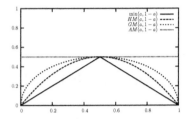

**Fig. 7.1.** Graphical representation of the functions $\mathcal{C}_{\min}$, $\mathcal{C}_{AM}(x) = 1/2$, $\mathcal{C}_{GM}(x) = \sqrt{x(1-x)}$, $\mathcal{C}_{HM}(x) = 2(1-x)x$

- $\mathcal{C}_{HM}(x) = 2(1-x)x$

Figure 7.1 illustrates the functions. It can be seen that min always returns the lowest value, followed by $HM$, $GM$, and, finally, $AM$.

Now, we will study some of the indices that can be applied to analyze the parameters of the aggregation operators and their influence on the outcome of the aggregation method. In particular, we review some of the power indices (e.g., the Shapley and Banzhaf values), interaction indices, and dispersion. Later, we consider evaluation methods that take into account the operator. At that point, we consider the average value and the degree of disjunction (orness).

## 7.1 Indices of Power: Shapley and Banzhaf Power Indices

When fuzzy measures on $X$ are restricted to take values in $\{0, 1\}$ (i.e., 0-1 fuzzy measures), they can be used to model coalitions of the individuals $x_i \in X$. In this case, for a given set $A \subseteq X$, the value $\mu(A)$ represents whether or not the set $A$ has a winning position when making a decision; for example, in a given vote, whether a coincident vote for all $x_i$ in the set $A$ ensures $A$'s opinion being selected. This interpretation, born in game theory, leads to several indices to measure the power of a particular $x_i$ in $X$ with respect to the winning positions. They are the indices of power, or power indices.

Power indices are not only applied to 0-1 fuzzy measures, as in the above example, but they can also be applied to general fuzzy measures. In such cases, a fuzzy measure is interpreted as the value of a coalition. Then, power indices stand for a measure of the worth of $x_i$, or how much $\mu$ increases when $x_i$ is included in a coalition.

Therefore, as fuzzy measures are used as parameters for fuzzy integrals (see Chapter 6), power indices are of interest in information fusion. This link is tightened by the fact that the application of fuzzy integrals to characteristic functions of a crisp set $A$ yields the measure of $A$. For example, $CI_\mu(A) = \mu(A)$ and $SI_\mu(A) = \mu(A)$ (see Propositions 6.30 and 6.44). According to this, power indices for a fuzzy measure $\mu$ give a summarization of the influence of the $x_i$ in the outcome of fuzzy integrals when the values to be aggregated are restricted to 0 or 1, or, in general, they give a trend on the output for values in $[0, 1]$.

Several power indices have been defined. We review and compare some of them (e.g., the Shapley and Banzhaf values) in the following sections. Typically, the indices are a function of the measure and of a particular $x_i$ in $X$. We will denote them by $\phi_{x_i}(\mu)$, where $\phi$ is the term used for a particular index name.

### 7.1.1 Shapley Value

When applied to 0-1 fuzzy measures, the Shapley value is a function that, for each $x_i$ in $X$, counts the number of times that an $x_i$ changes a losing coalition into a winning one, i.e., the number of sets $S$ such that, when including $x_i$ in the set, we have $\mu(S \cup \{x_i\}) = 1$ while $\mu(S) = 0$. When $\mu$ is not restricted to $\{0, 1\}$, the Shapley value measures a variation on the $\mu$ when $x_i$ enters into a set (or coalition).

Formally, the Shapley value is defined for arbitrary fuzzy measures, while for 0-1 fuzzy measures it is known as the Shapley-Shubik index. The formal definition of the Shapley value is given below.

**Definition 7.2.** *Given a fuzzy measure $\mu$, the Shapley value of $\mu$ for $x_i$, denoted by $\varphi_{x_i}(\mu)$, is defined as follows:*

$$\varphi_{x_i}(\mu) := \sum_{S \subseteq X \backslash \{x_i\}} \frac{|S|!}{(N - |S| - 1)! N!} \left( \mu(S \cup \{x_i\}) - \mu(S) \right) \qquad (7.1)$$

There are several expressions for the Shapley value that are equivalent to this. We present some of them in a proposition below.

The first expression is based on considering all possible total orders on the set $X$. Naturally, there are $|X|!$ such orders. We use $\rho_X$ to denote such a set. Then, we need to consider the set of $x_j$ that precedes $x_i$ in a particular ordering $r \in \rho_X$. We will denote such a preceding set by $r_{x_i}$. The example below illustrates $\rho_X$ and $r_{x_i}$ for $|X| = 3$.

*Example 7.3.* Let $X = \{x_1, x_2, x_3\}$; then, $\rho_X$, the set of all possible total orders on $X$, is defined as

$$\rho_X = \{(x_1, x_2, x_3), (x_1, x_3, x_2), (x_2, x_1, x_3), (x_2, x_3, x_1), (x_3, x_1, x_2), (x_3, x_2, x_1)\}.$$

Naturally, there are $|X|! = 3! = 6$ different orders. Given $r = (x_2, x_1, x_3)$, $r_{x_1} = \{x_2\}$, $r_{x_2} = \emptyset$, and $r_{x_3} = \{x_2, x_1\}$.

Now, we consider the alternative expressions for the Shapley value.

**Proposition 7.4.** *Let $\mu$ be a fuzzy measure on $X$ with Möbius transform $m$, let $N = |X|$, let $\rho_X$ be the set of orderings for $X$, and let $\varphi_{x_i}(\mu)$ be the Shapley value of $\mu$ for $x_i$ in $X$; then, the following equalities hold:*

1. *For all $x_i \in X$,*

$$\varphi_{x_i}(\mu) = \frac{1}{N!} \sum_{r \in \rho_X} \left( \mu(r_{x_i} \cup \{x_i\}) - \mu(r_{x_i}) \right), \tag{7.2}$$

*where $r_{x_i}$ denotes the set of $x \in X$ that precede $x_i$ in $r$.*

2. *For all $x_i \in X$,*

$$\varphi_{x_i}(\mu) = \frac{1}{N} \sum_{s=0}^{N-1} \frac{1}{\binom{N-1}{s}} \sum_{\substack{S \subseteq X \setminus \{x_i\} \\ |S| = s}} \left( \mu(S \cup \{x_i\}) - \mu(S) \right). \tag{7.3}$$

3. *For all $x_i \in X$,*

$$\varphi_{x_i}(\mu) = \sum_{\{S : x_i \in S \subseteq X\}} \frac{1}{|S|} m(S). \tag{7.4}$$

Expression 7.2 shows that the Shapley value for $x_i$ is an average of the gain for adding $x_i$ all possible *positions* in an order $r$. When the measure is a 0-1 fuzzy measure, the value counts the number of times the element $x_i$ provokes the change from 0 to 1 with its inclusion. Expression 7.3 also considers the average. The average is considered on the basis of the sizes of the sets (i.e., from size $s = 0$ to $s = N - 1$) and then considering all sets $S$ with size $s$. The last alternative expression is based on the Möbius transform of $\mu$ instead of on $\mu$ itself.

Characterizations of the Shapley value have been built showing that it is the only operator that summarizes a fuzzy measure satisfying some basic properties. We consider below one of such characterizations.

### 7.1.2 Characterization of the Shapley Value

Given a fuzzy measure $\mu$ on a set $X = \{x_1, \ldots, x_N\}$, the Shapley value is a function of $\mu$ for each $x_i$ in $X$. Accordingly, Shapley values can be seen as a vector $\varphi = (\varphi_1(\mu), \ldots, \varphi_N(\mu))$. The characterization is based on the following concepts: symmetry, efficiency, and additivity. We detail the concepts below.

**Symmetry:** The names of the elements $x_i$ do not play any relevant role in the computation of $\varphi_i(\mu)$. This property is stated in terms of permutations on $X$: Let $\pi$ be a permutation of $X$ (i.e., $\pi(x) \in X$ and $\pi(x) \neq \pi(y)$ if and only if $x \neq y$); then, $\mu_\pi$ is defined as $\mu_\pi(A) = \mu(\pi(A))$, where $\pi(A) = \cup_{a \in A} \{\pi(a)\}$.

Then, symmetry is fulfilled when, for all permutations $\pi$ on $X$,

$$\varphi_x(\mu) = \varphi_{\pi(x)}(\mu_\pi).$$

**Efficiency or carrier axiom:** The value is a summarization of the whole fuzzy measure. That is, not only is the individual value $\mu(\{x_i\})$ taken

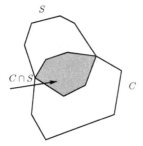

**Fig. 7.2.** Carrier axiom: subset $C$ is the carrier. In this case, $\mu(S) = \mu(C \cap S)$

into account to compute $\varphi_{x_i}(\mu)$, but also the other values to which the $x_i$ contribute. This is formalized in terms of the *carrier* concept.

A subset $C$ of $X$ is a carrier of $\mu$ if

$$\mu(S) = \mu(C \cap S)$$

for all $S \subseteq X$. Therefore, elements $x \in X$ that are outside any carrier have no influence on the measure (they contribute nothing to any coalition). That is, only the filled region in Figure 7.2 is relevant for computing $\mu(S)$.

The efficiency axiom is formalized using the carrier concept as follows. For any carrier $C$ in $X$,

$$\sum_{x_i \in C} \varphi_{x_i}(\mu) = \mu(C).$$

**Additivity:** When two independent fuzzy measures are combined, their values must be added element by element. This is formalized as follows. For any two fuzzy measures $\mu_1$ and $\mu_2$, for all $x_i$ in $X$,

$$\phi_{x_i}(\mu_1) + \phi_{x_i}(\mu_2) = \phi_{x_i}(\mu_1 + \mu_2),$$

where the fuzzy measure $\mu_1 + \mu_2$ is defined as $\mu(A) = \mu_1(A) + \mu_2(A)$ for all $A \subseteq X$.

Now, we can formulate the characterization of the Shapley value as follows.

**Theorem 7.5.** *A unique value function $\varphi$ exists satisfying symmetry, efficiency, and additivity. It is the Shapley value.*

That is, the Shapley value is the only index that satisfies the three conditions considered above.

### 7.1.3 Banzhaf Value

The application of the Shapley value shows that in some circumstances there are sets that are counted twice. The following example illustrates this situation.

*Example 7.6.* Let $\mu$ be a 0-1 fuzzy measure on $X = \{x_1, x_2, x_3\}$ such that $\mu(\{x_1, x_2\}) = 0$ and $\mu(\{x_1, x_2, x_3\}) = 1$. Then, using Expression 7.2, we can observe that the Shapley value for $x_3$ counts twice the fact that $x_3$ changes $\mu(\{x_1, x_2\})$, equal to 0 to $\mu(\{x_1, x_2, x_3\})$ equal to 1. This is so because both orderings, $r_1 = (x_1, x_2, x_3)$ and $r_2 = (x_2, x_1, x_3)$, will be considered when computing $\varphi_{x_3}(\mu)$.

The Banzhaf value, another power index, permits us to consider only once the influence of a source $x_i$ in such a situation. This is achieved considering the pairs $S$ and $S \setminus \{x_i\}$. Note that, for the particular set $S_1 = \{x_1, x_2, x_3\}$ in the example above, only $S_1$ and $S_1 \setminus \{x_3\} = \{x_1, x_2\}$ would be considered. We define below two variants of the Banzhaf index, the *unnormalized* or *non-standardized* Banzhaf index, and the *normalized* one. Both indices rely on the concept of an $x$ in $X$ being *essential* in a set $S$ or, equivalently, in the concept of the swing voter.

**Definition 7.7.** *Let $\mu$ be a 0-1 fuzzy measure on $X$; then, for any $S \subseteq X$, it is said that $x$ is an* essential *member in $S$ or, equivalently, that $x$ is a swing voter if removing $x$ from $S$ changes the measure from 1 to 0. In other words, $x$ is essential if $\mu(S) - \mu(S \setminus \{x_i\}) = 1$.*

Equivalently, in the context of game theory and coalitions, $x$ is a swing voter when the measure moves from a winning situation to a losing one.

Both normalized and unnormalized Banzhaf indices for a source $x$ count the number of sets in which $x$ is essential. Differences are due to the way the proportion is computed. The unnormalized index divides the count by the total number of coalitions in which $x_i$ is a member. The normalized Banzhaf index divides the count by the total number of distinct sets in which a member is essential. Both definitions are given below.

**Definition 7.8.** *Let $\mu$ be a fuzzy measure on $X$; then,*

1. *The* unnormalized *(or* nonstandardized *or* absolute*) Banzhaf index of $\mu$ for $x_i$ is defined by*

$$\beta'_{x_i}(\mu) := \frac{\sum_{S \subseteq X} \left( \mu(S) - \mu(S \setminus \{x_i\}) \right)}{2^{N-1}}.$$

2. *The* Penrose index *(or* normalized Banzhaf index *or* relative Banzhaf index*) of $\mu$ for $x_i$ is defined by*

$$\beta_{x_i}(\mu) := \frac{\sum_{S \subseteq X} \left( \mu(S) - \mu(S \setminus \{x_i\}) \right)}{\sum_{i=1}^{N} \sum_{S \subseteq X} \left( \mu(S) - \mu(S \setminus \{x_i\}) \right)}.$$

The normalized Banzhaf index can be expressed in terms of the unnormalized one as follows:

$$\beta_{x_i}(\mu) = \frac{\beta'_{x_i}(\mu)}{\sum_{x_j \in X} \beta'_{x_j}}.$$

### 7.1.4 Properties

A probability model for interpreting the power indices has been defined considering winning coalitions. The main idea is to consider the individual effect. This effect is defined as follows.

*Question of individual effect:* What is the probability that my vote will affect the outcome of the vote on a bill? In other words, what is the probability that a bill will pass if I vote for it, but fail if I vote against it ? (Straffin, [380])

Models about individual effects can be built by considering the probability with which each $x_i$ in $X$ either votes for the bill or against it. Let $p_k$ the probability of $x_k$ voting for the bill; then, for $X = \{x_1, \ldots, x_N\}$, we have a probability vector $(p_1, \ldots, p_N)$. Different assumptions can be considered with respect to the values $p_k$. In particular, the following two seem natural in this framework.

**Assumption 1 (Homogeneity assumption)** *A number $p$ is chosen from the uniform distribution on $[0,1]$, and $p_k = p$ for all $k$.*

**Assumption 2 (Independence assumption)** *Each $p_k$ is chosen independently from the uniform distribution on $[0,1]$.*

Then, the following two theorems can be proved.

**Theorem 7.9.** *The individual effect of $x_i$ under the homogeneity assumption is given by the Shapley value for $x_i$: $\varphi_{x_i}$.*

**Theorem 7.10.** *The individual effect of $x_i$ under the independence assumption is given by the (unnormalized) Banzhaf value for $x_i$: $\beta'_{x_i}$.*

According to these theorems, the Shapley and Banzhaf indices for 0-1 fuzzy measures can be interpreted in terms of a probabilistic model. In this model, differences stand for the voting probabilities $p_k$ for each voter $x_k$.

## 7.2 Interaction

Both Shapley and Banzhaf indices are computed for individual sources $x \in X$. Interaction indices have been developed to measure to what extent two or more elements interact in a given measure. These new indices can be seen as a generalization of the former, where interaction, understood as either a complementarity or a redundacy, is measured. The definition of the interaction index is based on the $S$-derivative. Such a derivative is defined as follows.

**Definition 7.11.** *Let $\mu$ be a fuzzy measure; then, given $S, T \subseteq X$, $\Delta_S[\mu(T)]$ is the $S$-derivative of $\mu$ at $T$, and it is recursively defined as*

$$\Delta_x[\mu(T)] := \mu(T) - \mu(T \setminus \{x\}) \qquad (7.5)$$

$$\Delta_S[\mu(T)] := \Delta_x\left[\Delta_{S \setminus \{x\}}[\mu(T)]\right] \ \text{for } x \in S, \qquad (7.6)$$

where $\Delta_x[\Delta_S(T)]$ is defined as

$$\Delta_x[\Delta_S[T]] := \Delta_S[\mu(T \cup S \cup \{x\})] - \Delta_S[\mu((T \cup S) \setminus \{x\})].$$

In fact, the equation for computing $\Delta_S[\mu(T)]$ holds for all $x \in S$, as the order of selecting $x$ in $S$ is not relevant in the computation of the derivative.

Induction on the cardinality of the set $S$ permits us to obtain a nonrecursive expression for the $S$-derivative. This expression is

$$\Delta_S[\mu(T)] = \sum_{K \subset S} (-1)^{|S| - |K|} \mu(T \cup K)$$

for all $T \subset X \setminus S$. For all $S$ that are subsets of a carrier $C$ (i.e., $\mu(S) = \mu(C \cap S)$ for all $S \subseteq X$), and for all $T \subseteq X \setminus S$, we have $\Delta_S[\mu(T)] = 0$.

Now, we define the interaction index for two sources $x_i$ and $x_j$ in a carrier $C$ of $X$ as follows:

$$I(\mu, \{x_i, x_j\}) := \sum_{T \subseteq C \setminus \{x_i, x_j\}} \frac{(|N| - |T| - 2)! |T|!}{(|N| - 1)!} \Delta_{\{x_i, x_j\}}[\mu(T)].$$

This is generalized for an arbitrary number of sources as follows:

$$I(\mu, S) := \sum_{T \subseteq C \setminus S} \frac{(|N| - |T| - |S|)! |T|!}{(|N| - |S| + 1)!} \Delta_S[\mu(T)].$$

This index is an extension of the Shapley value, as $I(\mu, \{x_i\})$ is equivalent to the Shapley value of $x_i$ in $\mu$ for all $\mu$ and all $x_i \in X$. That is, $I(\mu, \{x_i\}) = \varphi_{x_i}(\mu)$ (where $\varphi_{x_i}(\mu)$ follows Definition 7.2).

## 7.3 Dispersion

In most aggregation operators (all other indices being equal), it is often considered inappropriate to accumulate the weights or importances into a single source. Instead, the weights are distributed among the sources to maximize dispersion. This is done to reduce the influence of a particular source (see Section 2.2.6 about the use of influence functions for this purpose).

Entropy is a measure that can be used for evaluating the dispersion. As a weighting vector $\mathbf{p}$ is equivalent to a probability distribution, the standard definition of entropy is appropriate.

**Definition 7.12.** *Let* $\mathbf{p} = (p_1, \ldots, p_N)$ *be a weighting vector; then, its entropy (dispersion) $E$ is defined as*

$$E(p) := - \sum_{i=1}^{N} p_i \log p_i,$$

*with* $0 \log 0$ *defined as $0$ (to allow for zero weights).*

Alternatively, with the function $h$,

$$h(x) := \begin{cases} -x \ln x & \text{if } x > 0 \\ 0 & \text{if } x = 0, \end{cases} \tag{7.7}$$

the entropy can be defined as follows.

**Definition 7.13.** *Let* $\mathbf{p} = (p_1, \ldots, p_N)$ *be a weighting vector; then, its entropy (dispersion) $E$ is defined as*

$$E(\mathbf{p}) := \sum_{i=1}^{N} h(p_i).$$

For positive weights adding 1 (weighting vectors following Definition 6.1), the expression $E$ is maximal when all the weights are equal (i.e., $p_i = 1/N$). The maximal value obtained is $E(\mathbf{p}) = \log N$. In contrast, the minimal value $E(\mathbf{p}) = 0$ is obtained when $p_i = 1$ for one $i$.

A concept related to dispersion is variability. The variability is defined as the variance of the weights.

**Definition 7.14.** *Let* $\mathbf{p} = (p_1, \ldots, p_N)$ *be a weighting vector; then, its variance (variability) is defined as*

$$\sigma^2(\mathbf{p}) := E[(\mathbf{p} - E[\mathbf{p}])^2].$$

So, the variability can be computed as follows:

$$\sigma^2(\mathbf{p}) = \frac{1}{N} \sum_{i=1}^{N} (p_i - E[\mathbf{p}])^2 = \frac{1}{N} \sum_{i=1}^{N} (p_i - \frac{1}{N})^2 = \sum_{i=1}^{N} \frac{p_i^2}{N} - \frac{1}{N^2}.$$

### 7.3.1 Entropy for Fuzzy Measures

There exist two alternative definitions for entropy concerning fuzzy measures. They are the lower and upper entropies. We define them below using the function $h$ given above.

**Definition 7.15.** *Let $\mu$ be a fuzzy measure on $X = \{x_1, \ldots, x_N\}$; then, the lower entropy $E_l$ of $\mu$ is defined by*

$$E_l(\mu) := \sum_{i=1}^{N} \sum_{T \subseteq X \setminus \{x_i\}} \gamma_{|T|}(N) h[\mu(T \cup \{x_i\}) - \mu(T)], \qquad (7.8)$$

*where $h$ is defined as in Equation 7.7, and*

$$\gamma_t(n) := \frac{(n - t - 1)! t!}{n!}. \qquad (7.9)$$

**Definition 7.16.** *Let $\mu$ be a fuzzy measure on $X = \{x_1, \ldots, x_N\}$; then, the upper entropy $E_u$ of $\mu$ is defined by*

$$E_l(\mu) := \sum_{i=1}^{N} h \left( \sum_{T \subseteq X \setminus \{x_i\}} \gamma_{|T|}(N) [\mu(T \cup \{x_i\}) - \mu(T)] \right), \qquad (7.10)$$

*with $\gamma_t(n)$ defined as above.*

In this second definition, the entropy of a fuzzy measure $\mu$ corresponds to the entropy of the Shapley value (see Section 7.1.1) of the measure $\mu$.

The following properties have been proved for the entropies.

**Proposition 7.17.** *Let $E_l$ and $E_u$ be defined as above; then, the following hold*

1. *$E_l(\mu)$ and $E_u(\mu)$ are symmetric with respect to the permutation of the sources (the permutation of a measure follows the definition in Section 7.1.2):*
$$E_l(\mu_\pi) = E_l(\mu) \quad \text{and} \quad E_u(\mu_\pi) = E_u(\mu)$$
*for all permutation $\pi$.*
2. *$E_l(\mu) \leq E_u(\mu)$ for all $\mu$.*
3. *$E_l(\mu) = E_u(\mu)$ if and only if $\mu$ is additive.*
4. *When $\mu$ is additive, if $\mu$ is inferred from $\mathbf{p}$, then, $E_l(\mu) = E_u(\mu) = E(\mathbf{p})$. This implies that the entropy of a fuzzy measure reduces to the entropy of a weighting vector, and is consistent with the fact that the Choquet integral with respect to $\mu$ corresponds to the weighted mean with respect to $\mathbf{p}$.*
5. *Given a symmetric fuzzy measure generated from a weighting vector $\mathbf{w}$ (using Proposition 6.25), its lower entropy is*

$$\sum_{i=1}^{N} h(w_i),$$

*and its upper entropy is*

$$\log N.$$

*This roughly corresponds to the lower and upper entropies of the OWA operator with weights $\mathbf{w}$.*

## 7.4 Average Values

A simple index for any aggregation operator is its average value, i.e., the integral for all possible inputs.

**Definition 7.18.** *Let* $\mathbb{C}$ *be an aggregation operator in* $[0,1]^N$ *with parameter* $P$; *then, the* average value *of* $\mathbb{C}_P$ *is defined as*

$$AV(\mathbb{C}_P) := \int_0^1 \ldots \int_0^1 C_P(a_1, \ldots, a_N) \, da_1 \ldots da_N.$$

We give below the average value for some aggregation operators.

**Proposition 7.19.** *The average value for the minimum (min), the maximum (max), and the arithmetic mean (AM) is as follows:*

- $AV(\min) = N/(N+1)$
- $AV(\max) = 1/(N+1)$
- $AV(AM) = 1/2$

## 7.5 Orness or the Degree of Disjunction

Aggregation operators (as explained in Section 1.1, Equation 1.1) yield values between the minimum and the maximum. However, their behavior with respect to minimum and maximum is not the same for all of them. While some operators always yield values near the minimum, others yield values near the maximum; and some yield values that can be near either the minimum or the maximum (depending on the operator parameterization). Due to this, one way to evaluate the behavior of an operator is to measure its similarity to the maximum (or minimum) operator. To compute the similarity between operators, the average value introduced in Definition 7.18 can be used.

The index that computes the similarity with the maximum is the *degree of disjunction* or *orness*. This name is after the use in fuzzy logic of the maximum to model disjunction or the "or" connective (see Section 2.3.1). In an analogous way, the similarity with the minimum corresponds to the *degree of conjunction* or *andness*, as in fuzzy logic the minimum is used to model conjunction or the "and" connective.

We now define the degree of disjunction or orness.

**Definition 7.20.** *Let* $\mathbb{C}$ *be an aggregation operator with parameters* $P$; *then, the* orness *of* $\mathbb{C}_P$ *is defined by*

$$\text{orness}(\mathbb{C}_P) := \frac{AV(\mathbb{C}_P) - AV(\min)}{AV(\max) - AV(\min)}. \tag{7.11}$$

The degree of conjunction or andness can be defined either with an expression measuring similarity with maximum or, equivalently, in terms of the orness.

**Definition 7.21.** *The* andness *of* $\mathbb{C}_P$ *is defined by*

$$\text{andness}(\mathbb{C}_P) := 1 - \text{orness}(\mathbb{C}_P) = \frac{AV(\max) - AV(\mathbb{C}_P)}{AV(\max) - AV(\min)}$$

Naturally, these definitions are such that the following equations hold:

**Proposition 7.22.** *Orness and andness satisfy the following properties:*
$$\text{orness}\,(\max) = 1 \quad \text{andness}\,(\max) = 0$$
$$\text{orness}\,(\min) = 0 \quad \text{andness}\,(\min) = 1$$
$$\text{orness}\,(AM) = 1/2 \quad \text{andness}\,(AM) = 1/2$$

Simplified expressions for the orness can be found for particular aggregation operators. In the next proposition we give the corresponding expressions for the OWA operator (see Definition 6.4) and the Choquet integral (see Definition 6.17).

**Proposition 7.23.** *Let orness be defined as in Definition 7.20; then, the following equations hold:*

$$\text{orness}(WM_{\mathbf{p}}) = 1/2$$

$$\text{orness}(GM_{\mathbf{p}=(p_1,\dots,p_N)}) = \tfrac{N+1}{N-1}\left(\tfrac{N}{N+1}\right)^N - \tfrac{1}{N-1}$$

$$\text{orness}(OWA_{\mathbf{w}}) = \tfrac{1}{N-1}\sum_{i=1}^{N}(N-i)w_i$$

$$\text{orness}(HM_{\mathbf{p}=(p_1,p_2)}) = 0.2274 \text{ and } \text{orness}(HM_{\mathbf{p}=(p_1,p_2,p_3)}) = 0.2257$$

$$\text{orness}(CI_\mu) = \tfrac{1}{N-1}\sum_{A\subseteq X}\tfrac{N-|A|}{|A|+1}m(A),$$

*where m is the Möbius transform of $\mu$ (see Definition 5.14).*

Here, $GM$ is the Geometric Mean and $HM$ is the harmonic mean (see Section 4.2). It can be proved that

$$\text{orness}(GM)_{\mathbf{p}=(p_1,\dots,p_N)} < \text{orness}(GM)_{\mathbf{p}=(p_1,\dots,p_{N+1})}.$$

For example, for $N = 2$, $\text{orness}(GM) = 1/3 = 0.3333$, and for $N = 3$ $\text{orness}(GM) = 11/32 = 0.3437$.

The following property can be proved for the orness of the OWA operator.

**Proposition 7.24.** *Given a weighting vector* $\mathbf{w}$ *and a weighting vector* $\mathbf{w}'$ *satisfying* $w'_i = w_{N-i+1}$ *for all i,*

$$\text{orness}(OWA_{\mathbf{w}}) = 1 - \text{orness}(OWA_{\mathbf{w}'}).$$

## 7.5.1 Orness for Fuzzy Quantifiers

The orness definition given above for the OWA weighting vectors can yield different values when applied to weighting vectors extracted from the same quantifier but with different dimensions. That is, the orness of the OWA operator depends on the dimension when weights are extracted from the same quantifier. To make comparisons among quantifiers easier, and to make their evaluation possible, the orness of a quantifier has been defined. The measure is defined in such a way that, when weighting vectors of increasing dimension are extracted, their orness tends to be the orness of the quantifier. So, the orness of weighting vectors approximates the orness of the quantifier.

The definition for the quantifier is based on a rewriting of the expression of the orness for the OWA in Proposition 7.23. Note that

$$
\begin{aligned}
orness(\mathbf{w}) &= \sum_{i=1}^{N}(\frac{N-i}{N-1})w_i \\
&= \frac{N-1}{N-1}w_1 + \frac{N-2}{N-1}w_2 + ... + \frac{N-N}{N-1}w_N \\
&= \sum_{i=1}^{N-1}(1/N-1)Q(i/N),
\end{aligned}
\tag{7.12}
$$

where $Q$ is the quantifier associated with weights $\mathbf{w}$. That is, $Q$ is the quantifier that has been used to extract the $w_i$ or, alternatively, the quantifier that interpolates the points $\{(i/n, \sum_{j \leq i} w_j)\}$.

To generalize the orness measure, we consider the following expression instead of the previous ones:

$$
\sum_{i=1}^{N}(1/N)Q(i/N).
\tag{7.13}
$$

While it is clear that $\sum_{i=1}^{N}(1/N)Q(i/N)$ and $\sum_{i=1}^{N-1}(1/N-1)Q(i/N)$ tend to be equal for large values of $N$, the latter expression can be easily generalized by means of an integral. We will refer to this generalization as the continuous orness. Its definition is as follows.

**Definition 7.25.** *Given a fuzzy quantifier $Q$, the continuous orness measure for $Q$ is defined as*

$$
orness(Q) := \int_0^1 Q(x) \, dx.
\tag{7.14}
$$

This expression permits to study the orness for some families of quantifiers without focusing on particular dimensions. Additionally, it permits us to visualize that all fuzzy quantifiers with the same area in $[0,1]$ are equivalent with respect to orness. We consider some families of fuzzy quantifiers below, giving analytical expressions for their orness.

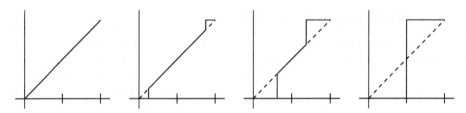

**Fig. 7.3.** $\alpha$-trimmed fuzzy quantifiers

| $\lambda$ | cont. orness | $N = 5$ Yager's | $N = 10$ Yager's | $N = 20$ Yager's |
|---|---|---|---|---|
| −0.99 | 0.7930 | 0.8473 | 0.8213 | 0.8074 |
| −0.90 | 0.6768 | 0.7114 | 0.6943 | 0.6856 |
| −0.50 | 0.5573 | 0.5687 | 0.5630 | 0.5602 |
| 0.00 | 0.5 | 0.5 | 0.5 | 0.5 |
| 0.50 | 0.4663 | 0.4596 | 0.4629 | 0.4646 |
| 1.00 | 0.4427 | 0.4312 | 0.4370 | 0.4398 |
| 1000 | 0.1437 | 0.0826 | 0.1105 | 0.1265 |

**Table 7.1.** Comparison of orness measures

*Example 7.26.* The $\alpha$-trimmed quantifier $Q_\alpha^t$ is defined as

$$Q_\alpha^t(x) := \begin{cases} 0 & \text{if } x < \alpha \\ x & \text{if } \alpha \leq x < 1 - \alpha \\ 1 & \text{if } x \geq 1 - \alpha \end{cases} \quad (7.15)$$

for $\alpha \leq 0.5$.

Note that the orness of these fuzzy quantifiers equals 0.5, and that the OWA operator with these fuzzy quantifiers corresponds to the $\alpha$-trimmed mean. Figure 7.3 represents some quantifiers of this family.

We now consider two other families of fuzzy quantifiers:

**Definition 7.27.** *Let Sugeno $\lambda$-quantifiers and Yager $\alpha$-quantifiers be defined as follows.*

*Sugeno $\lambda$-quantifier: for $\lambda > -1$, when $\lambda = 0$, $Q_\lambda(x) = x$ and when $\lambda \neq 0$,*

$$Q_\lambda(x) = (e^{x \ln(1+\lambda)} - 1)/\lambda.$$

*Yager $\alpha$-quantifier: for $\alpha > 0$,*

$$Q_\alpha(x) = x^\alpha.$$

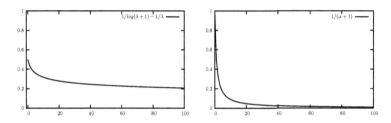

**Fig. 7.4.** Orness of the Sugeno $\lambda$-quantifiers for values of $\lambda \in (-1, 100]$ (left) and of the Yager $\alpha$-quantifiers for values of $\alpha \in (0, 100]$ (right)

$Q_\lambda$ is called a Sugeno $\lambda$-quantifier because $Q_\lambda$ generates a distorted probability with any probability distribution $\mathbf{p}$. That is, $\mu = Q_\lambda \circ \mathbf{p}$ is a Sugeno $\lambda$-measure for any probability distribution $\mathbf{p}$ and $\lambda > 1$. $Q_\lambda$ corresponds to the distortion function in Corollary 5.68.

**Proposition 7.28.** *The orness of the Sugeno $\lambda$-quantifier is*

$$orness(Q_\lambda) = \frac{1}{ln(1+\lambda)} - \frac{1}{\lambda}.$$

*When $\lambda = 0$, $orness(Q_\lambda) = 1/2$, which corresponds to the left and right limit.*

*The orness of the Yager $\alpha$-quantifier is*

$$orness(Q_\alpha) = \frac{1}{\alpha + 1}.$$

Table 7.1 gives the orness for several Sugeno $\lambda$-quantifiers. The table includes the continuous orness and the orness for dimensions $N = 5$, 10, and 20. It can be observed the convergence of the measure when $N$ increases with respect to the continuous orness.

A graphical representation of the orness for $Q_\lambda$, with $\lambda \in (-1, 100)$, is given in Figure 7.4 (left). This figure shows that, for large values of $\lambda$, we need a large variation in the parameter to obtain relevant changes in the orness. Figure 7.4 (right) represents the orness for $Q_\alpha$ quantifiers for $\alpha \in (0, 100)$. In this case, the orness moves rapidly from 1 to 0.

An important result of the two fuzzy quantifiers is the following.

**Proposition 7.29.** *The orness of the Sugeno $\lambda$-quantifier $(Q_\lambda(x))$ and of the Yager $\alpha$-quantifier $(Q_\alpha(x))$ are strictly monotone decreasing functions with respect to the parameters $\lambda$ and $\alpha$.*

This proposition will be used in Chapter 8.2, as it is useful for determining the weights in the OWA operator.

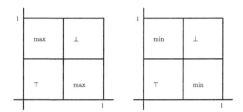

**Fig. 7.5.** Two families of uninorms

### 7.5.2 Pointwise Orness: Orness Distribution Function

The consideration that the orness is a pointwise property instead of a global one yields to the definition of the orness distribution function. This is motivated by the fact that an aggregation operator can be defined in such a way that in some subdomains it behaves like an and operator, while in other subdomains it behaves like an or operator. Then, the orness is defined for each point in $[0,1]^N$ as the similarity between the outcome of the operator and the maximum. In this way, an orness distribution is defined for all $[0,1]^N$, and then, if needed, average values can be computed to summarize the information.

Uninorms are examples of operators that do not have such a uniform behavior on the entire domain. As we saw in Section 4.1.1, uninorms have a conjunctive region and a disjunctive region. For example, in the case of Figure 4.1 (reproduced here in Figure 7.5), we have the region with a t-norm, the region with a t-conorm, and the regions with minimum or maximum. So, there are regions with a low orness (the regions with a t-norm) and regions with a high orness (the regions with a t-conorm).

This pointwise orness definition resembles the previous expression (see Expression 7.11). The main difference is that now no average value is computed, but the orness distribution function is defined on the space of inputs.

**Definition 7.30.** *Let* $\mathbb{C}$ *be an aggregation operator with parameters* $P$*; then, the* orness distribution function *of* $\mathbb{C}_P$ *is defined by*

$$\mathrm{odf}(\mathbb{C}_P, \mathbf{a}) = \frac{\mathbb{C}_P(\mathbf{a}) - \min(\mathbf{a})}{\max(\mathbf{a}) - \min(\mathbf{a})} \tag{7.16}$$

*for all* $\mathbf{a} \neq (a', \ldots, a')$.

Definition 7.30 is properly defined, as Expression 7.16 can be properly computed (i.e., it does not diverge to infinity) when the input vector $\mathbf{a}$ tends to a situation in which all the inputs are equal.

Now, we can consider the computation of some indices for this distribution so that its information is summarized. We will consider an orness average value.

**Definition 7.31.** *Let* $\mathbb{C}$ *be an aggregation operator with parameters* $P$, *and let* $orness(\mathbb{C}_P)$ *be its orness distribution function; then, its orness average value is defined by*

$$\overline{odf}(\mathbb{C}_P) = \int_0^1 \ldots \int_0^1 odf(\mathbb{C}_P,(a_1,\ldots,a_N))da_1 \ldots da_N.$$

Simplified expressions are known for some aggregation operators.

**Proposition 7.32.** *The following relations hold:*

1. $\overline{odf}(WM_\mathbf{p}) = 0.5$
2. $\overline{odf}(OWA_\mathbf{w}) = orness(OWA_\mathbf{w})$
3. $\overline{odf}(UN_e) = (1-e)^N$
4. $\overline{odf}(UN^e) = 1 - e^N$
5. $\overline{odf}(GM) = 0.385$ *(for $N = 2$)*
6. $\overline{odf}(HM) = 0.306$ *(for $N = 2$)*

Here, $UN_e$ and $UN^e$ correspond to the uninorms defined in Section 4.1.1, and $GM$ and $HM$ correspond, respectively, to the geometric and harmonic means (Section 4.2).

Although this proposition shows that the orness average value of the OWA operators are equivalent to the orness of the OWA operators, this is not the usual case. Note that equality does not hold for $HM$ and $GM$ (see Proposition 7.23).

Note also that equivalence between the orness and the average value of the orness distribution corresponds to the equivalence between

$$\overline{odf}(\mathbb{C}_P) = \int_0^1 \ldots \int_0^1 odf(\mathbb{C}_P,(a_1,\ldots,a_N))da_1 \ldots da_N =$$

$$\int_0^1 \ldots \int_0^1 \frac{\mathbb{C}_P(\mathbf{a}) - \min(\mathbf{a})}{\max(\mathbf{a}) - \min(\mathbf{a})}da_1 \ldots da_N$$

and

$$orness(\mathbb{C}_P) = \frac{AV(\mathbb{C}_P) - AV(\min)}{AV(\max) - AV(\min)}.$$

### 7.5.3 Interpretation

An important point of the orness measure is that it can be interpreted beyond its definition in terms of maximum and minimum. Some of the existing interpretations are given below. Interpretations are rooted in some examples.

**Similarity to maximum:** The simplest interpretation is that orness corresponds to the similarity to the maximum operator. In this case, similarity can be computed from a global perspective (as in Definition 7.20) or as the average of a pointwise property (as in Definitions 7.31).

**Compensation:** In this case, orness measures to what extent a *bad* score (a small value) for one of the inputs influences the output. When compensation is allowed (orness is maximum and equal to one), any *good* value (a large value) overcomes any *bad* one. That is, the maximum is selected. In contrast, when no compensation is allowed (orness is equal to zero) *bad* values (the smaller ones) overcome all *good* values, no matter how many good values there are.

**Optimism:** Under this interpretation, the orness corresponds to a measure of the optimism of a decision maker, while the andness corresponds to pessimism. With regard to the aggregation of utilities, the larger the orness, the larger the risk the decision maker is accepting (the aggregated utility will be larger for the same inputs with larger ornesses). In contrast, when the orness is small, the decision maker is giving more importance to small utilities, and, thus, is more concerned about risk. Therefore, the decision maker, with regard to risk, is more pessimistic (less optimistic).

**Fuzzy quantifier:** The orness is interpreted in terms of the orness of a fuzzy quantifier. It is known that the "for all" quantifier has an orness equal to 0. When this quantifier is applied, all criteria are required to be satisfied. This corresponds to a conjunction of all criteria. In contrast, the orness of the "there exists" is equal to 1. In this case, only the satisfaction of one of the criteria is required. Therefore, this is disjunctive behavior. Other quantifiers can also be used, and the corresponding orness is interpreted in terms of such quantifiers (e.g., about 50%, a third).

## 7.6 Bibliographical Notes

1. **Power indices:** Power indices have been developed in game theory, the von Neumann and Morgenstern book [425] being the preeminent reference in the area. For essays on the history of game theory, see [433]. See [135, 136] for an outline on the history of indices. For a recent book on game theory, see [285].

   The Shapley value was first defined in [357] (in 1953), and then specialized into the Shapley-Shubik power index in [359] (1954). The Banzhaf index was introduced in [31] (in 1965), and its standardized definition was proposed by Dubey and Shapley in [108] (in 1979) as a modification of the standardized one. Nevertheless, in 1946, L. S. Penrose had already published the index in [319] (which was introduced into mainstream research by Morriss in 1987 [274]). The Banzhaf index was characterized by Dubey (see [107, 108]). [27] studies indices for non-atomic games. That is, $N$-player games for non finite $N$, or, equivalently, fuzzy measures over a non finite set. Appendix A in [27] reviews finite games and their values (the Shapley value for fuzzy measures). For a detailed account of the Shapley value and related issues, see the book edited by Roth [336]. [357] and [359] are reprinted there.

At present, a large number of power indices have been defined in the literature of game theory, e.g., the Deegan-Packel index [90], the Coleman index [81] (reprinted in [82]), and the Executive power index (by Colomer and Martinez [83, 84]). While these indices only rely on the measure/game $\mu$, other indices have been defined in game theory that consider some additional information (e.g., whether the individuals in a set are connected). This is the case with the Garret and Tsebelis index [159, 160, 161]. See [19] for a review of the indices. Felsenthal and Machover (1998) [135] give a critical perspective of the area (see also Morris, 1987 [274]).

Similarities and relationships between the indices have also been studied, and paradoxes have been analyzed. For example, [44] and [45] proved that in some circumstances the indices lead to unsatisfactory results. Straffin in [379] and [380] compares the Shapley-Shubik and Banzhaf power indices. He gives an example showing that the two indices can yield different results, and even rank the power of the voters differently. Straffin argues that the comparison of the two indices in terms of the order in which winning coalitions are formed is misleading. He introduces the interpretation based on probability models and the homogeneity and independence assumptions. So, the order plays no part in the interpretations. He also studies alternative situations, such as partial homogeneity assumptions, which can model the situation where voters can be partitioned into groups (e.g., parties), and there is homogeneity between two members of the same group but not between two members of two different groups. Examples of paradoxes of the indices can be found in Felsenthal and Machover [135].

For Web resources, see [311] (it includes software for computing power indices) and [310].

The Shapley value for fuzzy coalitions has been studied by several researchers. See [53] and [420]

2. **Interaction index:** The interaction index for sets of two elements was introduced by Owen (1972) [309] (Section 5), and later rediscovered by Murofushi and Soneda [278]. The extension for an arbitrary number of sources is due to Grabisch [169]. The interaction index presented here generalizes the Shapley value. Accordingly, it can be cited as the Shapley interaction index. Other interaction indices are also possible. For example, Roubens in [337] introduced the Banzhaf interaction index that extends the Banzhaf value. Some characterizations of interaction indices are given in [152, 153, 154].

3. **Average value, degree of conjunction, degree of disjunction or orness:** The average values for minimum and maximum were given by Dujmović in [101]. Orness and andness concepts were introduced by Dujmović in 1973 and 1974. The orness distribution function (called *local orness*) and the orness average value (called *mean local orness*) were introduced in [100]. Then, the orness (as defined in Definition 7.20) was introduced in [102]. [102] gives, among other results, the orness of the geometric mean. Numerical computations for some root-mean-powers are

also included. [103] includes a few additional results. The orness for the quasi-arithmetic mean is studied in [230]. [230] presents an alternative definition for measuring orness that uses the generator of the quasi-arithmetic mean.

Independently, Yager defined orness and andness in [442] for the OWA operator. Yager and Rybalov defined in 1996 the orness and andness of uninorms $UN_e$ and $UN^e$ in [457]. Marichal, 1998, gives in [245] the general definition of the degree of disjunction and finds an expression (in Proposition 7.23) for the orness of the Choquet integral. Marichal also proves the consistency of Yager's expression for the OWA orness with Dujmović's degree of disjunction. Before Marichal, in 1994, Fodor and Roubens [146] had already drawn attention to Dujmović's work. More recently Fernández Salido and Murakami [138] reintroduced the orness distribution functions and the orness average value. We use their terminology in this book.

The orness for fuzzy quantifiers was first defined by Yager in [449]. Torra [405] uses this expression to study the orness of Sugeno $\lambda$-quantifiers. [59] introduced an alternative expression for the orness of quantifiers that was proved to be equivalent to Yager's in [453]. See also [57]. The equivalent expression is:

$$orness(Q) = 1 - \int_0^1 xQ'(x)dx = 1 - \int_0^1 xq(x)dx,$$

where $Q$ is the fuzzy quantifier and $q'$ its generating function.

[443] defines an ordinal OWA and a nonnumerical orness (orness on an ordinal scale). [133] discusses this ordinal orness.

4. **Orness interpretation:** Orness as compensation is considered in [396]. A similar concept, used in [245], is orness as a degree of tolerance. [301] and [444] interpret orness as optimism and andness as pessimism, which is used to define the pessimism-optimism index criterion of Hurwicz. See [131, 132] or Luce and Raiffa [239] for a description of this criteria. [22] gives an interpretation of the orness in terms of the number of disjunctions in a statement. Dujmović [102] and Fernandez-Salido and Murakami [138] use dissimilarity with the minimum as orness. [138] and [245] review some interpretations of orness.

5. **Entropy and dispersion:** For the definition of entropy and its properties, see [356] and [25]. Some characterizations of these measures can be found in [10] and [114]. The use of entropy [356] for evaluating the dispersion of the weights appears, among others, in [301, 302, 60, 442, 400]. The first use with OWA weights was in Yager's 1988 paper [442]. In these works, entropy is used for measuring dispersion of a weighting vector either for the weighted mean or for the OWA operator.

For fuzzy measures, original definitions for entropy were given by Yager in [448] and Marichal in [245] (see also [248]). Yager's definition corresponds to the upper entropy, and Marichal's corresponds to the lower entropy. The lower and upper probabilities are due to Marichal and Roubens.

In [254], they compare the two definitions and prove, among other things, that the upper entropy always leads to a value larger than the lower one (and that equality is obtained only for additive fuzzy measures). The names for the upper and lower entropies naturally follows from this property.

Entropy for fuzzy measures as defined in this chapter assumes values in $[0, 1]$. Marichal and Roubens defined, also in [254], entropy suitable for ordinal fuzzy measures. The authors affirm that this definition is linked to the Sugeno integral (as both the entropy and the integral are specially suited for working in ordinal scales).

Let $\mu$ be defined in $L = \{l_0, \ldots, l_r\}$, with $l_0 <_L l_1 <_L \cdots <_L l_r$; then, the ordinal entropy $E_L$ of $\mu$ is defined by $E_L(\mu) = l_{|R|} - 2$, where $R = \{\mu(A)|A \subset X\}$. The rationale for this definition is that the entropy $E_L(\mu)$ measures the diversity of the coefficients used in the fuzzy measure $\mu$. Note that the measure is based on the number of different terms used in $\mu$, less two. Among the properties of $E_L$, we have that its minimum is $l_0$ and its maximum is $l_{r-1}$ (or $l_{2^N-1}$ if $2^N < r$). Moreover, the entropy for WMax, WMin, OWMax, and OWMin using a weighting vector $\mathbf{u} = (u_1, \ldots, u_N)$ is $E_L(\mu) := l_{|\{u_1,\ldots,u_N\}|-2}$ (again, the number of different terms now in $\mathbf{u}$).

The use of variance to measure variability can be found in [446], but the concept is slightly different. Fullér and Majlender [157] use the same approach as defined here.

# 8

# Selection of the Model

*The moment of truth is a running program.*

H. A. Simon, [362] (p. 96)

When an application needs a fusion mechanism, the developer has to solve an essential problem: the construction of the appropriate model. This corresponds to (i) the selection of an aggregation operator and (ii) the determination of its parameters. This process should take into account several factors. Some of them are highlighted here.

Mathematical properties: The operator should be selected taking into account the desired properties of the model. Characterizations can help in the selection. Indices and any other behavioral analysis, e.g., orness or breakdown points (if we are interested in robust behavior), can also be useful.

Interpretability: An expert or user needs to understand the model. As the model consists of an operator and its parameters, interpretability is applicable to both operators and parameters. In the case of complex operators such as fuzzy integrals, tools for analyzing the parameters (as with power indices) can improve the interpretability of the model.

Adaptability: The environment of any system changes with respect to time. Selection should consider whether the operator will still be valid after some time of operation (say, with a readjustment of its parameters).

An additional factor to be taken into account is *simplicity*. We have described in previous chapters a broad variety of operators. The most simple ones have a limited descriptive capability (e.g., the arithmetic mean), while the more complex ones (e.g., the Sugeno integral) have additional capabilities. Such additional capabilities come with the price of more parameters (e.g., $2^{|N|} - 2$ values for any fuzzy integral).

In general, other factors being equal, the simpler the model, the better. Too simple operators might produce unsuccessful results, whatever the parameter used. However, unnecessarily complex operators might cause overfitting to a particular situation or to a particular data.

In the rest of this chapter we will consider the problem of parameter determination. At present, there exist several methods for this purpose, and they can be roughly classified into the following two classes.

Methods based on an expert:  Two main alternatives exist in which the help of an expert is fundamental for fixing the parameter. One alternative is that the expert (almost) directly supply the required parameters. The Analytic Hierarchy Process is an example of this approach. The method permits us to elicit the weights for operators like the weighted mean. In short, an expert is interviewed, and from pairwise comparison of the sources the weights are calculated. This method permits inconsistencies to some extent. Another alternative is that the expert supply some relevant information that is later used for parameter determination. This is the case of parameter determination from orness or compensation.

Methods based on data:  Parameters are learned from a set of examples. One alternative consists of having preferences or a partial order of the examples (or of the outcomes of the aggregation); for example, we may prefer example 1 to example 5, or the outcome of example 1 may be larger than that of example 5. The order is not required to be a total order. Another alternative consists of having examples where each is defined in terms of the inputs of the model as well as the intended output for the inputs.

In this chapter we will review methods of the two classes. First, we will consider the Analytic Hierarchy Process. Then, we will describe the method to determine OWA weights from orness (or compensation) and dispersion. Finally, we will review some methods to determine parameters from examples.

## 8.1 Analytic Hierarchy Process

The Analytic Hierarchy Process (AHP) was designed to derive ratio scales. Therefore, AHP is a valid methodology for weight determination for operators with weights represented in ratio scales.

Weighted means and OWA operators are an example of such operators. In their weighting vectors, the relevant aspect is the relationship (the ratio) between weights; that is, whether one weight is, for example, two times or three times another weight. This is equivalent to expressing that one information source (sensor or expert) is two times or three times more relevant or important than another. Besides, there is an absolute zero for weights. Note also that the exact value for the weight is not much relevant for expressing importance, and that any transformation that keeps the ratio would be, in principle, valid. The requirement of all weights adding 1 permits unanimity to hold. The difference between weights $\mathbf{w}$ adding 1 and weights $\mathbf{w}'$ adding

| Importance | Meaning |
|:----------:|:-------:|
| 1 | Equal importance |
| 3 | Moderate importance |
| 5 | Essential or strong importance |
| 7 | Very strong importance |
| 9 | Extreme importance |

**Table 8.1.** Scale for intensity of importances

$r$ is a multiplication factor of $r$ in the outcome. All these properties imply a ratio scale.

In a given problem, the first step in the AHP is to formalize the structure of the objects under consideration, establishing their relationships. Typically, such a structure corresponds to a hierarchy or a network. In the case of simple weight determination, we will only consider a single set $X = \{x_1, \ldots, x_N\}$ consisting of the information sources. So, no hierarchical structure or network is needed. More complex situations arise when AHP is applied in a multicriteria decision making problem. In that case, we have several criteria as well as several alternatives. Both, the criteria and alternatives are used to define a network.

The next step is to compare each pair of objects. That is, in our case, to compare the pairs of sources with respect to their importance. So, for each pair $x_i, x_j$ in $X$, we assign to it a value representing whether the importance of $x_i$ is larger than, equal to or lower than that of $x_j$. Let $a_{ij}$ be this value. Then, $a_{ij}$ corresponds to the ratio of the importance of $x_i$ to that of $x_j$. It is usual to use the 1-9 scale (see Table 8.1) to express importance. All comparisons $a_{ij}$ define a square matrix, where $a_{ij} = 1/a_{ji}$ because the comparison satisfies reciprocity.

*Example 8.1.* Let us consider the problem of selecting the most suitable student for a particular task in school. Selection will be based on his or her marks in the five subjects considered in Example 5.46. To rate the weight of each subject in the selection, we will use the AHP.

Therefore, we need first to compare each pair of subjects using the 1-9 scale. To do so, we define $X$ such that, according to the requirements, it corresponds to the set $\{ML, P, M, L, G\}$, where $ML$ stands for Mathematical Logic, $P$ for Physics, $M$ for Mathematics, $L$ for Literature, and $G$ for Greek. Second, we consider the pairwise comparison of the subjects. According to the particular task to be assigned to the student, we establish that

- $ML$ is considered three times more relevant than $P$ and two times more relevant than $M$. In relation to Humanities, $ML$ is considered seven times more important than $L$ and nine times more important than $G$.
- $M$ is the second most important subject, and its importance is two times that of $P$ and three times that of $L$ and $G$.

|    | ML  | P   | M   | L   | G   |
|----|-----|-----|-----|-----|-----|
| ML | 1   | 3   | 2   | 7   | 9   |
| P  | 1/3 | 1   | 1/2 | 2   | 3   |
| M  | 1/2 | 2   | 1   | 3   | 3   |
| L  | 1/7 | 1/2 | 1/3 | 1   | 2   |
| G  | 1/9 | 1/3 | 1/3 | 1/2 | 1   |

**Table 8.2.** Pairwise comparison matrix for $\{ML, P, M, L, G\}$

The pairwise comparisons, as well as all the other required comparisons, are given in Table 8.2.

When the values of $a_{ij}$ are in a ratio scale, they should satisfy $a_{ik} = a_{ij}a_{jk}$. A matrix satisfying this equality is said to be consistent. In a consistent matrix, all rows are equal except for a multiplicative factor. That is, for two rows $r$ and $s$, we have that $a_{sj} = a_{rj}k$ for a certain value $k$. Specifically, $a_{sj} = a_{1j}/a_{1s}$. For such consistent matrices, weights can be obtained by taking any row and normalizing it. That is, when $N$ sources are under consideration, we define the weights as $w_i = a_{sj}/\sum_{i=1}^{N} a_{sj}$, for any row $s$. It is simple to prove that the result does not depend on the row selected.

Nevertheless, AHP permits us to obtain weighting vectors even in the case of inconsistent matrices. This is common in real practice. The weights are obtained from the principal eigenvector once it is normalized. We will illustrate the process with the matrix defined in the previous example. Note that the matrix is inconsistent.

*Example 8.2.* Let us consider the pairwise comparison matrix in Example 8.1. The weighting vector associated with this matrix (Table 8.2) is obtained as follows.

1. Obtain the principal eigenvector of the matrix. This vector corresponds to
$$(8.8920 \quad 2.7254 \quad 4.2718 \quad 1.4899 \quad 1).$$

2. Normalize the vector. That is, divide each term by $(8.8920 + 2.7254 + 4.2718 + 1.4899 + 1)$:

$$(0.4838104 \quad 0.148288 \quad 0.23242705 \quad 0.0810649 \quad 0.054409627).$$

Therefore, the weights are $p(ML) = 0.4838104$, $p(P) = 0.148288$, $p(M) = 0.23242705$, $p(L) = 0.0810649$, and $p(G) = 0.054409627$. Note that the weights roughly satisfy the constraints specified in the matrix.

AHP offers a measure of consistency to evaluate the quality of the matrix. This measure is based on the principal eigenvalue. This is so because it is

| $N$ | 1 | 2 | 3 | 4 | 5 | 6 | 7 | 8 | 9 | 10 |
|------|---|---|------|------|------|------|------|------|------|------|
| $RI$ | 0 | 0 | 0.58 | 0.90 | 1.12 | 1.24 | 1.32 | 1.41 | 1.45 | 1.49 |

**Table 8.3.** Random consistency indices $(RI)$ for different dimensions $N$

known that for a positive reciprocal matrix of dimension $N$, its principal eigenvalue is $N$, and that the value differs from $N$ for nonreciprocal matrices.

When the principal eigenvalue of a matrix is $\lambda_{\max}$, its consistency index is defined as

$$CI = \frac{\lambda_{\max} - N}{N - 1}.$$

Additionally, there is a consistency ratio that compares the consistency index with an average of consistency indices computed from random matrices of different dimensions. Formally, the index is defined by

$$CR = CI/RI,$$

where the $RI$ is the Random Consistency Index, which depends on the dimension of the matrix $N$. Table 8.3 gives such indices $RI$ for different values of $N$.

The $CR$ is a normalized value. For values of $CR$ larger than 0.10, revising the original matrix is recommended.

*Example 8.3.* The consistency index for the matrix in Example 8.1 is computed from the principal eigenvalue (that is $\lambda_{max} = 5.0654$), as follows:

$$CI = \frac{5.0654 - 5}{5 - 1} = 0.01635003.$$

The consistency ratio is

$$CR = \frac{0.01635003}{1.12} = 0.014598242.$$

## 8.2 OWA Weights from Orness

In Section 7.5, we described some measures and indices for aggregation operators. Orness or degree of disjunction is one such measure. This measure, defined as a similarity to the maximum operator, can be understood as compensation or optimism.

One approach for defining the parameters of an operator is to fix some index or measure, and then to select a parameter satisfying it. The orness, due to its simple interpretation, is specially adequate for this purpose. We illustrate the process by considering how to define the parameter for the OWA operator when the weights are determined from a fuzzy quantifier. Nevertheless, the same approach can be applied for other operators as well.

In the particular case of OWA with a fuzzy quantifier $Q$, we have that the orness of the OWA operator corresponds to (Equation 7.25)

$$orness(Q) = \int_0^1 Q(x) \, dx. \tag{8.1}$$

Then, the determination of the parameter corresponds to finding the quantifier $Q$ such that $orness(Q) = \delta$ for a degree of orness equal to $\delta$ (supplied by the expert). The following example illustrates this fact.

*Example 8.4.* The problem consists of determining a quantifier of the Yager family $(Q_\alpha(x) = x^\alpha)$ for the OWA operator when an expert supplies a degree of orness or compensation equal to 0.2. That is, $\delta = 0.2$.

According to Proposition 7.28, the orness of $Q_\alpha$ is given by

$$orness(Q_\alpha) = \frac{1}{\alpha + 1}.$$

Therefore, the problem is to find $\alpha$ such that

$$\delta = \frac{1}{\alpha + 1}.$$

So, $\alpha = (1 - \delta)/\delta = (1 - 0.2)/0.2 = 4$.

Proposition 7.29, which establishes that the orness of Sugeno $\lambda$-quantifiers and of Yager $\alpha$-quantifiers are decreasing functions, is useful when defining the parameters of the OWA operator, because it implies that, for a single degree of orness $\delta$, there is a single quantifier.

However, in general, the degree of disjunction does not determine the parameters uniquely. For example, if we consider OWA operators with weighting vectors, there exist several weighting vectors with orness equal to 0.5. In this case, solutions can be further constrained by adding additional requirements. A usual constraint is to require maximum dispersion among the weights. Other constraints might be also of interest. For example, selecting weights so that the robustness of the operator is maximized.

## 8.2.1 Orness and Dispersion

The rationale of this approach is to assume that, when several weighting vectors are equally good, it is better to distribute the weights as much as posible among the sources instead of accumulating them in a single source. For example, in the case of the OWA operator the two weighting vectors $(0, 1/2, 1/2, 0)$ and $(1/4, 1/4, 1/4, 1/4)$ have the same orness (equal to 0.5). We prefer the second vector, as all sources contribute to the output. As seen in Section 7.3, the dispersion of a weighting vector can be measured using entropy. So, the constraint of maximum dispersion corresponds to maximum entropy. The same can be applied to operators other than the OWA operator. In the case where fuzzy measures are used, the fuzzy entropy for fuzzy measures defined in Section 7.3.1 can be used.

With regard to a degree of disjunction $\delta$ and maximum dispersion, the problem of weight determination is formulated as follows:

$$Maximize\ dispersion$$
$$Subject\ to$$
$$\delta = orness$$

$$w\ is\ a\ weighting\ vector$$

For the OWA operator, it is formulated as follows:

$$\text{Minimize } -\sum_{i=1}^{N} w_i \ln w_i$$
$$\text{Subject to}$$
$$\delta = \frac{1}{N-1} \sum_{i=1}^{N} (N-i)w_i$$

$$\sum_{i=1}^{N} w_i = 1$$

$$w_i \geq 0$$

The following result permits us to compute the weights in the case of the OWA operator.

**Proposition 8.5.** *Given an OWA operator of dimension $N > 2$, and an orness value equal to $\delta \in [0,1]$, the mathematical programming problem given above has a unique solution given by the weighting vector $\mathbf{w}$ with*

$$w_j = \frac{t^j}{\sum_{i=1}^{N} t^i}$$

*for all $j = 1, ..., N$, where $t$ is the only real positive zero of the equation*

$$ax^{N-1} + (a+1)x^{N-2} + \cdots (a+N-2)x + (a+N-1) = 0,$$

*with $a = -\delta(N-1)$.*

In the case of $N = 2$, the problem formulated above corresponds to $w_1 = \delta$ and $w_2 = 1-\delta$. An OWA operator with weights of this form is called *Maximum Entropy* OWA (ME-OWA).

$$
\begin{array}{cccc|c}
a_1^1 & a_2^1 & \dots & a_N^1 & b^1 \\
a_1^2 & a_2^2 & \dots & a_N^2 & b^2 \\
\vdots & \vdots & & \vdots & \vdots \\
a_1^M & a_2^M & \dots & a_N^M & b^M
\end{array}
$$

**Table 8.4.** Data examples

## 8.3 Extracting Parameters from Examples: Expected Outcome

Now, we consider parameter determination when there is a set of examples for fixing the model and the examples are defined by *(input, output)* pairs. This problem can either be expressed as an optimization problem or as a learning problem.

We will present in this section a mathematical formulation of the problem. Then, in the following sections, we will describe some particular methods that can be used for some particular aggregation operators.

### Formulating the optimization problem

As stated, we consider examples defined by *(input, output)* pairs. Therefore, examples follow the structure described in Table 8.4. That is, there are $M$ different examples, each consisting of the values supplied by $N$ information sources and the *correct* outcome that we are intended to approximate from the values. Therefore, each example consists of $N+1$ values, with $(a_1^j a_2^j \dots a_N^j | b^j)$ denoting the values for the $j$th example. Here, $a_i^j$ is the value supplied by the $i$th information source (say, $x_i$), and $b^j$ is the ideal outcome for the same example. $\mathbf{A}$ denotes the matrix $\mathbf{A} = \{a_i^j\}$, and $\mathbf{b}$ is the vector such that $\mathbf{b}' = (b^1 \dots b^M)$. We use $\mathbf{b}'$ to denote the transpose of the vector $\mathbf{b}$. Again, we use $X = \{x_1, \dots, x_N\}$ to denote the information sources, and the function $f^j$ to denote that $a_i^j = f^j(x_i)$ is the value supplied by $x_i$ for the $j$th example.

Given a set of examples, and assuming that the aggregation function $\mathbb{C}$ is known, the goal is to determine the parameters of $\mathbb{C}$ given $\mathbf{A}$. When $\mathbb{C}$ is the Weighted Mean (WM), problem is to find a weighting vector $\mathbf{p}$ for the $WM$ so that the difference between $b^j$ and the estimated value for the $j$th example is minimum. Similarly, if $\mathbb{C}$ is the Choquet integral, then the goal is to find a fuzzy measure $\mu$ that also minimizes such a difference.

Naturally, to establish this problem properly we need a way to measure the difference between the estimated value and the *correct* outcome. In the case of the $j$th example, this corresponds to the difference between $b^j$ and $\mathbb{C}(f^j(x_1), \dots, f^j(x_N))$. One alternative (and the one most used) is to take the squared difference (i.e., $distance(x, y) = (x - y)^2$). In this case, the best model is the one that minimizes

$$D_{\mathbb{C}}(parameters(\mathbb{C})) = \sum_{j=1}^{M}(\mathbb{C}(f^j(x_1),\ldots,f^j(x_N)) - b^j)^2. \qquad (8.2)$$

We will simplify this notation, denoting the *parameters* of $\mathbb{C}$ by $P$, and then using $\mathbb{C}_P$ to express that the aggregation operator depends on $P$. Thus, instead of the last expression, we will use

$$D_{\mathbb{C}}(P) = \sum_{j=1}^{M}(\mathbb{C}_P(f^j(x_1),\ldots,f^j(x_N)) - b^j)^2. \qquad (8.3)$$

Typically, as $\mathbb{C}$ is assumed to be known and fixed, the goal is to obtain the minimum $D_{\mathbb{C}}(P)$ over the possible parameters $P$. The minimization of this equation results in a least sum of squares (see Section 2.2.5). Nevertheless, this is a constrained least square problem, because parameters have to satisfy particular constraints that depend on their nature. For example, weights in weighting vectors have to add to 1 and be positive. So, in general, the problem to be solved takes the following form:

$$\begin{aligned} &\text{Minimize } D_{\mathbb{C}}(P) = \sum_{j=1}^{M}(\mathbb{C}_P(a_1^j,\ldots,a_N^j) - b^j)^2 \\ &\text{Subject to logical constraints on } P \end{aligned} \qquad (8.4)$$

When the aggregation operator is the weighted mean with a weighting vector $\mathbf{p} = (p_1,\ldots,p_N)$, we have $\sum_i p_i = 1$ and $p_i \geq 0$, and, thus, the problem corresponds to

$$\begin{aligned} &\text{Minimize } D_{WM}(\mathbf{p} = (p_1,\ldots,p_N)) = \sum_{j=1}^{M}(WM_{\mathbf{p}}(a_1^j,\ldots,a_N^j) - b^j)^2 \\ &\text{Subject to} \\ &\qquad \sum_{i=1}^{N} p_i = 1 \\ &\qquad p_i \geq 0 \end{aligned} \qquad (8.5)$$

Let us consider an example of weight determination for the weighted mean.

*Example 8.6.* Let us consider the problem of giving a global score to students in a school in terms of their marks in the five subjects considered in Example 5.46; that is, the marks in Mathematical Logic ($ML$), Physics ($P$), Mathematics ($M$), Literature ($L$), and Greek ($G$).

Then, let us consider a set of 10 students $\{s_1,\ldots,s_{10}\}$, for who the marks in all five subjects are known, and let us consider a subjective overall rate. Such marks and overall rates are given in Table 8.5.

Now, if we use the weighted mean as a model for our subjective rating, then we can use the problem formalized in Equation 8.5 for determining the weights. In this case, with $\mathbf{p} = (p_{ML}, p_P, p_M, p_L, p_G)$, and using the data in Table 8.5, the previous problem is rewritten as

| Student | $ML$ $P$ $M$ $L$ $G$ | Subjective evaluation |
|---|---|---|
| $s_1$ | 0.8 0.9 0.8 0.1 0.1 | 0.7 |
| $s_2$ | 0.7 0.6 0.9 0.2 0.3 | 0.6 |
| $s_3$ | 0.7 0.7 0.7 0.2 0.6 | 0.6 |
| $s_4$ | 0.6 0.9 0.9 0.4 0.4 | 0.8 |
| $s_5$ | 0.8 0.6 0.3 0.9 0.9 | 0.8 |
| $s_6$ | 0.2 0.4 0.2 0.8 0.1 | 0.3 |
| $s_7$ | 0.1 0.2 0.4 0.1 0.2 | 0.1 |
| $s_8$ | 0.3 0.3 0.3 0.8 0.3 | 0.4 |
| $s_9$ | 0.5 0.2 0.1 0.2 0.1 | 0.3 |
| $s_{10}$ | 0.8 0.2 0.2 0.5 0.1 | 0.5 |

**Table 8.5.** Marks given to ten students, and their subject evaluation

$$\text{Minimize } D_{WM}(\mathbf{p}) = \sum_{j=1}^{10}((p_{ML}a_{ML}^j + p_P a_P^j + p_M a_M^j + p_L a_L^j + p_G a_G^j) - b^j)^2$$
*Subject to*
$$p_{ML} + p_P + p_M + p_L + p_G = 1$$
$$p_{ML} \geq 0, \quad p_P \geq 0, \quad p_M \geq 0, \quad p_L \geq 0, \quad p_G \geq 0$$

$$(8.6)$$

The optimal solution of this quadratic problem subject to linear constraints gives the following weights:

$$p_{ML} = 0.4244, \ p_P = 0.4108, \ p_M = 0.0000, \ p_L = 0.1249 \text{ and } p_G = 0.0399.$$

So, this model implies that there are two main marks that are considered relevant in our subjective evaluation: the one for $ML$ and the one for $P$.

### Dealing with multiple solutions

Before going into the details on the computation of a solution for the optimization problem, it is important to note that the minimization problem does not always lead to a single solution. Instead, several solutions can exist with the same distance $D_{\mathbb{C}}(P)$.

This situation is similar to the one we found in the case of parameter determination from a given orness. Again, we can consider an additional measure such as dispersion. So, given several parameters $P$ with the same $D_{\mathbb{C}}(P)$, we select the one that maximizes dispersion. In this case, the problem to be solved is formalized as follows.

(i) Find a solution $P$ of the minimization problem in Expression 8.4, and then define $\Delta$ as the error for this parameter $P$. That is, $\Delta = D(P)$
(ii) Solve the following problem:

$$\text{Maximize } dispersion(P)$$
$$\text{Subject to}$$
$$\text{logical constraints on } P$$
$$\text{and}$$
$$\Delta = D_{\mathbb{C}}(P),$$

where *dispersion* is measured in terms of entropy.

### 8.3.1 Weighted Mean

The simplest problem for learning weights is when the aggregation operator corresponds to the weighted mean. In this case, using the Euclidean norm $||x|| = \sqrt{x'x}$, we can express $D_{WM}(\mathbf{p})$ as $D_{WM}(p) = ||Ap - b||^2 = (Ap - b)'(Ap - b)$. Due to the fact that this expression gets its minimum when we have the minimum for $(1/2)p'Q'p + r'p$, where $Q = A'A$ and $r = A'b$, we can reformulate the problem given above in Expression 8.5 as follows:

$$\text{Minimize } (1/2)p'Qp + r'p$$
$$\text{Subject to}$$
$$\sum_{i=1}^{N} p_i = 1$$
$$p_i \geq 0$$

From now on, we consider this problem as the one to minimize. Nevertheless, for the sake of clarity, we use $D_{WM}(\mathbf{p})$ to denote the difference between the model using $p$ and the examples, instead of using the new minimization function $(1/2)p'Qp + r'p$.

The new formulation of the problem shows clearly that this is a typical optimization problem: a quadratic program with linear inequality constraints. There exist several algorithms to solve this class of problems. We consider here two different approaches. First, a method that is a variation of the gradient descent adapted for the weighted mean, and, second, a method that is generic and can be applied to any kind of optimization problem with a quadratic function and linear inequality constraints. The latter method can be applied to the case of the weighted mean.

Example 8.6 given above is an example of the optimization problem.

### Using gradient descent

The key element of this approach is to reformulate the problem so that no equality or inequality constraint is needed. So, we drop constraints $p_i \geq 0$ and $\sum_i p_i = 1$. This is achieved considering an unconstrained vector $\mathbf{\Lambda} = (\lambda_1 \ldots \lambda_N)$ from which the constrained weights $p_i$ are extracted as follows:

$$\mathbf{p_\Lambda} = \left( \frac{e^{\lambda_1}}{\sum_{j=1}^{N} e^{\lambda_j}} \quad \cdots \quad \frac{e^{\lambda_N}}{\sum_{j=1}^{N} e^{\lambda_j}} \right). \tag{8.7}$$

Note that, in this way, any vector $\mathbf{\Lambda} \in \mathbb{R}^N$ leads to a valid weighting vector. That is, for all $\mathbf{\Lambda}$, we have that all $\mathbf{p_\Lambda}$ constructed using Equation 8.7 satisfy $p_i \geq 0$ and $\sum_i p_i = 1$. Therefore, the problem of learning weights for the WM operator is equivalent to the following unconstrained minimization problem:

$$\text{Minimize } D_{WM}(\Lambda) = \sum_{j=1}^{M} \left(WM_{\mathbf{P_\Lambda}}(a_1^j, \ldots, a_N^j) - b^j\right)^2. \tag{8.8}$$

That is,

$$\text{Minimize } D_{WM}(\Lambda) = \sum_{j=1}^{M} \left(\sum_{i=1}^{N} \frac{e^{\lambda_i}}{\sum_{j=1}^{N} e^{\lambda_j}} a_i^j - b^j\right)^2. \tag{8.9}$$

To apply gradient descent to this problem, an iterative process is applied where one of the examples is considered at each step, and the parameters $\lambda_i$ are updated according to the error for the example. So, given the $j$th example

$$(a_1^j \ a_2^j \ \ldots \ a_N^j \ b^j),$$

we define the error $e^j$ as

$$e^j = \left(\sum_{i=1}^{N} \frac{e^{\lambda_i}}{\sum_{j=1}^{N} e^{\lambda_j}} a_i^j - b^j\right)^2.$$

Then, given the parameters $\lambda_i$ at time $t$ (such parameters are expressed by $\lambda_i(t)$), the new parameters $\lambda_i(t+1)$ are defined as follows:

$$\lambda_i(t+1) := \lambda_i(t) - \beta \left. \frac{\partial e^j}{\partial \lambda_i} \right|_{\lambda_i = \lambda_i(t)}$$

for $i = 1, \ldots, N$, where $\beta$ is a learning rate (a small value $0 \leq \beta \leq 1$). This expression is equivalent to

$$\lambda_i(t+1) := \lambda_i(t) - \beta \frac{e^{\lambda_i(t)}}{\sum_{j=1}^{N} e^{\lambda_j(t)}} (a_i^j - \hat{b}^j)(\hat{b}^j - b^j), \tag{8.10}$$

where $\hat{b}^j$ is the estimate of $b^j$ at time $t$. That is, $\hat{b}^j = WM_{\Lambda(t)}(a_1^j, \ldots, a_N^j)$. Algorithm 1 describes this algorithm.

**Using active set methods**

Another approach for solving the optimization problem is to use an algorithm for quadratic programming. We consider an algorithm based on active set methods. These methods exploit the fact that the computation of the solution of quadratic problems with linear equality constraints is simple. Active set methods are iterative methods in which, at each step, inequality constraints are partitioned into two groups, those that are to be treated as active (considered as equality constraints) and those that are treated as inactive (essentially

---

**Algorithm 1** Gradient descent

**Algorithm** GradientDescent $(A, b$: Examples) **returns** weighting vector **is**
**begin**
    int t:=0;
    define $\Lambda(t) := (1...1)$;
    **while** no convergence **do**
        Select example $j$
        $\hat{b}^j := WM_{\lambda(t)}(a_1^j, \ldots, a_N^j)$
        Compute $\lambda_i(t+1)$ for $i = 1, \ldots, N$ using Equation 8.10
        t:=t+1;
    **end while**
    Use Equation 8.7 to find weights **p** from $\Lambda$.
    return **p**;
**end**

---

ignored). Once the partition is known, the algorithm proceeds by moving on the surface defined by the working set of constraints (the set of active constraints) to an improved point. In this movement, some constraints are added to the working set and others are removed. Then, the algorithm computes a new movement on the surface. This process is repeated until the minimum is reached.

In our case, inequality constraints are the ones that restrict the weights to be positive. Due to this, when the constraint corresponding to the weight $p_i$ is active, the weight is forced to be zero. In contrast, when a constraint is not active, the value of the corresponding weight is not restricted. Thus, at a certain step, the working set is defined by the initial equality constraint (all weights add to 1) and the active ones.

Assuming that, at the $k$th step, $p^k$ is a nonoptimal solution found in the previous step, the movement on the surface described above consists of modifying $p^k$ so that a better approximation (more minimal than the previous one) is obtained. That is, at time $k$, a vector $d^k$ is computed that corresponds to the step to perform from the last solution $p^k$. Therefore, we have that the new solution $p^{k+1}$ is computed as $p^{k+1} = p^k + d^k$. As $p^{k-1}$ is a weighting vector (i.e., $\sum_{i=1}^N p_i^{k+1} = 1$), and as $p^k$ is also, the vector $d^k$ satisfies $\sum_{i=1}^N d_i^k = 0$.

Hence, the vector $d^k$ is obtained as the solution of the following optimization problem

$$\text{Minimize } (1/2)(d^k)'Qd^k + (g^k)'d^k$$
$$\text{Subject to } c_i'd^k = 0 \quad for \ all \ i \in W_k,$$

where $g^k = r + Qp^k$ and $c_i'$ is a vector with the coefficients of the $i$th equality constraint in the working set at the $k$th step $(W_k)$.

This problem can be solved with the following system of linear equations:

$$Qd^k + C\lambda = -g^k$$
$$c'd^k = 0,$$

where $\lambda$ are the Lagrange multipliers, and $C$ is the matrix formed with all the coefficients of the active constraints.

To complete the method, we need to specify the procedures for adding a constraint to the working set and for removing a constraint from it. Both are considered in the algorithm that describes the whole method. The algorithm requires an initial weighting vector $p^0$ that satisfies all the constraints (feasible solution). This is easy to find, as $p^0 = (1/N \ldots 1/N)$, $N$ being the number of weights, is a suitable weighting vector. The algorithm is given as Algorithm 2. The input data to this algorithm are the matrix $Q$, the vector $r$, and the initial $p^0$. The result is the optimal weighting vector and the corresponding $D_{WM}(p)$.

### One or multiple solutions for the weighted mean

Now, we study the optimization problem to establish for which conditions there is a single solution. In the case of multiple solutions, we can consider the optimization of another property, such as dispersion, for selecting one weighting vector from among those with the same optimal $D_{WM}(\mathbf{p})$.

Multiple solutions can be caused by redundant information. Note that redundant information might imply that the data in some information sources is deduced from the data in the other sources. In this case, the matrix $Q$ might be singular, and, in some cases, the singularity causes the system to have several solutions. We consider these issues in some detail below.

First, we should consider the case of independent sources. If they are independent, then there is a single solution of the minimization problem. This is established below.

**Proposition 8.7.** *Let $A$ be a matrix of examples of dimension $M \times N$ (number of examples $\times$ number of sources); then, if the columns of $A$ are linearly idependent, the minimization problem in Equation 8.3.1 has a single solution.*

So, there is a single vector $\mathbf{p}$ that minimizes the distance $D_{WM}(\mathbf{p})$.

In the case of dependencies, we have three different situations. The first one is as follows.

**Proposition 8.8.** *If there is a column $a_k$ in $A$ that is a linear combination of the other columns $a_i$ in such a way that*

$$a_k = \sum_{i \neq k} \alpha_i a_i,$$

*with $\sum_i \alpha_i = 1$ and $\alpha_i \geq 0$, then, if the vector $\mathbf{p}$ is an optimal solution of the minimization of $D_{WM}(\mathbf{p})$ with $p_k \neq 0$, there is at least one other optimal solution $\mathbf{p}^*$ such that $p_k^* = 0$. Naturally, as both are optimal solutions, $D(\mathbf{p}^*) = D(\mathbf{p})$.*

---

**Algorithm 2** Learning weights for the weighted mean

---

**Algorithm** LearningWeightsWeightedMean ($p$: weighting vector;
$Q$: Quadratic matrix; $r$: vector) **returns** (weighting vector, optimal distance) **is**
**begin**
    $k:=0$; $p^k := p$;
    $W_k :=$ the equality constraint corresponding to have $a = (1 \ldots 1)$;
    boolean exit := false;
    **while** not exit **do**
        boolean check-lagrange := true;
        $g^k := r + Qp^k$;
        Compute $d^k$ and $\lambda$ (lagrange multipliers) as a solution of:
            Minimize $(1/2)(d^k)'Qd^k + (g^k)'d^k$ subject to $c_i'd^k = 0$ for all $i \in W_k$
        **if** $d^k \neq 0$ **then**
            $\alpha^k := min\{1, min\{(b_i - c_i'p^k)/(c_i'd^k)|c_i'd^k > 0\}\}$;
            $p^{k+1} := p^k + \alpha^k d^k$;
            **if** $\alpha^k < 1$ **then**
                /* add restriction (the equality constraint) corresponding to $p^k\alpha^k = 0$
                */
                $W_{k+1} = W_k+$ the constraint of the index of $\alpha^k$;
                check-lagrange := false;
            **end if**
        **end if**
        **if** check-lagrange **then**
            $\lambda_q := min \{\lambda_i|i \in I \cap W_k\}$;
            **if** $\lambda_q \geq 0$ **then**
                exit:= true;
            **else**
                /* drop restriction (equality constraint) corresponding to the $q$th
                weight*/
                $W_{k+1} := W_k$ - the $q$th equality constraint
            **end if**
        **end if**
        k:=k+1;
    **end while**
    return $< p^k, D(p^k) >$;
**end**

---

So, we can eliminate the $k$th column, as this does not change the optimal value of $D_{WM}(\mathbf{p})$.

The second case of dependency is where, for any $a_k$ that can be written as $a_k = \sum_{i \neq k} \alpha_i a_i$, we have some $\alpha_i < 0$. In this case, the following result is known.

**Proposition 8.9.** *Let us consider the case where, for all $a_k$ that can be written as $a_k = \sum_{i \neq k} \alpha_i a_i$ with $\sum_{i \neq k} \alpha_i = 1$, there exists at least one $\alpha_j$ such that $\alpha_j < 0$. In this case, one of the weights will be zero.*

Nevertheless, it is not known which of the weights will be zero.

---

**Algorithm 3** Optimal solution for the weighted mean

---

**Algorithm** optimalSolutionWMn ($p$: weighting vector; $A$: Data;
$\qquad\qquad\qquad\qquad\qquad\qquad$ $b$: expectedResults) **returns** weighting vector **is**
**begin**
$\quad$ **if** $Q$ has not linear dependent columns **then**
$\quad\quad$ Compute $Q = A'A$ and $r = A'b$
$\quad\quad$ $< p, D(p) > = $ LearningWeightsWeightedMean $(p,Q,r)$;
$\quad\quad$ return p;
$\quad$ **else**
$\quad\quad$ **if** exists $a_k$ such that $a_k = \sum_{i \neq k} \alpha_i a_i$ with $\sum_{i \neq k} \alpha_i = 1$ and $\alpha_i \geq 0$ for all
$\quad\quad$ $i \neq k$ **then**
$\quad\quad\quad$ Remove column $a_k$ from matrix $A$
$\quad\quad\quad$ Compute $Q'$ and $r'$ accordingly
$\quad\quad\quad$ $< p, D(p) > = $ LearningWeightsWeightedMean $(p,Q',r')$;
$\quad\quad\quad$ build $p'$ from $p$ and with $p'_i = 0$
$\quad\quad\quad$ return p;
$\quad\quad$ **else**
$\quad\quad\quad$ **for** $i = 1$ in $1 \cdots N$ **do**
$\quad\quad\quad\quad$ Remove column $a_i$ from matrix $A$
$\quad\quad\quad\quad$ Compute $Q'$ and $r'$ accordingly
$\quad\quad\quad\quad$ $< p(i), D(p(i)) > = $ LearningWeightsWeightedMean $(p,Q',r')$;
$\quad\quad\quad\quad$ build $p'(i)$ from $p(i)$ and with $p'(i)_i = 0$
$\quad\quad\quad$ **end for**
$\quad\quad\quad$ return the weighting vector p(i) with minimal $D(p(i))$;
$\quad\quad$ **end if**
$\quad$ **end if**
**end**

---

The last case occurs when there is dependency, but $\sum a_i \neq 1$.

**Proposition 8.10.** *When, for all linearly dependent columns $a_k$ in $A$ (i.e., $a_k = \sum_{i \neq k} \alpha_i a_i$), we have $\sum \alpha_i \neq 1$, the optimization problem has a single solution.*

$\quad$ Algorithm 3 returns the best solution for any kind of problem, either with dependencies or without them. The inputs of the algorithm are the data to be used to determine the parameters, that is, the matrix $A$ and the vector $b$. Additionally, we should supply a first initial weighting vector $p^0$ to start the iteration process. The outcome of the algorithm is the optimal $p$.

$\quad$ In Section 8.3 we considered the case with multiple solutions, arguing that an additional property can be optimized. We will illustrate this requiring maximal dispersion. Let $\Delta$ be the error for the optimal solution; then, the minimization problem for the optimal weighting vector with respect to dispersion is as follows:

$$\text{Minimize } - E(\mathbf{p}) = \sum_{i=1}^{N} p_i \log p_i$$
$$\text{Subject to}$$
$$D_{WM}(\mathbf{p}) = \Delta \qquad (8.11)$$
$$\sum_{i=1}^{N} p_i = 1$$
$$p_i \geq 0$$

We now consider this problem restricted to the case of a single dependent source $x_k$ such that $a_k = \sum_{i \neq k} \alpha_i a_i$ and $\sum_i \alpha_i = 1$. The set of solutions is characterized in the following proposition.

**Proposition 8.11.** *Let $A$ be such that there is an $a_k = \sum_{i \neq k} \alpha_i a_i$ with $\sum_{i \neq k} \alpha_i = 1$, and its removal leads to a matrix $A'$ with independent columns. Then, all optimal solutions of the optimization problem for the weighted mean (Equation 8.3.1) are of the form*

$$p^* = p - \tau_k \alpha,$$

*with $\alpha$ being the $N$ dimensional vector defined from the values $\alpha_i$ and $\alpha_k = -1$. That is,*

$$\alpha = (\alpha_1, \ldots, \alpha_{k-1}, -1, \alpha_{k+1}, \ldots, \alpha_N)$$

*and*

$$\tau_k \in \left[ 0, \min \left( 1, \min_{\alpha_i < 0} (p_i - 1)/\alpha_i, \min_{\alpha_i > 0} p_i/\alpha_i \right) \right].$$

Naturally, the solution of the problem stated in Equation 8.11 is of this form. Therefore, as the entropy is a convex function, finding the optimal solution with maximum dispersion corresponds to finding the maximum of a one-variable convex function. The variable of this function is $\tau_k$.

### 8.3.2 OWA Operators

The algorithms studied in the previous section for the weighted mean can be applied without difficulty to the case of the OWA operator. This is due to the fact that, from a computational point of view, the only difference between the weighted mean and the OWA is the ordering step in the OWA.

Accordingly, the application of the previous algorithms to the OWA only requires us to reorder the data for each example in decreasing order. Then, the first weight will correspond to the largest value, the second weight to the second largest value, and so on.

For illustration, we reconsider the data in Example 8.6, and learn a model based on the OWA operator.

*Example 8.12.* Let us consider the data in Table 8.5, and assume that our subjective model is based on the OWA operator. This data is reproduced in Table 8.6. Then, first we order each record in decreasing order to obtain Table 8.7. Naturally, the subjective evaluation does not change. Next, we apply the algorithm based on active set methods described in Section 8.3.1, and get

| Student | $ML$ | $P$ | $M$ | $L$ | $G$ | Subjective evaluation |
|---------|------|-----|-----|-----|-----|----------------------|
| $s_1$ | 0.8 | 0.9 | 0.8 | 0.1 | 0.1 | 0.7 |
| $s_2$ | 0.7 | 0.6 | 0.9 | 0.2 | 0.3 | 0.6 |
| $s_3$ | 0.7 | 0.7 | 0.7 | 0.2 | 0.6 | 0.6 |
| $s_4$ | 0.6 | 0.9 | 0.9 | 0.4 | 0.4 | 0.8 |
| $s_5$ | 0.8 | 0.6 | 0.3 | 0.9 | 0.9 | 0.8 |
| $s_6$ | 0.2 | 0.4 | 0.2 | 0.8 | 0.1 | 0.3 |
| $s_7$ | 0.1 | 0.2 | 0.4 | 0.1 | 0.2 | 0.1 |
| $s_8$ | 0.3 | 0.3 | 0.3 | 0.8 | 0.3 | 0.4 |
| $s_9$ | 0.5 | 0.2 | 0.1 | 0.2 | 0.1 | 0.3 |
| $s_{10}$ | 0.8 | 0.2 | 0.2 | 0.5 | 0.1 | 0.5 |

**Table 8.6.** Marks given to ten students, and their subject evaluation

| Student | $a_{\sigma(1)}$ | $a_{\sigma(2)}$ | $a_{\sigma(3)}$ | $a_{\sigma(4)}$ | $a_{\sigma(5)}$ | Subjective evaluation |
|---------|------|------|------|------|------|----------------------|
| $s_1$ | 0.9 | 0.8 | 0.8 | 0.1 | 0.1 | 0.7 |
| $s_2$ | 0.9 | 0.7 | 0.6 | 0.3 | 0.2 | 0.6 |
| $s_3$ | 0.7 | 0.7 | 0.7 | 0.6 | 0.2 | 0.6 |
| $s_4$ | 0.9 | 0.9 | 0.6 | 0.4 | 0.4 | 0.8 |
| $s_5$ | 0.9 | 0.9 | 0.8 | 0.6 | 0.3 | 0.8 |
| $s_6$ | 0.8 | 0.4 | 0.2 | 0.2 | 0.1 | 0.3 |
| $s_7$ | 0.4 | 0.2 | 0.2 | 0.1 | 0.1 | 0.1 |
| $s_8$ | 0.8 | 0.3 | 0.3 | 0.3 | 0.3 | 0.4 |
| $s_9$ | 0.5 | 0.2 | 0.2 | 0.1 | 0.1 | 0.3 |
| $s_{10}$ | 0.8 | 0.5 | 0.2 | 0.2 | 0.1 | 0.5 |

**Table 8.7.** Marks given to ten students, reordered for learning OWA weights

the following weights: $w_1 = 0.1245$, $w_2 = 0.6385$, $w_3 = 0.0531$, $w_4 = 0.0000$, $w_5 = 0.1839$. Therefore, when using the OWA model, we have that the most relevant mark is the second one. This mark accounts for 63% of the final mark. The first and the last marks have a similar importance (0.1245 and 0.1839), and the fourth has no importance.

### 8.3.3 The WOWA Operator

In the case of the WOWA operator, the optimization problem is not quadratic. Therefore, the active set methods described in Section 8.3.1 cannot be applied. Nevertheless, the gradient descent method (described in Section 8.3.1) can be applied, as the WOWA operator uses two weighting vectors as parameters.

As the WOWA operator generalizes both weighted mean and OWA, we can bootstrap the gradient descent with the best solution obtained with either the OWA or the weighted mean. So, we apply a mixed approach combining the two methods described above. This is detailed below. We will use **p** to denote the WOWA weights used in the weighted mean, and $\mathbf{w}$ to denote the weights used in the OWA operator.

The method for the WOWA is as follows.

1. Solve the optimization problem for the weighted mean and determine $\mathbf{p}$. Let $D_{WM}(\mathbf{p})$ be the optimal distance achieved. Solve this problem using an optimal solver (such as the active set methods described in Section 8.3.1).
2. Solve the optimization problem for the OWA operator. Let $\mathbf{w}$ be its solution and let $D_{OWA}(\mathbf{w})$ be the optimal distance achieved.
3. Compute $D_{WOWA}$ for the pairs $(\mathbf{p}, (1/N, \ldots, 1/N))$, $((1/N, \ldots, 1/N), \mathbf{w})$ and $(\mathbf{p}, \mathbf{w})$. Select the pair with minimal $D_{WOWA}$.
4. Define $\Lambda^p$ from $\mathbf{p}$ and $\Lambda^w$ from $\mathbf{w}$ so that

$$\mathbf{p} = (e^{\lambda_1^p} / \sum_{j=1}^{N} e^{\lambda_j^p} \ldots e^{\lambda_N^p} / \sum_{j=1}^{N} e^{\lambda_j^p})$$

$$\mathbf{w} = (e^{\lambda_1^w} / \sum_{j=1}^{N} e^{\lambda_j^w} \ldots e^{\lambda_N^w} / \sum_{j=1}^{N} e^{\lambda_j^w})$$

Unless there is one $p_i$ or $w_i$ equal to zero, such vectors can be obtained defining $\lambda_i^p = \log p_i$ and $\lambda_i^w = \log w_i$. Zero weights can only be approximated with a large enough negative value for $\lambda$.
5. Apply gradient descent for the WOWA operator. This corresponds to the algorithm in Section 8.3.1, where $e^j$ has been defined in terms of the WOWA operator. The computation of

$$\lambda_i(t+1) := \lambda_i(t) - \beta \left. \frac{\partial e^j}{\partial \lambda_i} \right|_{\lambda_i = \lambda_i(t)}$$

uses a numerical approximation of the derivative. The gradient descent is applied until some convergence criterion is met.
6. Once the $\Lambda$ parameters are known, obtain the corresponding weights $\mathbf{p}$ and $\mathbf{w}$.

This learning approach is based on the definition of WOWA with two weighting vectors. Alternatively, we can consider the use of a parametric fuzzy quantifier. Similar approaches can be applied in this case.

### 8.3.4 Choquet Integral

Now, we consider the problem of parameter determination in the case of the Choquet integral. As we did with previous methods, we consider again the minimization of the least sum of squares.

Let $k$ be an integer; then, $\delta_N^k \delta_{N-1}^k \ldots \delta_1^k$ is the dyadic representation of $k$ when

$$k = 2^{N-1}\delta_N^k + 2^{N-2}\delta_{N-1}^k \ldots 2^0 \delta_1^k.$$

For example, 101 is the dyadic representation of 5 and 11111 is the dyadic representation of 31.

---

**Fig. 8.1.** Dyadic representation of an integer

We start by considering the most general case of a Choquet integral with an unconstrained fuzzy measure $\mu$. The problem in Equation 8.4 can be rewritten as

$$\text{Minimize } D_{\mathbb{C}}(\mu) = \sum_{j=1}^{M}(CI_\mu(a_1^j, \ldots, a_N^j) - b^j)^2$$
Subject to
$$\mu(\emptyset) = 0 \qquad\qquad (8.12)$$
$$\mu(X) = 1$$
$$\mu(A) \leq \mu(B) \text{ when } A \subseteq B$$

The problem can be solved as a quadratic problem with linear constraints. Therefore, the active set methods described in Section 8.3.1 are also suitable here. We give below the details on how to formulate the problem.

The basic idea is to determine the Möbius transform of the measure instead of the measure itself. Then, the optimization function should be rewritten using the Möbius transform, and appropriate constraints should be added to the problem.

We consider some notation. As usual, let $X = \{x_1, \ldots, x_N\}$ and $a_i^j = f^j(a_i)$. Now, instead of using $\mu(A)$ to denote the measures of subsets of $X$ we will use $\mu_k$, with $k \in \{0, \ldots, 2^N - 1\}$. To do so, we need a mapping between $\mu_k$ and the subsets $A \subseteq X$. This will be achieved using the dyadic representation of integers (see Figure 8.1). In particular, let $\delta_N^k \delta_{N-1}^k \ldots \delta_1^k$ be the dyadic representation of $k$; then, $\mu_k$ denotes the measure of the following set:

$$\mu_k = \mu(\{x_l \in X | \delta_l^k = 1 \ \ for \ \ l = 1, \ldots, N\}).$$

Additionally, we use $\mu_{k(A)}$ to denote the measure of the set $\mu(A)$.

*Example 8.13.* Let $X = \{x_1, x_2, x_3, x_4, x_5\}$; then, $\mu(\{x_1, x_3, x_4\})$ is represented by $\mu_{13}$, because the dyadic representation of 13 is 01101. In particular,

$$\delta_5^{13}\delta_4^{13}\delta_3^{13}\delta_2^{13}\delta_1^{13} = 01101.$$

Therefore, $\delta_4^{13} = 1$, $\delta_3^{13} = 1$ and $\delta_1^{13} = 1$, but $\delta_5^{13} = 0$ and $\delta_2^{13} = 0$. So, we have $\mu_{13} = \mu(\{x_4, x_3, x_1\})$. Similarly, $\mu_1 = \mu(\{x_1\})$, $\mu_2 = \mu(\{x_2\})$, $\mu_3 = \mu(\{x_1, x_2\})$, $\mu_4 = \mu(\{x_3\})$, $\mu_5 = \mu(\{x_1, x_3\})$, $\mu_6 = \mu(\{x_2, x_3\})$, $\mu_7 = \mu(\{x_1, x_2, x_3\})$, $\ldots$, $\mu_{2^N-1} = \mu(X)$.

Using this notation, we have that a fuzzy measure can be represented by the vector $(\mu_0, \mu_1, \ldots, \mu_{2^N-1})$. As $\mu_0$ should not be learned, because it is always zero, we only consider the vector $(\mu_1, \ldots, \mu_{2^N-1})$, denoted by $\mu^+$. In a similar way, we will consider the vector $\mathbf{m}^+ = (m_1, \ldots, m_{2^N-1})$ that corresponds to the Möbius transform of $\mu$. Then, for each example, we can rewrite its Choquet integral as the product of a vector $\mathbf{a}^+$ by a vector $\mathbf{m}^+$. Such a product is the result of rewritting the expression of the Choquet integral.

First, let us recall the definition of the Choquet integral (see Equation 6.1):

$$CI_\mu(a_1, \ldots, a_N) = \sum_{i=1}^{N} \left(a_{s(i)} - a_{s(i-1)}\right)\mu(A_{s(i)}), \qquad (8.13)$$

where $a_{s(i)} = f(x_{s(i)})$ is defined from the permutation $s$ so that $0 \leq f(x_{s(1)}) \leq \cdots \leq f(x_{s(N)}) \leq 1$, where $A_{s(i)} = \{x_{s(i)}, \ldots, x_{s(N)}\}$ and $f(x_{s(0)}) = 0$.

Using the Möbius transform, this expression can rewritten for each example $j$ as follows:

$$CI_\mu(a_1^j, \ldots, a_N^j) = \sum_{i=1}^{N}\left(\left(a_{s(i)}^j - a_{s(i-1)}^j\right)\left(\sum_{A \subset A_{s(i)}^j} m(A)\right)\right) =$$
$$= \sum_{i=1}^{N} \sum_{A \subset A_{s(i)}^j} \left(\left(a_{s(i)}^j - a_{s(i-1)}^j\right)m(A)\right),$$

where $a_{s(i)}^j$ denotes the $i$th lowest value in $(a_1 \ldots a_N)$. Note that the permutation $s(i)$ depends on the $j$th example. So, formally, we have a permutation $s^j$ for each example $j = 1, \ldots, M$. Similarly, the sets $A_{s(i)}$ also depend on the example. That is why we use $A_{s(i)}^j$ to denote them.

Now, using a proper ordering of the terms, the expression above can be rewritten as follows:

$$CI_\mu(a_1^j, \ldots, a_N^j) = \sum_{k=1}^{2^N-1} a_i^{+,j} m_k = \mathbf{a}^{+,j}\mathbf{m}^+.$$

So, the minimization problem established in Equation 8.12 can be rewritten as

Minimize $D_\mathbb{C}(P) = \sum_{j=1}^{M}(\mathbf{a}^{+,j}\mathbf{m}^+ - b^j)^2$
Subject to

$$\sum_{k=1}^{2^N-1} m_k = 1$$
$$\sum_{B' \subset B} m_{k(B')} - \sum_{A' \subset A} m_{k(A')} \geq 0 \text{ for all } A \subset B$$

$$(8.14)$$

Note that the condition $m(\emptyset) = 0$ is not needed, as $m_0$ is not included in the model, and that the constraints $\mu(X) = 1$ and $\mu(A) \leq \mu(B)$ when $A \subseteq B$ are replaced by the appropriate constraints on $m$.

This optimization problem has the same structure as the one of the weighted mean. It is a quadratic optimization problem with linear constraints. Therefore, it can be solved using the same methods used for the weighted mean. We illustrate below its application with one example.

*Example 8.14.* Let us reconsider Example 8.6; but, in this case, we consider a Choquet integral as the appropriate aggregation operator. Then, using the model in Equation 8.14, we have that the fuzzy measure learned, as well as its corresponding Möbius transform, is given in Table 8.8. The Möbius transform permits us to distinguish the most relevant sets; that is, the sets with no null Möbius value. They are the following ones

- $m(\{L\}) = 0.106946016$
- $m(\{L, G\}) = 0.058218250823$
- $m(\{P\}) = 0.40697540673$
- $m(\{ML\}) = 0.4278602983$

It can be observed that these most revelant subjects are related with the subjects distinguished in Example 8.6, where the weighted mean was used for aggregation. In that case we had $p_{ML} = 0.4244$, $p_P = 0.4108$, $p_M = 0.0000$, $p_L = 0.1249$, and $p_G = 0.0399$. Thus, again, $ML$ and $P$ have larger weights, followed by $L$. While $G$ had a non-null weight in the case of the weighted mean, we observe that in the case of the Choquet integral $G$ is only relevant when used in conjunction with $L$.

### Learning constrained fuzzy measures

The model based on the Choquet integral can be easily extended for learning two types of fuzzy measures: $k$-order additive fuzzy measures and belief functions. We detail how the optimization problem above should be modified so that the fuzzy measure learned is of this type.

$k$-order additive fuzzy measures: A fuzzy measure is $k$-order additive when the Möbius transform is zero for all subsets of $X$ with cardinality larger than $k$. In order to obtain a fuzzy measure that is of this class for a given $k$, the optimization problem should constrain all sets $A$ where $|A| > k$ to have $m(A) = 0$. This can be done by adding such constraints to the model, or, alternatively, removing terms $m_{k(A)}$ with $|A| > k$ from the set of variables in the optimization problem.

Belief functions: A fuzzy measure is a belief function when the Möbius transform is always positive. Therefore, in order to obtain such fuzzy measures from the model, we need to add $m_k \geq 0$, for all $k = 1, \ldots, 2^N - 1$ to the optimization problem.

In addition to these two types of fuzzy measures, two other types are worth mentioning: additive and symmetric. Note that the Choquet integral with respect to these measures are, respectively, a weighted mean and an OWA operator. Therefore, it is simpler to solve the optimization problem established in Sections 8.3.1 and 8.3.2. Optimal solutions are also obtained in this case. If the general algorithm described here is used, then the constraints for additive fuzzy measures are, of course, $m(A) = 0$ for all $|A| > 1$. In the

| $X = \{ML, P, M, L, G\}$ | $\mu_{\mathbf{p,w}}$ | Möbius transform |
|---|---|---|
| { 0 0 0 0 0 } | 0.0 | 0.0 |
| { 0 0 0 0 1 } | 3.4022290872E-10 | 3.402229E-10 |
| { 0 0 0 1 0 } | 0.106946016 | 0.106946016 |
| { 0 0 0 1 1 } | 0.1651642671632229 | 0.058218250823 |
| { 0 0 1 0 0 } | 8.3601024454E-9 | 8.36010244E-9 |
| { 0 0 1 0 1 } | 8.83795255408E-9 | 1.3762720999999943E-10 |
| { 0 0 1 1 0 } | 0.1069460245331454 | 1.7304299659848255E-10 |
| { 0 0 1 1 1 } | 0.1651642760308463 | 1.9685075791642248E-10 |
| { 0 1 0 0 0 } | 0.40697540673 | 0.40697540673 |
| { 0 1 0 0 1 } | 0.4069754072782796 | 2.0805673850432527E-10 |
| { 0 1 0 1 0 } | 0.5139214228473248 | 1.1732481652870774E-10 |
| { 0 1 0 1 1 } | 0.5721396745940198 | 3.754153654611514E-10 |
| { 0 1 1 0 0 } | 0.4069754253555859 | 1.0265483518789864E-8 |
| { 0 1 1 0 1 } | 0.4069754261791199 | 1.3762718742427182E-10 |
| { 0 1 1 1 0 } | 0.5139214418189969 | 1.7304313537636062E-10 |
| { 0 1 1 1 1 } | 0.5721396942346477 | 1.9685064689412002E-10 |
| { 1 0 0 0 0 } | 0.4278602983 | 0.4278602983 |
| { 1 0 0 0 1 } | 0.4278602992562150 | 6.159921461801332E-10 |
| { 1 0 0 1 0 } | 0.5348063178062644 | 3.5062643899408386E-9 |
| { 1 0 0 1 1 } | 0.5930245705338666 | 9.48387146593177E-10 |
| { 1 0 1 0 0 } | 0.4278603068168202 | 1.5671774988845755E-10 |
| { 1 0 1 0 1 } | 0.4278603080796658 | 1.6900336685665707E-10 |
| { 1 0 1 1 0 } | 0.5348063266691706 | 1.7304302435405816E-10 |
| { 1 0 1 1 1 } | 0.5930245800971049 | 1.9685075791642248E-10 |
| { 1 1 0 0 0 } | 0.8348357051643919 | 1.3439194201936289E-10 |
| { 1 1 0 0 1 } | 0.8348357066267909 | 2.981271896018711E-10 |
| { 1 1 0 1 0 } | 0.9417817250059003 | 2.179191271878267E-10 |
| { 1 1 0 1 1 } | 0.9999999789905173 | 3.754155875057563E-10 |
| { 1 1 1 0 0 } | 0.8348357240771254 | 1.3042977808908063E-10 |
| { 1 1 1 0 1 } | 0.8348357261527856 | 1.6900347787895953E-10 |
| { 1 1 1 1 0 } | 0.9417817446108061 | 1.7304324639866309E-10 |
| { 1 1 1 1 1 } | 0.9999999999960869 | 1.9685031382721263E-10 |

**Table 8.8.** Fuzzy measure $\mu$ and its Möbius transform. The first column denotes the subsets of $X = \{x_1, \ldots, x_5\}$ (a 0 in the $i$th column means that $x_i$ is not included, while a 1 in the $i$th column means that $x_i$ is included)

case of symmetric measures, we will add the constraint $m(A) = m(B)$ for all subsets of $X$ where $|A| = |B|$.

In the case that the fuzzy measure is a distorted probability, the problem cannot be solved with a similar approach. Nevertheless the method described in Section 8.3.3 is appropriate as the WOWA operator is equivalent to a Choquet integral with a distorted probability.

| Student | ML  P   M   L   G | Subjective evaluation |
|---------|-------------------|-----------------------|
| $s_1$ | 0.8  0.9 0.8 0.1 0.1 | 3rd |
| $s_2$ | 0.7  0.6 0.9 0.2 0.3 | 4th |
| $s_3$ | 0.7  0.7 0.7 0.2 0.6 | 4th |
| $s_4$ | 0.6  0.9 0.9 0.4 0.4 | 1st |
| $s_5$ | 0.8  0.6 0.3 0.9 0.9 | 1st |
| $s_6$ | 0.2  0.4 0.2 0.8 0.1 | 8th |
| $s_7$ | 0.1  0.2 0.4 0.1 0.2 | 10th |
| $s_8$ | 0.3  0.3 0.3 0.8 0.3 | 7th |
| $s_9$ | 0.5  0.2 0.1 0.2 0.1 | 8th |
| $s_{10}$ | 0.8  0.2 0.2 0.5 0.1 | 6th |

**Table 8.9.** Marks given to ten students, and their subject evaluation using preferences

## 8.4 Extracting Parameters from Examples: Preferences or Partial Orders

Another case where learning approaches can be considered is when there are examples (e.g., alternatives) and a (partial) order is defined over them according to our preferences. The following example illustrates this situation.

*Example 8.15.* Let us consider again the students in Example 8.6, and let us assume that our subjective overall rate is replaced by our preference on the students. Table 8.9 includes the marks of the students and our subjective preference (right column).

The subjective preference now gives an ordering of the students. The order is partial, as there are groups of students that are indistinguishable. This is the case of the two 1st students in the class ($s_4$ and $s_5$).

To formulate this problem we consider the examples and a (partial) order relation $<$. The set $S$ that represents this relation is defined by the pairs $(r, t)$ such that $s_r > s_t$ (student $s_r$ is preferred to student $s_t$). Then, given a model defined by an aggregation operator $\mathbb{C}$ with parameter $P$, the goal is to find $P$ such that, for all $(r, t) \in S$, it follows that

$$\mathbb{C}_P(\text{evaluation-student } r) > \mathbb{C}_P(\text{evaluation-student } t)$$

or, following the notation in Section 8.3, where $f^r$ denotes example $r$,

$$\mathbb{C}_P(f^r(x_1), \ldots, f^r(x_N)) > \mathbb{C}_P(f^t(x_1), \ldots, f^t(x_N)).$$

Naturally, this equation can be rewritten as

$$\mathbb{C}_P(f^r(x_1), \ldots, f^r(x_N)) - \mathbb{C}_P(f^t(x_1), \ldots, f^t(x_N)) > 0.$$

Although we want this equation to hold for all $(r, t) \in S$, data might be inconsistent; so, for each $(r, t)$, we consider a variable $y_{(r,t)} \geq 0$ that we expect to be as small as possible so that the following equation holds:

$$\mathbb{C}_P(f^r(x_1), \ldots, f^r(x_N)) - \mathbb{C}_P(f^t(x_1), \ldots, f^t(x_N)) + y_{(r,t)} > 0.$$

Considering all $(r,t) \in S$, we are interested in the parameter $P$ that minimizes the number of violations:

$$\sum_{(r,t) \in S} y_{(r,t)}.$$

Therefore, the problem to minimize is as follows:

Minimize $\sum_{(r,t) \in S} y_{(r,t)}$
Subject to
$$\mathbb{C}_P(f^r(x_1), \ldots, f^r(x_N)) - \mathbb{C}_P(f^t(x_1), \ldots, f^t(x_N)) + y_{(r,t)} > 0$$
$$y_{(r,t)} \geq 0$$
logical constraints on $P$

$$(8.15)$$

In the particular case where the model uses the weighted mean, this problem is formulated as

Minimize $\sum_{(r,t) \in S} y_{(r,t)}$
Subject to
$$\sum_{i=1}^{N} p_i(f^r(x_i) - f^t(x_i)) + y_{(r,t)} > 0$$
$$y_{(r,t)} \geq 0 \qquad (8.16)$$
$$\sum_{i=1}^{N} p_i = 1$$
$$p_i \geq 0$$

Other models have been proposed with similar objectives. One of them assumes that the difference between two preferred examples should be larger than a certain threshold (defined as constant), and then maximizes the minimum difference between the two alternatives.

## 8.5 Analysis

We focus the analysis to the case of parameter determination in the case of using examples.

The main two approaches considered (gradient descent and quadratic programming) have advantages and disadvantages. In general, quadratic programming, when applicable, is faster, and enables us to obtain a global optimum in reasonable time. For this kind of problems the gradient descent is not adequate, because of its slow convergence and because the initial weighting vector influences the final result. Besides, there is no one-to-one correspondence between $\Lambda$ vectors and weighting vectors. Therefore, it is more efficient to use this approach for the weighted mean and the OWA operator. The same is true for the Choquet integral for unconstrained fuzzy measures, as well as for $k$-order additive fuzzy measures or belief functions.

Nevertheless, in other types of problems, when the function to minimize is not quadratic (as is the case with the WOWA operator), active set methods are complex and difficult to implement. Other approaches are more suitable. The gradient descent and genetic algorithms are two of such approaches.

An additional problem to be considered when learning parameters from examples is that of missing data. When an example has missing data, we can either drop the example or adapt the optimization function so that it can be applied to the remaining data. Some operators can be easily adapted for this purpose. For example, the OWA operators with quantifiers can be applied to an arbitrary number of parameters. Nevertheless, such transformations cause the problems to be more complex, and, no longer quadratic.

## 8.6 Bibliographical Notes

1. **Analytic Hierarchy Process:** The Analytic Hierarchy Process was developed by Saaty between 1971 and 1975. The classical reference is [344]. A short paper describing the main insights of the approach is [343]. [390] is a Web page that allows us to compute the AHP for a single matrix. The consistency ratio described here is not the only one possible, and other definitions exist in the literature. See [15] for a recent analysis. The paper includes some historical notes on this issue.

2. **Parameters from orness:**   The selection of parameters from orness or degree of disjunction was first proposed by Dujmović in [102] with respect to root-mean-powers for two inputs. [103] considered additional inputs. The method to determine weights for the OWA as a maximization of dispersion given a particular orness value was introduced by O'Hagan in [301]. This corresponds to the ME-OWA operator (the first use of this term seems to be in [303]). O'Hagan solve the optimization problem using geometric programming formulation (see [301] and [302]). In [302], he gives the weighting vectors for all solutions of dimensions 3, 4, 5, and 6 when the orness is one of $\{0.95, 0.9, 0.85, 0.8, 0.75, 0.7, 0.65, 0.6, 0.55\}$. For orness equal to 0.5, the optimal solution is $p_i = 1/N$. For orness below 0.5, Proposition 7.24 can be used. Proposition 8.5, concerning OWA weights, was proved by Carbonell, Mas, and Mayor in [60] (1997). This case was independently studied later by Fullér and Majlender in [156]. The same authors considered the problem of learning weights for OWA given orness, assuring minimal variability (Definition 7.14) for the weights. See [157] for details. [438] considered the problem of learning weights for OWA with constraints on the weights.

3. **Optimization methods:** Algorithms for solving optimization problems can be found in [162, 240, 300]. Our first implementation (in Java) for

quadratic problems with linear constraints using active set methods followed [240]. There exist (free and commercial) software for solving this problem. We are currently using BPMPD by Mészáros [267]. The Kappalab package [171] ( *"laboratory for capacities"*) developed for the language and environment for statistical computing R [327] implements some learning methods, as well as some aggregation operators and indices.

4. **Parameters from preferences or partial orders:** Srinivasan and Schoker (1973) [372, 373] were the first to propose linear programming for learning the weights of a linear model. Weights are found from a set of examples and preferences, as in Example 8.15. They show that the quality of the model (*goodness of fit*) is not influenced by whether the linear model is a weighted mean with weights $\mathbf{p}$ (that is, $\sum_i p_i = 1$) or a linear model with $\mathbf{p}$ such that $\sum_i p_i = K \neq 1$. Nevertheless, they mainly concentrate on a third approach, where *goodness of fit* (G) and *poorness of fit* (B) have a difference of $h$ (that is, $G - B = h$). Their model is known as LINMAP (LINear programming techniques for Multidimensional Analysis of Preferences). Pekelman and Sen (1974) [317] consider in some detail the same problem with a weighted mean. They define objective functions to minimize the amount of violation (as in the previous works by Srinivasan and Schoker), and to minimize the number of violations.

While in the previous works a single set of preferences $S$ is considered, Horsky and Rao (1984) [192] considered the case of different sets of preferences. This is modeled, considering a partition of $S$ into $\{s_1, \ldots, s_s\}$, where $(r, t) \in S_3$ and $(r', t') \in S_1$ means that the difference between $r$ and $t$ is much larger than the one between $r'$ and $t'$. Formally,

$$[\mathbb{C}(f^r) - \mathbb{C}(f^t)] > [\mathbb{C}(f^{r'}) - \mathbb{C}(f^{t'})]$$

for $(r, t) \in S_a$, and $(r', t') \in S_b$ if $a > b$

[432] studies the case of a single set of preferences $S$ when there is a group of experts, each with his or her own opinions. Another related problem was studied in [79], where weights are learned in a two-class classification problem. Although this approach to learning parameters from preferences has been mainly restricted to the weighted mean, Meyer and Roubens considered the case for the Choquet integral [268].

5. **Parameters from outcomes.**

   a) **Weighted mean and OWA operators:** The determination of parameters for the OWA operators were initially studied by Filev and Yager using the gradient descent with the transformation $e^{\lambda_i} / \sum_j e^{\lambda_j}$ in [140, 140]. This problem was latter studied by Torra using active set methods [400] for both OWA and the weighted mean. Results on linear dependencies on the data can be found in [406]. This paper also considers learning parameters for quasi-arithmetic means.

   b) **WOWA operators:** An algorithm for learning the parameters of the WOWA operator is given in [401] and [408]. This problem is equivalent

to the one for solving a Choquet integral with a distorted probability.
The consideration of genetic algorithms for learning the parameters
of the WOWA was considered in [298].

c) **The Choquet integral:** The earliest approaches for using learning parameters for the Choquet integral are due to Mori and Murofushi [273] and Tanaka and Murofushi [391]. [273] was the first to
consider quadratic programming for learning fuzzy measures. [199]
and [200] present a method based on genetic algorithms. Mathematical results for this optimization problem are given in [195, 196]. A
related problem was considered in [428], a regression model based on
a Choquet integral, where the fuzzy measure as well as a few additional parameters are learned using genetic algorithms.

As stated above, the case of learning WOWA weights (studied
in [298, 401, 408]) corresponds to determining distorted probabilities.
In this case, both the weights and the functions should be learned.
[298, 401, 408] learn the function through the weighting vector $\mathbf{w}$,
later interpolating the function from $\mathbf{w}$. Identification of Sugeno $\lambda$-
measures with genetic algorithms was studied in [72, 222].

Learning $k$-order additive fuzzy measures has been studied for the
particular case of $k = 2$ by Marichal and Roubens [253]. They consider
additional constraints on the interaction between information sources;
that is, whether interaction (Möbius transform) is positive or negative.

d) **The Sugeno integral:** Learning models for the Sugeno integral have
not been studied much. [198] seems to be the first publication to use
gradient descent for learning the measure in a Sugeno integral. [423]
uses genetic algorithms for learning fuzzy measures (the $k$-maxitive)
for a Sugeno integral.

e) **Other operators:** Learning algorithms have been applied to several
other families of aggregation operators. For example, [38] considers
uninorms, weighted quasi-arithmetic means, and weighted root-mean-
powers; [39] studies methods for the Generalized OWA, the Generalized Choquet integral, and the Geometric OWA; [197] presents an
algorithm for the twofold integral.

6. **Unsupervised learning methods:** Research on fuzzy measures and
integrals mainly uses measures either defined heuristically or learned using supervised approaches. Nevertheless, there is also a trend toward using unsupervised learning approaches. Soria-Frisch [370, 371] has developed one such approach, applied to the field of computer vision. Kojadinovic [212, 213] has also considered this type of learning method.

7. **Reinforcement learning methods:** The method proposed by Keller
and Osborn in [208] to learn the fuzzy measure for a Sugeno integral can
be classified as a reinforcement learning approach. The method uses a
reward and punishment scheme based on the performance of the model in
a classification task.

8. **Other approaches:** Besides the heuristic approach for parameter determination, in the case of fuzzy measures for fuzzy integrals, Proposition 5.40, which permits us to determine a Sugeno $\lambda$-measure from the measures on the singletons, has also been used. This is the case in the paper by Cho and Kim [77], which uses this approach with a Sugeno integral. The proposition can be applied to with other methods. This is the case in [429], where genetic algorithms are used to determine the measures on the singletons (for a Sugeno $\lambda$-measure) in a pattern recognition application. The aggregation model in that paper uses a Sugeno integral.

# A

## Properties

We list below the main properties used in this book. The properties are listed in alphabetical order. The list is not exhaustive.

Associativity:
$$(x \circ y) \circ z = x \circ (y \circ z)$$

Internality:
$$\min_i a_i \leq \mathbb{C}(a_1, \ldots, a_N) \leq \max_i a_i$$

Neutral element $e$:
$$x \circ e = x$$

Positive Homogeneity:
$$\mathbb{C}(ra_1, \ldots, ra_N) = r\mathbb{C}(a_1, \ldots, a_N)$$

for $r > 0$

Reciprocity:
$$\mathbb{C}(1/a_1, \ldots, 1/a_N) = 1/\mathbb{C}(a_1, \ldots, a_N)$$

Sensitive: For all $k = 1, \ldots, N$, $a_k \neq a'_k$,
$$\mathbb{C}(a_1, \ldots, a_{k-1}, a_k, a_{k+1}, \ldots, a_N) \neq \mathbb{C}(a_1, \ldots, a_{k-1}, a'_k, a_{k+1}, \ldots, a_N)$$

Separable: $\mathbb{C}(a_1, \ldots, a_N)$ is separable if there exist functions $g_1, \ldots, g_N$ and $\circ$ (continuous, associative, and cancellative) such that
$$\mathbb{C}(a_1, \ldots, a_N) = g_1(a_1) \circ g_2(a_2) \circ \cdots \circ g_N(a_N)$$

Symmetry: For any permutation $\pi$ of $\{1, \ldots, N\}$,
$$\mathbb{C}(a_1, \ldots, a_N) = \mathbb{C}(a_{\pi}(1), \ldots, a_{\pi}(N))$$

Unanimity:
$$\mathbb{C}(a, \ldots, a) = a$$

This property is also known as identity, reflexivity, or agreement.

Comparable: $B_1$ and $B_2$ are said to be comparable when either $B_1 \preceq B_2$ or $B_2 \preceq B_1$. $B_1 \preceq B_2$ if, for all $x, y \in [0, 1]$, we have $B_1(x, y) \leq B_2(x, y)$ (a similar definition applies for operators on $[0, 1]^N$).

# B

## Some Aggregation Operators

For reference, we list below some of the aggregation operators studied in this book. The operators are listed in alphabetical order. The definitions use the following notation: $X$ is a set of reference or information sources, $f(x_i)$ is the value supplied by $x_i$ (with $a_i = f(x_i)$), $\mu$ is a fuzzy measure, $\mathbf{p}$ and $\mathbf{w}$ are (probabilistic) weighting vectors, $\mathbf{u}$ is a possibilistic weighting vectors, $\sigma$ is a permutation such that $a_{\sigma(i)} \geq a_{\sigma(i+1)}$, $s$ is a permutation such that $a_{s(i)} \leq a_{s(i+1)}$, $A_{\sigma(k)} = \{x_{\sigma(j)} | j \leq k\}$, and $A_{s(k)} = \{x_{\sigma(j)} | j \geq k\}$

Arithmetic mean (AM):

$$\sum_{i=1}^{N} a_i / N$$

Choquet integral (CI):

$$\sum_{i=1}^{N} [f(x_{s(i)}) - f(x_{s(i-1)})] \mu(A_{s(i)})$$

Geometric mean (GM):

$$\sqrt[N]{\prod_{i=1}^{N} a_i}$$

Harmonic mean (HM):

$$\frac{N}{\sum_{i=1}^{N} \frac{1}{x_i}}$$

OWA:

$$\sum_{i=1}^{N} w_i a_{\sigma(i)}$$

Root-mean-power:

$$\sqrt[\alpha]{\frac{\sum_{i=1}^{N} x_i^{\alpha}}{N}}$$

Sugeno integral (SI):

$$\max_{i=1,N} \min(f(x_{s(i)}), \mu(A_{s(i)}))$$

Twofold integral (TI):

$$TI_{\mu_S,\mu_C}(f) = \sum_{i=1}^{n} \left( \left( \bigvee_{j=1}^{i} f(x_{s(j)}) \wedge \mu_S(A_{s(j)}) \right) \left( \mu_C(A_{s(i)}) - \mu_C(A_{s(i+1)}) \right) \right)$$

Weighted maximum (WMax):

$$\max_i \min(u_i, a_i)$$

Weighted mean (WM):

$$\sum_{i=1}^{N} p_i a_i$$

Weighted minimum (WMin):

$$\min_i \max(neg(u_i), a_i)$$

WOWA:

$$\sum_{i=1}^{N} \omega_i a_{\sigma(i)}$$

with $\omega_i = w^*(\sum_{j \leq i} p_{\sigma(j)}) - w^*(\sum_{j < i} p_{\sigma(j)})$
$w^*$ nondecreasing function from $\{(i/N, \sum_{j \leq i} w_j)\}_{i=1,...,N} \cup \{(0,0)\}$

# References

1. Abidi, M. A., Gonzalez, R. C. (ed.) (1992) Data Fusion in Robotics and Machine Intelligence, Academic Press.
2. Aczél, J. (1948) Sur les opérations définies pour nombres réels, Bulletin de la Société Mathématique de France 76 59-64.
3. Aczél, J. (1961) Vorlesungen über Funktionalgleichungen und ihre Anwendungen, Birkhäuser Verlag.
4. Aczél, J. (1966) Lectures on Functional Equations and their Applications, Academic Press.
5. Aczél, J. (1984) On weighted synthesis of judgements, Aequationes Mathematicae 27 288-307.
6. Aczél, J. (1987) A Short Course on Functional Equations, D. Reidel Publishing Company (Kluwer Academic Publishers Group).
7. Aczél, J., Alsina, C. (1984) Classification of some classes of quasilinear functions with applications to triangular norms and to synthesizing judgements, Meth. Opl. Res. 48 3-22.
8. Aczél, J., Alsina, C. (1986) On synthesis of judgements, Socio-Econ. Plann. Sci 20:6 333-339.
9. Aczél, J., Alsina, C. (1987) Synthesizing judgements: A functional equations approach, Mathematical Modelling 9:3/5 311-320.
10. Aczél, J., Daróczy, Z. (1975) On Measures of Information and their Characterizations, Academic Press.
11. Aczél, J., Dhombres, J. (1989) Functional Equations in Several Variables: with Applications to Mathematics, Information Theory and to the Natural and Social Sciences, Cambridge University Press.
12. Aczél, J., Roberts, F. S. (1989) On the possible merging functions, Math. Social Sci 17 205-243.
13. Aczél, J., Saaty, T. L. (1983) Procedures for synthesizing ratio judgements, Journal of Mathematical Psychology 27 93-102.
14. Agustí, J., Esteve, F., Garcia, P., Godo, L., Sierra., C. (1991) Combining multiple-valued logics in modular expert systems, Proc. 7th Conference on Uncertainty in AI, Los Angeles, CA.
15. Alonso, J. A., Lamata, M. T. (2006) Consistency in the analytic hierarchy process: A new approach, Int. J. of Unc., Fuzz. and Knowledge Based Systems 14:4 445-459.

16. Alsina, C., Feliu, G., Marquet, L. (1996) Diccionari de mesures catalanes, Manuals Curial.
17. Alsina, C., Frank, M. J., Schweizer, B. (2006) Associative Functions: Triangular Norms and Copulas, World Scientific.
18. Alsina, C., Trillas, E., Valverde, L. (1983) On some logical connectives for fuzzy sets theory, Journal of Mathematical Analysis and Applications 93 15-26.
19. Andjiga, N.-G., Chantreuil, F., Lepelley, D. (2003) La mesure du pouvoir de vote, Working Paper INRA - Unité ESR Rennes 03-01.
20. Anonymous (1821) Dissertation sur la recherche du milieu le plus probable, entre les résultats de plusieurs observations ou expériences, Annales de Mathématiques Pures et Appliquées, 12 181-204. According to Harter [123] (see p. 157), the author of this paper is Svanberg.
21. Arnold, B. C., Balakrishnan, N., Nagaraja, H. N. (1992) A First Course in Order Statistics, John Wiley and Sons.
22. Arnould, T., Ralescu, A. (1992) From "and" to "or," Proc. of the IPMU 1992 Conference, 323-328.
23. Arrow, K. J. (1951) Social Choice and Individual Values, Wiley (2nd edition 1963).
24. Arrow, K. J., Sen, A. K., Suzumura, K. (eds.) (2002) Handbook of Social Choice and Welfare, Elsevier.
25. Ash, R. B. (1965) Information Theory, Wiley.
26. Ash, R. B. (1972) Real Analysis and Probability, Academic Press.
27. Aumann, R. J., Shapley, L. S. (1974) Values of Non-Atomic Games, Princeton University Press.
28. Bajraktarević, M. (1958) Sur une équation fonctionnelle aux valeurs moyennes, Glasnik. Mat. Fiz. I Astr 13:4 243-248.
29. Bajraktarević, M. (1963) Sur une généralisation des moyennes quasilineaire, Publ. Inst. Math. Beograd. 3:17 69-76.
30. Bajraktarević, M. (1969) Über die Vergleichbarkeit der mit Gewichtsfunktionen gebildeten Mittelwerte, Stud. Sci. Math. Hungar. 4 3-8.
31. Banzhaf, J. F., III (1965) Weighted voting does not work: a mathematical analysis, Rutgers Law Review 19 317-343.
32. Banzhaf, J. F., III (1968) One man, ? votes: A mathematical analysis of political consequences and judicial choice, George Washington Law Review 36 808-823.
33. Barnard, G. A. (1949) Statistical inference, Journal of the Royal Statistical Society, Series B XI:2 115-149.
34. Barnett, V., Lewis, T. (1984) Outliers in Statistical Data, 2nd Edition, Wiley.
35. Barral Souto, J. (1938) El modo y otras medias, casos particulares de una misma expression matemática, Boletín Matemático 11:3 29-41.
36. Barthelemy, J.-P., McMorris, F. R. (1986) The median procedure for n-trees, Journal of Classification 3 329-334.
37. Beliakov, G. (2001) Shape preserving splines in constructing WOWA operators, Fuzzy Sets and Systems 121:3 549-550.
38. Beliakov, G. (2003) How to build aggregation operators from data, Int. J. of Intel. Syst. 18 903-923.
39. Beliakov, G. (2005) Learning weights in the generalized OWA operators, Fuzzy Optimization and Decision Making 4 119-130.

40. Bemporad, G. (1926) Sul principio della media aritmetica, Atti Accad. Naz. Lincei 6:3 87-91.
41. Benvenuti, P., Mesiar, R., Vivona, D. (2002) Monotone set functions-based integrals, in E. Pap (ed.) Handbook of Measure Theory, North-Holland, 1329-1379.
42. Billingsley, P. (1995) Probability and Measure, Wiley, 3rd Edition (1st Edition, 1979; 2nd Edition, 1986).
43. Bouyssou, D., Marchant, T., Pirlot, M., Perny, P., Tsoukiàs, A., Vincke, P. (2000) Evaluation and Decision Models: A Critical Perspective, Kluwer's International Series, Kluwer Academic Publishers.
44. Brams, S. (1975) Game Theory and Politics, Free Press.
45. Brams, S. J., Affuso, P. J. (1976) Power and size: a new paradox, Theory and Decision 7 29-56.
46. Brecht, T., Arjomandi, E., Li, C., Pham, H. (2001) Controlling garbage collection and heap growth to reduce the execution time of Java applications, Proc. of the ACM Conference on Object-Oriented Programming, Systems, Languages, and Applications (OOPSLA), 353-366.
47. Brooks, R. R., Iyengar, S. S. (1998) Multi-Sensor Fusion: Fundamentals and Applications with Software, Prentice Hall, New-Jersey.
48. Buchanan, B., Shortliffe, E. (1984) Rule-based expert systems, The MYCIN experiments of the Stanford heuristic programming project, Addison-Wesley, Reading, MA.
49. Bull, J. M., Smith, L. A., Ball, C., Pottage, L., Freeman, R. (2003) Benchmarking Java against C and Fortran for scientific applications, Concurrency and Computation: Practice and Experience 15 417-430.
50. Bullen, P. S., Mitrinović, D. S., Vasić, P. M. (1988) Means and their Inequalities, D. Reidel Publishing Company.
51. Bullen, P. S. (2003) Handbook of Means and Their Inequalities, Kluwer Academic Publishers.
52. Bustince, H., Herrera, F., Montero, J. (eds.) (2007) Fuzzy Sets and their Extensions: Representation, Aggregation and Models, Springer-Verlag, forthcoming.
53. Butnariu, D., Klement, E. P. (1993) Triangular Norm-Based Measures and Games with Fuzzy Coalitions, Kluwer Academic Publishers.
54. Calders, P. (1984) De teves a meves, Laia.
55. Calvo, T., Kolesárová, A., Komorníková, M., Mesiar, R. (2002) Aggregation operators: properties, classes and construction methods, in T. Calvo, G. Mayor, R. Mesiar (eds.) Aggregation Operators, Physica-Verlag, 3-104.
56. Calvo, T., Mayor, G., Mesiar, R. (eds.) (2002) Aggregation Operators, Physica-Verlag.
57. Calvo, T., Mayor, G., Torrens, J., Suñer, J., Mas, M., Carbonell, M. (2000) Generation of weighting triangles associated with aggregation functions, Int. J. of Unc., Fuzz. and Knowledge Based Systems 8:4 417-451.
58. Calvo, T., Mesiarová, A., Valášková, L. (2003) Construction of aggregation operators – new composition method. Kybernetika 39 643-650.
59. Calvo, T., Suñer, J. (2001) On the degree of orness of a class of weighting triangles, manuscript.

60. Carbonell, M., Mas, M., Mayor, G. (1997) On a class of monotonic extended OWA operators, Proc. of the 6th IEEE Int. Conf. on Fuzzy Systems, 1695-1699.
61. Carlson, B. C. (1970) Problem 70-10, An inequality of mixed arithmetic and geometric means, SIAM Review 12:2 287-288.
62. Cartan, H. (1945) Theories du potentiel newtonien: energie, capacite, suites de potentiels, Bulletin de la Societe Mathematique de France, 73 74-106.
63. Castillo, E., Cobo, A., Gutiérrez, J. M., Pruneda, R. E. (1999) Introducción a las redes funcionales con aplicaciones, Paraninfo.
64. Castillo, E., Cobo, A., Gutiérrez, J. M., Pruneda, R. E. (1999) Functional Networks with Applications. A Neural-Based Paradigm, Kluwer Academic Publishing.
65. Castillo, E., Ruiz-Cobo, M. R. (1992) Functional Equations and Modelling in Science and Engineering, Marcel Dekker, Inc.
66. Cauchy, A. L. (1821) Cours d'Analyse de l'Ecole Royale Polytechnique, $I^{re}$ Partie, Analyse Algébraique, Paris.
67. Chan, S.C., Wong, K.C., Chiu, D.K.Y. (1992) A survey of multiple sequence comparison methods, Bulletin of Mathematical Biology 54:4 563-598.
68. Chateauneuf, A. (1994) Combination of compatible belief functions and relation of specificity, in Advances in Dempster-Shafer Theory of Evidence, Wiley, 97-114.
69. Chateauneuf, A. (1996) Decomposable measures, distorted probabilities and concave capacities, Mathematical Social Sciences 31 19-37.
70. Chateauneuf, A., Jaffray, J.-Y. (1989) Some characterizations of lower probabilities and other monotone capacities through the use of Möbius inversion, Mathematical Social Sciences 17:3 263-283.
71. Chen, J. E., Otto, K. N. (1995) Constructing membership functions using interpolation and measurement theory, Fuzzy Sets and Systems 73:3 313-327.
72. Chen, T.-Y., Wang, J.-C. (2001) Identification of $\lambda$-measures using sampling design and genetic algorithms, Fuzzy Sets and Systems 123 321-341.
73. Cheng, Y., Kashyap, R. L. (1992) Recursive fusion operators: Desirable properties and illustrations, in M. Al Abidi, R. C. Gonzalez (eds.) Data Fusion in Robotics and Machine Intelligence, Academic Press, 245-265.
74. Chiclana, F., Herrera, F., Herrera-Viedma, E. (2000) The ordered weighted geometric operator: properties and applications, Proc. of the IPMU 2000 Conference, 985-991.
75. Chiclana, F., Herrera-Viedma, E., Herrera, F., Alonso, S. (2004) Induced ordered weighted geometric operators and their use in the aggregation of multiplicative preference relations, Int. J. of Intel. Syst. 19 233-255.
76. Chisini, O. (1929) Sul concetto di media, Periodico di Matematiche, Serie 4 Vol. 9:2 106-116.
77. Cho, S.-B., Kim, J. H. (1995) Multiple network fusion using fuzzy logic, IEEE Trans. on Neural Networks, 6:2 497-501.
78. Cholewa, W. (1985) Aggregation of fuzzy opinions – An axiomatic approach, Fuzzy Sets and Systems 17 249-258.

79. Choo, E. U., Wedley, W. C. (1985) Optimal criterion weights in repetitive multicriteria decision-making, The Journal of the Operational Research Society, 36:11 983-992.

80. Choquet, G. (1953/54) Theory of capacities, Ann. Inst. Fourier, 5 131-295.

81. Coleman, J. S. (1971) Control of collectivities and the power of a collectivity to act, in B. Lieberman (ed.) Social Choice, Gordon and Breach, 269-300.

82. Coleman, J. S. (1986) Individual interests and collective action, Cambridge University Press.

83. Colomer, J. M. (1996) Measuring parliamentary deviation, European Journal of Political Research 30 87-101.

84. Colomer, J. M., Martínez, F. (1995) The paradox of coalition trading, Journal of Theoretical Politics 7 41-63.

85. Cotes, R. (1722) Aestimatio errorum in mixta mathesi, per variationes partium trianguli plani et sphaerici. Appended to Harmonia Mensurarum (New York: Readex Microprint, 1968).

86. Day, W. H. E., McMorris, F. R. (1992) Critical comparison of consensus methods for molecular sequences, Nucleic Acids Research 20:5 1093-1099.

87. Day, W. H. E., McMorris, F. R. (1992) Consensus sequences based on plurality rule, Bulletin of Mathematical Biology 54:6 1057-1068.

88. De Baets, B., Fodor, J. (1999) Residual operators of uninorms, Soft Computing 3 89-100.

89. De Baets, B., Fodor, J. (1999) Van Melle's combining function in MYCIN is a representable uninorm: An alternative proof, Fuzzy Sets and Systems 104 133-136.

90. Deegan, J., Packel, E. W. (1978) A new index of power for simple n-person games, Int. J. of Game Theory 7 113-123.

91. De Finetti, B. (1931) Sul concetto di media, Giorn. Ist. Ital. Attuari 3:2 369-396.

92. Delgado, M., Verdegay, J. L., Vila, M. A. (1993) On aggregation operations of linguistic labels, Int. J. of Intel. Syst. 8 351-370.

93. Dellacherie, C. (1971) Quelques commentaires sur les prolongements de capacités, Séminaire de Probabilités 1969/1970, Strasbourg, Lecture Notes in Mathematics 191 77-81.

94. Dempster, A. P. (1967) Upper and lower probabilities induced by a multivalued mapping, Ann. Math. Stat. 38 325-339.

95. de Soto, A. R., Trillas, E. (1999) On antonym and negate in fuzzy logic, Int. J. of Intel. Syst. 14 295-303.

96. de Weerdt, M. (2003) Plan merging in multi-agent systems, PhD Dissertation, Technische Universiteit Delft, The Netherlands.

97. Dombi, J. (1982) Basic concepts for the theory of evaluation: The aggregative operator, European Journal of Operations Research 10 282-293.

98. Driankov, D., Hellendoorn, H., Reinfrank, M. (1993) An introduction to fuzzy control, Springer-Verlag, USA.

99. Duda, R., Hart, P., Nilsson, N. (1976) Subjective bayesian methods for rule-based inference systems, AFIPS Conference Proceedings of the 1976 Nat. Computer Conference, 45 1075-1082.

100. Dujmović, J. J. (1973) A generalization of some functions in continous mathematical logic – evaluation function and its applications (in Serbo-Croatian), Proc. of the Informatica Conference, paper d27, Bled, Yugoslavia.

101. Dujmović, J. J. (1973) Two integrals related to means, Journal of the University of Belgrade EE Dept., Series Mathematics and Physics 412-460 231-232.

102. Dujmović, J. J. (1974) Weighted conjunctive and disjunctive means and their application in system evaluation, Journal of the University of Belgrade, EE Dept., Series Mathematics and Physics 483 147-158.

103. Dujmović, J. J. (1975) Extended continuous logic and the theory of complex criteria, Journal of the University of Belgrade, EE Dept., Series Mathematics and Physics 498 197-216.

104. Dujmović, J. J. (1996) A method for evaluation and selection of complex hardware and software systems, The 22nd International Conference for the Resource Management and Performance Evaluation of Enterprise Computing Systems, CMG 96 Proceedings, Vol. 1, 368-378.

105. Dujmović, J. J. (1998) Evaluation and design of benchmark suites, Chapter 12 in K. Bagchi, G. Zobrist, K. Trivedi (ed.) State-of-the-art in Performance Modeling and Simulation: Theory, Techniques and Tutorials, Gordon and Breach Publishers, 287-323.

106. Dujmović, J. J., Larsen, H. L. (2004) Properties and modeling of partial conjunction / Disjunction, Current Issues in Data and Knowledge Engineering, Warszawa, Poland, 215-224.

107. Dubey, P. (1975) On the uniqueness of the Shapley value, Int. J. of Game Theory 4 131-40.

108. Dubey, P., Shapley, L. S. (1979) Mathematical properties of the Banzhaf power index, Mathematics of Operations Research 4 99-131.

109. Dubois, D. (1983) Modèles Mathématiques de l'Imprécis et de l'Incertain, en Vue d'Applications aux Techniques d'Aide à la Décision, Thèse d'Etat, Univ. de Grenoble.

110. Dubois, D., Prade, H. (1982) A class of fuzzy measures based on triangular norms: a general framework for the combination of uncertain information, Int. J. of General Systems 8 43-61.

111. Dubois, D., Prade, H. (1985) A review of fuzzy set aggregation connectives, Information Sciences 36 85-121.

112. Dubois, D., Prade, H. (1986) Weighted minimum and maximum operations in fuzzy set theory, Information Sciences 39 205-210.

113. Dubois, D., Prade, H., Testemale, C. (1988) Weighted fuzzy pattern matching, Fuzzy Sets and Systems 28 313-331.

114. Ebanks, B., Sahoo, P., Sander, W. (1997) Characterizations of Information Measures, World Scientific.

115. Edwards, W. (1953) Probability-preferences in gambling, American Journal of Psychology 66 349-364.

116. Edwards, W. (1962) Subjective probabilities inferred from decisions, Psychological Review 69 109-135.

117. Einstein, A. (1922) The Meaning of Relativity, Princeton University Press (Spanish translation, El Significado de la Relatividad, Espasa-Calpe, 1984).

118. Halmos, P. R. (1974) Measure Theory, Springer.
119. Halpern, J. Y. (2003) Reasoning about Uncertainty, The MIT Press.
120. Halpern, J. Y., Fagin, R. (1992) Two views of belief: belief as generalized probability and belief as evidence, Artificial Intelligence 54 275-317.
121. Handa, J. (1977) Risk, probabilities, and a new theory of cardinal utility, Journal of Political Economy 85 97-122.
122. Hardy, G. H., Littlewood, J. E., Pólya, G. (1934) Inequalities, Cambridge University Press, 2nd Edition, 1952.
123. Harter, H. L. (1974) The method of least squares and some alternatives - Part I, Int. Stat. Rev. 42:2 147-174.
124. Harter, H. L. (1974) The method of least squares and some alternatives - Part II, Int. Stat. Rev. 42:3 235-264 and 282.
125. Harter, H. L. (1975) The method of least squares and some alternatives - Part III, Int. Stat. Rev. 43:1 1-44.
126. Harter, H. L. (1975) The method of least squares and some alternatives - Part IV, Int. Stat. Rev. 43:2 125-190.
127. Harter, H. L. (1975) The method of least squares and some alternatives - Addendum to Part IV, Int. Stat. Rev. 43:2 273-278.
128. Harter, H. L. (1975) The method of least squares and some alternatives - Part V, Int. Stat. Rev. 43:3 269-278.
129. Harter, H. L. (1975) The method of least squares and some alternatives - Part VI Subject and Author Indexes, Int. Stat. Rev. 44:1 113-159.
130. Hawkins, D. M. (1980) Identification of Outliers, Chapman and Hall, London.
131. Hurwicz, L. (1951) Optimality criteria for decision making under ignorance, Cowles Commission Discussion Paper, Statistics 370.
132. Hurwicz, L. (1951) Some specification problems and applications to econometric models, Econometrica 19 343-344 (abstract).
133. Fan, T., Ralescu, D. A. (1997) On the comparison of OWA operators and ordinal OWA operators, Int. J. of Unc., Fuzz. and Knowledge Based Systems 5:1 1-12.
134. Fechner, G. Th. (1878) Ueber den Ausgangswerth der kleinsten Abweichungssumme, dessen Bestimmung, Verwendung und Verallgemeinerung, Abhandlungen der Königliche Sächsische Gesellschaft der Wissenshaft zu Leipzig, XVIII 1-76.
135. Felsenthal, D. S., Machover, M. (1998) The measurement of voting power: theory and practise, problems and paradoxes, Edward Elgar.
136. Felsenthal, D. S., Machover, M. (2005) Voting power measurement: a story of misreinvention, Social Choice and Welfare 25 485-506.
137. Fremlin, D. H. (2006) Measure theory, Vols. I–V, Torres Fremlin. (See also http://www.essex.ac.uk/maths/staff/fremlin/mt.htm)
138. Fernández Salido, J. M., Murakami, S. (2003) Extending Yager's orness concept for the OWA aggregators to other mean operators, Fuzzy Sets and Systems 139 515-549.
139. Filev, D., Yager, R. R. (1994) Learning OWA operator weights from data, Proc. of the 3rd IEEE Int. Conf. on Fuzzy Systems (IEEE WCCI), Vol. 1 468-473.
140. Filev. D. P., Yager, R. R. (1998) On the issue of obtaining OWA operator weights, Fuzzy Sets and Systems 94 157-169.

141. Filkov, V., Skiena, S. (2004) Heterogeneous data integration with the consensus clustering formalism, Lecture Notes in Bioinformatics 2994 110-123.

142. Fishburn, P. C. (1969) Weak qualitative probability on finite sets, Ann. Math. Stat. 40 2118-2126.

143. Fishburn, P. C., Rubinstein, A. (1986) Aggregation of equivalence relations, Journal of Classification 3 61-65.

144. Fodor, J. C., Marichal, J.-L. (1997) On nonstrict means, Aequationes Mathematicae 54 308-327.

145. Fodor, J., Marichal, J.-L., Roubens, M. (1995) Characterization of the ordered weighted averaging operators, IEEE Trans. on Fuzzy Systems 3:2 236-240.

146. Fodor, J., Roubens, M. (1994) Fuzzy Preference Modelling and Multi-criteria Decision Support, Kluwer Academic Publishers.

147. Fodor, J. C., Roubens, M. (1995) Characterization of Weighted Maximum and some Related Operations, Information Sciences 84 173-180.

148. Fodor, J. C., Yager, R. R., Rybalov, A. (1997) Structure of uninorms, Int. J. of Unc., Fuzz. and Knowledge Based Systems 5 411-427.

149. Foster, R. M. (1922) Derivation of mode as special case of generalized mean, personal communication according to E. V. Huntington, Mathematical Memoranda, Handbook of Mathematical Statistics (ed. H. L. Rietz), p. 7, Boston, 1924.

150. Foulser, D., Li, M., Yang, Q. (1992) Theory and algorithms for plan merging, Artificial Intelligence 57:2-3 143-182.

151. Fujimoto, K. (1995) On Hierarchical Decomposition of Choquet Integral Model, Ph. D. Dissertation, Tokyo Institute of Technology.

152. Fujimoto, K. (2003) Cardinal-probabilistic interaction indices and their applications: A survey, J. of Advanced Computational Intelligence and Intelligent Informatics 7:2 79-85.

153. Fujimoto, K., Kojadinovic, I., Marichal, J.-L. (2004) Characterization of probabilistic and cardinal-probabilistic interaction indices, Proc. of the IPMU 2004 Conference, 45-52.

154. Fujimoto, K., Kojadinovic, I., Marichal, J.-L. (2006) Axiomatic characterizations of probabilistic and cardinal-probabilistic interaction indices, Games and Economic Behavior 55 72-99.

155. Fujimoto, K., Murofushi, T., Sugeno, M. (1998) Canonical hierarchical decomposition of Choquet integral over finite set with respect to null additive fuzzy measure, Int. J. of Unc., Fuzz. and Knowledge Based Systems 6:4 345-363.

156. Fullér, R., Majlender, P. (2001) An analytic approach for obtaining maximal entropy OWA operator weights, Fuzzy Sets and Systems 124 53-57.

157. Fullér, R., Majlender, P. (2003) On obtaining minimal variability OWA operator weights, Fuzzy Sets and Systems 136:2 203-215.

158. Fung, L., Fu, K. S. (1973) The k-th Optimal Policy Algorithm for decision making in fuzzy environments, Proc. 3rd IFAC Symposium on Identification and System Parameter Estimation.

159. Garret, G., Tsebelis, G. (1996) An institutional critique of intergovernmentalism, International Organization 50:2 269-299.

160. Garret, G., Tsebelis, G. (1996) An institutional critique of intergovernmentalism: Erratum, International Organization 50:3 539.
161. Garret, G., Tsebelis, G. (1999) Why resist the temptation to apply power indices to the European Union?, Journal of Theoretical Politics 11 291-308.
162. Gill, P. E., Murray, W., Wright, M. H. (1981) Practical Optimization, Academic Press.
163. Gini, C. (1958) Le Medie, Unione Tipografico-Editrice Torinese.
164. Godo, L., Torra, V. (2000) On aggregation operators for ordinal qualitative information, IEEE Trans. on Fuzzy Systems 8:2 143-154.
165. Goguen, J. A. (1967) L-fuzzy sets, J. of Math. Analysis and Applications 18:1 145-174.
166. Goodman, I. R. (1987) A general theory for the fusion of data, Triservice data fusion symposium, 254-270. Also in [345].
167. Grabisch, M. (1995) Fuzzy integral in multicriteria decision making, Fuzzy Sets and Systems 69 279-98.
168. Grabisch, M. (1996) $k$-order additive discrete fuzzy measures, Proc. of the IPMU 1996 Conference, 1345-1350.
169. Grabisch, M. (1997) $k$-order additive discrete fuzzy measures and their representation, Fuzzy sets and systems 92:2 167-189.
170. Grabisch, M. (2006) Capacities and games on lattices: a survey of results, Int. J. of Unc., Fuzz. and Knowledge Based Systems 14:4 371-392.
171. Grabisch, M., Kojadinovic, I., Meyer, P. (2006) Package "Kappalab," http://www.polytech.univ-nantes.fr/kappalab
172. Grabisch, M., Nguyen, H. T., Walker, E. A. (1995) Fundamentals of uncertainty calculi with applications to fuzzy inference, Kluwer Academic Publishers.
173. Grabisch, M., Murofushi, T., Sugeno, M. (1992) Fuzzy measures of fuzzy events defined by fuzzy integrals, Fuzzy Sets and Systems 50 293-313.
174. Grabisch, M., Murofushi, T., Sugeno, M. (eds.) (2000) Fuzzy Measures and Integrals: Theory and Applications, Physica-Verlag.
175. Greco, G. (1982) Sulla rappresentazione di funzionali mediante integrali, Rend. Sem. Mat. Univ. Padova 66 21-42.
176. Hägele, G., Pukelsheim, F. (2001) Llull's writings on electoral systems, Studia Lulliana 41 3-38
177. Hájek, P. (1985) Combining function for certainty degrees in consulting systems, Int. J. of Man-Machine Studies 22 59-67.
178. Hájek, P., Valdes, J. (1990) Algebraic foundations of uncertainty processing in rule-based expert systems (group theoretic approach), Computers and Artificial Intelligence 9:4 325-344.
179. Hall, D. L., Llinas, J. (1997) An introduction to multisensor Data Fusion, Proc. of the IEEE, 85:1 6-23.
180. Hamacher, H. (1975) Über logische Verknüpfungen unscharfer Aussagen und deren zugehörige Bewertungsfunktionen, Working Paper 75/14, RWTH, Aachen.
181. Hampel, F. R., Ronchetti, E. M., Rousseeuw, P. J., Stahel, W. A. (1986) Robust Statistics: The Approach based on Influence Functions, John Wiley and Sons.
182. Hastie, T., Tibshirani, R., Friedman, J. (2001) The Elements of Statistical Learning, Berlin: Springer

183. Hays, W. L. (1973) Statistics for the Social Sciences, Holt, Rinehart, Winston, New York.

184. Heckerman, D. (1986) Probabilistic interpretations for MYCIN's certainty factors, in L. N. Kanal, J. F. Lemmer (eds.) Uncertainty in Artificial Intelligence, Elsevier Science Publishers, 167-196.

185. Herrera, F., Herrera-Viedma, E., Chiclana, F. (2003) A study of the origin and uses of the ordered weighted geometric operator in multicriteria decision making, Int. J. of Intel. Syst. 18 689-707.

186. Herrera, F., Herrera-Viedma, E., Verdegay, J. L. (1995) A Sequential selection process in group decision making with a linguistic assessment approach, Information Science 85 223-239.

187. Herrera, F., Herrera-Viedma, E., Verdegay, J. L. (1996) Direct approach processes in group decision making using linguistic OWA operators, Fuzzy Sets and Systems 79 175-190.

188. Hollander, M., Wolfe, D. A. (1973) Nonparametric Statistical Methods, John Wiley & Sons.

189. Honda, A., Nakano, T., Okazaki, Y. (2002) Distortion of fuzzy measures, Proc. of the SCIS/ISIS conference.

190. Honda, A., Nakano, T., Okazaki, Y. (2002) Subjective evaluation based on distorted probability, Proc. of the SCIS/ISIS conference.

191. Honecker, M. (1937) Lullus-Handschriften aus dem Besitz des Kardinals Nikolaus von Cues - Nebst einer Beschreibung der Lullus-Texte in Trier und einem Anhang über den wiederaufgefundenen Traktak *De arte electionis*, Spanische Forschungen der Görresgesellschaft, Erste Reihe, 6 252-309

192. Horsky, D., Rao, M. R. (1984) Estimation of attribute weights from preference comparisons, Management Science 30:7 801-822.

193. Hovanessian, S. A. (1988) Introduction to Sensor Systems, Artech House.

194. Huber, P. J. (1981) Robust Statistics, John Wiley and Sons.

195. Imai, H., Asano, D., Sato, Y. (2003) An algorithm based on alternative projections for a fuzzy measure identification problem, in V. Torra (ed.) Information Fusion in Data Mining, Springer, 149-159.

196. Imai, H., Miyamori, M., Miyakosi, M., Sato, Y. (2000) An algorithm based on alternative projections for a fuzzy measures identification problem, Proc. Int. Conf. Soft Comp. (Iizuka), CD-ROM.

197. Imai, H., Torra, V. (2003) On a modeling of decision making with a twofold integral, Proc. EUSFLAT 2003, 714-717.

198. Ishii, K., Sugeno, M. (1985) A model of human evaluation process using fuzzy measure, Int. J. of Man-Machine Studies 22 19-38.

199. Ishii, Y. (2000) Identification of Multiple Choquet Integral Models by GA (in Japanese), Master Thesis, Tokyo Institute of Technology.

200. Ishii, Y., Murofushi, T. (2001) Identification of fuzzy measures using real valued GA and considering outliers, Proc. of 6th Workshop on Evaluation of Heart and Mind, 1-4.

201. Jackson, D. (1921) Note on the median of a set of numbers, Bulletin of the American Mathematical Society 27 160-164.

202. Jacod, J., Protter, P. (2003) Probability Essentials, Springer, 2nd Edition.

203. Jensen, J. L. W. V. (1905) Om konvekse funktioner og uligheder imellem middelvaerier, Mat. Tidsskr, B 49-68.
204. Jensen, J. L. W. V. (1906) Sur les Fonctions Convexes et les Inégalités entre les Valeurs Moyennes, Acta Math. 30 175-193.
205. Jessen, B. (1930) Über die Verallgeneinerung des arithmetischen Mittels, Acta Scientiarum Mathematicarum (Szeged), 5 108-116.
206. Kahneman, D., Tversky, A. (1979) Prospect theory: An analysis of decision under risk, Econometrica 47 263-291.
207. Kandel, A., Byatt, W. J. (1978) Fuzzy sets, fuzzy algebra and fuzzy statistics, Proc. of the IEEE, 68 1619-1639.
208. Keller, J. M., Osborn, J. (1996) Training the fuzzy integral, Int. J. of Approx. Reasoning 15 1-24.
209. Klement, E.-P., Mesiar, R., Pap, E. (1996) On the relationship of associative compensatory operators to triangular norms and conorms, Int. J. of Unc., Fuzz. and Knowledge Based Systems 4 129-144.
210. Klement, E. P., Mesiar, R., Pap, E. (2000) Triangular Norms, Kluwer Academic Publisher.
211. Klir, G., Yuan, B. (1995) Fuzzy Sets and Fuzzy Logic: Theory and Applications, Prentice Hall, UK
212. Kojadinovic, I. (2004) Estimation of the weights of interacting criteria from the set of profiles by means of information-theoretic functionals, European Journal of Operational Research 155 741-751.
213. Kojadinovic, I. (2004) Unsupervised aggregation by the Choquet integral based on entropy functionals: application to the evaluation of students, MDAI 2004, Lecture Notes in Artificial Intelligence 3131 163-174.
214. Kolmogorov, A. N. (1930) Sur la notion de la moyenne, Accad. Naz. Lincei Mem. Cl. Sci. Fis. Mat. Natur. Sez. 12 388-391.
215. Kolmogorov, A. (1933) Grundbegriffe der Wahrscheinlichkeitsrechnung, Berlin (English translation: Foundations of the Theory of Probability, Chelsea, New York, 1956).
216. König, H. (1997) Measure and Integration: an Advanced Course in Basic Procedures and Applications, Springer.
217. Krantz, D. H., Luce, R. D., Suppes, P., Tversky, A. (1971) Foundations of Measurement, Vol. 1: Additive and Polynomial Representations, Academic Press.
218. Kruse, R. (1983) Fuzzy integrals and conditional fuzzy measures, Fuzzy Sets and Systems 10 309-313.
219. Kruse, R., Schwecke, E., Heinsohn, J. (1991) Uncertainty and Vagueness in Knowledge Based Systems, Springer.
220. Kyburg Jr., H. E. (1988) Higher order probabilities and intervals, Int. J. of Approx. Reasoning 2 195-209.
221. Laplace, P. S. (1812) Théorie Analytique des Probabilités, Courcier, Paris; 2nd Edition 1818, containing two supplements; 3rd edition 1820, containing three supplements. Also in Vol. VII of Oeuvres Complètes de Laplace, Gauthier-Villars.
222. Lee, K.-M., Leekwang, H. (1995) Identification of λ-fuzzy measure by genetic algorithms, Fuzzy Sets and Systems 75 301-309.
223. Legendre, A. M. (1791) Eléments de Géométrie. Note IV. Didot, Paris, 1791 (New York: Readex Microprint, 1970).

224. Legendre, A. M. (1805) Nouvelles Méthodes pour la Détermination des Orbites des Cométes, F. Didot (New York: Readex Microprint, 1970).
225. Lehmann, E. L. (1975) Nonparametrics: Statistical Methods based on Ranks, Holden Day.
226. Leszczyski, K., Penczek, P., Grochulski, W. (1985) Sugeno's fuzzy measure and fuzzy clustering, Fuzzy Sets and Systems 15 147-158.
227. Ling, C. H. (1965) Representation of associative functions, Publ. Math. Debrecen 12 182-212.
228. Liu, X. (2005) On the properties of equidifferent RIM quantifier with generating function, Int. J. of General Systems 34:5 579-594.
229. Liu, X. (2006) Some properties of the weighted OWA operator, IEEE Trans. on Fuzzy Systems, part B 36:1 118-127.
230. Liu, X. (2006) An orness measure for quasi-arithmetic means, IEEE Trans. on Fuzzy Systems 14:6 837-848.
231. Liu, X., Chen, L. (2004) On the properties of parametric geometric OWA operator, Int. J. of Approx. Reasoning 35 163-178.
232. Liu, Y., Kerre, E. E. (1998) An overview of fuzzy quantifiers (I). Interpretations, Fuzzy Sets and Systems 95 1-21.
233. Liu, Y., Kerre, E. E. (1998) An overview of fuzzy quantifiers (II). Reasoning and applications, Fuzzy Sets and Systems 95 135-146.
234. Llull, R. (1982) Llibre d'Evast e Blanquerna, Edicions 62 (written c. 1283). English translation: Blanquerna, Londres: Jarrolds (1926) and reprinted by London: Dedalus and New York: Hippocrene (1987).
235. Llull, R. (2004) The Augsburg Web Edition of Llull's Electoral Writings, http://www.math.uni-augsburg.de/stochastik/lull/
236. López de Mántaras, R. (1990) Approximate Reasoning Models, Ellis Horwood Limited, England.
237. Losonczi, L. (1971) Über eine neue Klasse von Mittelwerte, Acta Sci. Math (Acta Univ. Szeged) 32 71-78.
238. Luce, R. D., Krantz, D. H., Suppes, P., Tversky, A. (1990) Foundations of Measurement, Vol. III: Representation, Axiomatization, and Invariance, Academic Press.
239. Luce, R. D., Raiffa, H. (1957) Games and decisions, John Wiley and Sons. Reprinted by Dover Publications Inc.
240. Luenberger, D. G. (1973) Introduction to Linear and Nonlinear Programming, Addison-Wesley, Menlo Park, California.
241. Luo, R.C., Kay, M.G. (1992) Data fusion and sensor integration: State-of-the-art 1990s, in M. Al Abidi, R. C. Gonzalez (eds.) Data Fusion in Robotics and Machine Intelligence, Academic Press, 7-135.
242. Magdalena, L. (2000) Hierarchical fuzzy control of a complex system using meta-knowledge, Proc. of the IPMU 2000 Conference, 630-637.
243. Maire, C., Boscovich, R. J. (1755) De litteraria Expeditione per Pontificiam ditionem ad dimetiendos duos Meridiani gradus, et corrigendam mappam geographicam. Jussu, et auspiciis Benedicti XIV. Pont. Max. Romae (New York: Readex, 1986).
244. Margush, T., McMorris, F.R. (1981) Consensus n-Trees, Bulletin of Mathematical Biology 43 239-244.
245. Marichal, J.-L. (1998) Aggregation Operators for Multicriteria Decision Aid, Ph. D. Dissertation, Institute of Mathematics, University of Liège, Liège, Belgium.

246. Marichal, J.-L. (2000) On an axiomatization of the quasi-arithmetic mean values without the symmetry axiom, Aequationes Mathematicae 59 74-83.
247. Marichal, J.-L. (2000) On comparison meaningfulness of aggregation functions, Journal of Mathematical Psychology 45 213-223.
248. Marichal, J.-L. (2002) Entropy of discrete Choquet capacities, European Journal of Operational Research 137:3 612-624.
249. Marichal, J.-L. (2002) On order invariant synthesizing functions, Journal of Mathematical Psychology 46 661-676.
250. Marichal, J.-L. (2004) $k$-intolerant capacities and Choquet integrals, Proc. of the IPMU 2004 Conference, 601-608.
251. Marichal, J.-L., Mathonet, P., Tousset, E. (1999) Characterization of some aggregation functions stable for positive linear transformations, Fuzzy Sets and Systems 102 293-314.
252. Marichal, J.-L., Roubens, M. (1993) Characterization of some stable aggregation functions, Proc. Int. Conf. on Industrial Engineering and Production Management, Mons, Belgium, 187-196.
253. Marichal, J.-L., Roubens, M. (2000) Determination of weights of interacting criteria from a reference set, European Journal of Operational Research 124 641-650.
254. Marichal, J.-L., Roubens, M. (2000) Entropy of discrete fuzzy measures, Int. J. of Unc., Fuzz. and Knowledge Based Systems 8:6 625-640.
255. Mayor, G. Torrens, J. (1993) On a class of operators for expert systems, Int. J. of Intel. Syst. 8 771-778.
256. McAllister, D. F., Roulier, J. A. (1981) An algorithm for computing a shape-preserving oscillatory quadratic spline, ACM Trans. Math. Software 7 331-347.
257. McLean, I. (1990) The Borda and Condorcet Principles: Three Medieval Applications, Social Choice Welfare 7 99-108.
258. McLean, I., London, J., Ramon Llull and the theory of voting, Studia Lulliana 32 21-37.
259. McLean, I., Lorrey, H. (2004) Voting in medieval universities and religious orders, manuscript. http://www.nuff.ox.ac.uk/Users/McLean/
260. Menger, K. (1942) Statistical metrics, Proc. Nat. Acad. Sci. 28 535-537.
261. Mesiar, R. (1997) $k$-order pan-additive discrete fuzzy measures, Proc. of the 7th IFSA Conference, 488-490.
262. Mesiar, R. (1999) Generalizations of $k$-order additive discrete fuzzy measures, Fuzzy Sets and Systems 102 423-428.
263. Mesiar, R., Mesiarová, A. (2004) Fuzzy integrals, MDAI 2004, Lecture Notes in Artificial Intelligence 3131 7-14.
264. Mesiar, R., Mesiarová, A., Fuzzy integrals – what they are?, Int. J. of Intel. Syst., in press.
265. Mesiar, R., Vivona, D. (1999) Two-step integral with respect to fuzzy measure, Tatra Mt. Math. Publ. 16 359-368.
266. Messafa, H. (1992) An Algorithm to maximize the agreement between partitions, Journal of Classification 9 5-15.
267. Mészáros, Cs. (2006) http://www.sztaki.hu/~meszaros/bpmpd
268. Meyer, P., Roubens, M. (2005) Choice, ranking and sorting in fuzzy multiple criteria decision aid, in J. Figueira, S. Greco, M. Ehrgott (eds.)

Multiple Criteria Decision Analysis: State of the Art Surveys, Springer, 471-506.

269. Miranda, P., Grabisch, M. (2002) *p*-symmetric fuzzy measures, Proc. of the IPMU 2002 Conference, 545-552.

270. Miranda, P., Grabisch, M., Gil, P. (2002) *p*-symmetric fuzzy measures, Int. J. of Unc., Fuzz. and Knowledge Based Systems 10 105-123.

271. Mitchell, H. B., Estrakh, D. D. (1997) A modified OWA operator and its use in lossless DPCM image compression, Int. J. of Unc., Fuzz. and Knowledge Based Systems 5:4 429-436.

272. Mitchell, H. B., Schaefer, P. A. (2000) Multiple priorities in an induced ordered weighted averaging operator, Int. J. of Intel. Syst. 15 317-327.

273. Mori, T., Murofushi, T. (1989) An analysis of evaluation model using fuzzy measure and the Choquet integral (in Japanese), Proc. of the 5th Fuzzy System Symposium, 207-212.

274. Morriss, P. (1987) Power: A Philosophical Analysis, Manchester University Press.

275. Muirhead, R. F. (1903) Some methods applicable to identities and inequalities of symmetric algebraic functions of *n* letters, Proc. Edinburgh Math. Soc., 21 144-157.

276. Murofushi, T., Narukawa, Y. (2002) A characterization of multi-step discrete Choquet integral, Abstracts of 6th Int. Conf. Fuzzy Sets Theory and Its Applications, 94.

277. Murofushi, T., Narukawa, Y. (2002) A characterization of multi-level discrete Choquet integral over a finite set (in Japanese). Proc. of 7th Workshop on Evaluation of Heart and Mind, 33-36.

278. Murofushi, T., Soneda, S. (1993) Techniques for reading fuzzy measures (iii): interaction index. Proc. 9th Fuzzy System Symposium, Sapporo, Japan. 693-696.

279. Murofushi, T., Sugeno, M. (1989) An interpretation of fuzzy measures and the Choquet integral as an integral with respect to a fuzzy measure, Fuzzy Sets and Systems 29 201-227.

280. Murofushi, T., Sugeno, M. (1991) Fuzzy t-conorm integral with respect to fuzzy measures: Generalization of Sugeno integral and Choquet integral, Fuzzy Sets and Systems 42 57-71.

281. Murofushi, T., Sugeno, M. (1992) Hierarchical decomposition of Choquet integral systems, Journal of the Japanese Fuzzy Society 4:4 749-752.

282. Murofushi, T., Sugeno, M., Fujimoto, K (1997) Separated hierarchical decomposition of the Choquet integral, Int. J. of Unc., Fuzz. and Knowledge Based Systems 5:5 563-585.

283. Sugeno, M., Fujimoto, K., Murofushi, T. (1995) A hierarchical decomposition of Choquet integral model, Int. J. of Unc., Fuzz. and Knowledge Based Systems 3 1-15.

284. Murofushi, T., Sugeno, M., Machida, M. (1994) Non-monotonic fuzzy measures and Choquet integral, Fuzzy Set and Systems 64 73-86.

285. Myerson, R. B. (1991) Game Theory, Harvard University Press.

286. Nagumo, M. (1930) Über eine Klasse der Mittelwerte, Japan. J. of Math 6 71-79.

287. Narukawa, Y. (1990) A Study of Fuzzy Measure and Choquet Integral (in Japanese), Master Thesis, Tokyo Institute of Technology.

288. Narukawa, Y., Murofushi, T. (2003) Choquet integral and Sugeno integral as aggregation functions, in V. Torra (ed.) Information Fusion in Data Mining, Springer.

289. Narukawa, Y., Murofushi, T., Sugeno, M. (2000) Regular fuzzy measure and Representation of comonotonically additive functionals, Fuzzy Sets and Systems 112:2 177-186.

290. Narukawa, Y., Torra, V. (2003) Choquet integral based models for general approximation, in I. Aguiló, L. Valverde, M. T. Escrig (eds.) Artificial Intelligence Research and Development, IOS Press, 39-50.

291. Narukawa, Y., Torra, V. (2004) On $m$-dimensional distorted probabilities and p-symmetric fuzzy measures, Proc. of the IPMU 2004 Conference, 1279-1284.

292. Narukawa, Y., Torra, V. (2004) Twofold integral and Multi-step Choquet integral, Kybernetika 40:1 39-50.

293. Narukawa, Y., Torra, V. (2005) Fuzzy measure and probability distributions: distorted probabilities, IEEE Trans. on Fuzzy Systems 13:5 617-629.

294. Narukawa, Y., Torra, V. (2005) Graphical interpretation of the twofold integral and its generalization, Int. J. of Unc., Fuzz. and Knowledge Based Systems 13:4 415-424.

295. Negoita, C. V., Flondor, P. (1976) On fuzziness in information retrieval, Int. J. of Man-Machine Studies 8 711-716.

296. Nelles, O. (2001) Nonlinear System Identification, Springer-Verlag.

297. Nelsen, R. B. (1993) Proofs Without Words: Exercises in Visual Thinking, Mathematical Association of America.

298. Nettleton, D., Torra, V. (2001) A comparison of active set methods and genetic algorithm approaches for learning weighting vectors in some aggregation operators, Int. J. of Intel. Syst. 16:9 1069-1083.

299. Newcomb, S. (1912) Researches on the motion of the moon: Part II. The mean motion of the moon and other astronomical elements derived from observations of eclipses and occultations extending from the period of the Babylonians until A.D. 1908, Astronomical Papers prepared for the Use of the Astronomical Ephemeris and Nautical Almanac, 9 1-249.

300. Nocedal, J., Wright, S. (2000) Numerical Optimization, Springer.

301. O'Hagan, M. (1987) Fuzzy decision aids, Proc. 21st Annual Asilomar Conf. on Signals, Systems, and Computers, IEEE and Maple Press, Vol. 2, 624-628 (Published in 1988).

302. O'Hagan, M. (1988) Aggregating template or rule antecedents in real-time expert systems with fuzzy set logic, in Proc. 22nd Ann. IEEE Asilomar Conf. Signals, Systems, Computers, Pacific Grove, CA, 681-689.

303. O'Hagan, M. (1990) A fuzzy neuron based on maximum entropy ordered weighted averaging, in Proc. 24th Ann. IEEE Asilomar Conf. Signals, Systems, Computers, 618-623.

304. Orlov, A. (1981) The connection between mean quantities and admissible transformations, Mathematical Notes 30 774-778.

305. Ovchinnikov, S. (1996) Means on ordered sets, Mathematical Social Sciences 32 39-56.

306. Ovchinnikov, S. (1998) An analytic characterization of some aggregation operators, Int. J. of Intel. Syst. 13 59-68.

307. Ovchinnikov, S. (1998) Invariant functions on simple orders, Order 14 365-371.
308. Ovchinnikov, S. (1999) Invariance properties of ordinal OWA operators, Int. J. of Intel. Syst. 14 413-418.
309. Owen, G. (1972) Multilinear extensions of games, Management Science 18:5 64-79.
310. Pajala, A. (2002) Voting power and power index website: a voting power WWW-Resource including POWERSLAVE voting body analyser, University of Turku, Turku. http://powerslave.val.utu.fi/
311. Pajala, A., Meskanen, T., Kause, T. (2002) POWERSLAVE power index calculator: A voting body analyser in the voting power and power index website, University of Turku, http://powerslave.val.utu.fi/
312. Pap, E. (ed.) (2002) Handbook of Measure Theory, Vols. I and II, North-Holland.
313. Pappus (1982) La Collection Mathématique, Librairie Scientifique et Technique Albert Blanchard. Translation and Notes by Paul Ver Eecke.
314. Pasupathy, S. (1989) Glories of Gaussianity (Light Traffic), IEEE Communications Magazine 27:8 (August 1989) p.38.
315. Paunić, D. (2002) History of measure theory, in E. Pap (ed.) Handbook of Measure Theory, North-Holland, 1-26.
316. Pedersen, O. (1974) A Survey of the Almagest, Odense University Press.
317. Peklman, D., Sen, S. K. (1974) Mathematical programming models for the determination of attribute weights, Management Science 20:8 1217-1229.
318. Penny, D., Foulds, L. R., Hendy, M. D. (1982) Testing the theory of evolution by comparing phylogenetic trees constructed from five different protein sequences, Nature 297 197-200.
319. Penrose, L. S. (1946) The elementary statistics of majority voting, Journal of the Royal Statistical Society 109:1 53-57.
320. Perez Martínez, L. (1959) El "Ars Notandi" y el "Ars Electionis" dos obras desconocidas de Ramón Llull, Estudios Lulianos 3 275-278.
321. Pham, T.D. (2002) Combination of multiple classifiers using adaptive fuzzy integral, Proc. IEEE Int. Conf. on Artificial Intelligence Systems 50-55.
322. Plackett, R. L. (1958) Studies in the history of probability and statistics: VII: The principle of the arithmetic mean, Biometrika 45:1/2 130-135.
323. Plackett, R. L. (1972) Studies in the history of probability and statistics XXIX: The discovery of the method of least squares, Biometrika 59 239-251.
324. Poincaré, H. (1912) Calcul des Probabilités, Paris.
325. Prelec, D. (1998) The probability weighting function, Econometrica 66:3 497-527.
326. Preston, M. G., Baratta, P. (1948) An experimental study of the auction-value of an uncertain outcome, American Journal of Psychology 61 183-93.
327. R (a language and environment for statistical computing and graphics), http://www.r-project.org/
328. Ralescu, A. L., Ralescu, D. A. (1997) Extensions of fuzzy aggregation, Fuzzy Sets and Systems 86 321-330.

329. Radojević, D. (1999) Logical interpretation of discrete Choquet integral defined by general measure, Int. J. of Unc., Fuzz. and Knowledge Based Systems 7:6 577-588.

330. Radojević, D. (1999) The logical representation of the discrete Choquet integral, Proc. of the EUROFUSE 1999 Workshop.

331. Rao, M. M. (1987) Measure Theory and Integration, Wiley.

332. Rapoport, A. (1989) Decision Theory and Decision Behavior, Kluwer Academic Publishers.

333. Rey, W. J. J. (1983) Introduction to Robust and Quasi-Robust Statistical Methods, Springer-Verlag.

334. Roberts, F. S. (1979) Measurement Theory, Addison-Wesley.

335. Rota, G.-C. (1964) On the foundations of combinatorial theory. I. Theory of Möbius functions, Z. Wahrscheinlichkeitstheorie 2 340-368.

336. Roth, A. E. (ed.) (1988) The Shapley Value, Cambridge University Press.

337. Roubens, M. (1996) Interaction between criteria and definition of weights in MCDA problems. Proc. 44th meeting of the European working group "Multicriteria Aid for Decisions", Brussels, Belgium, October 1996.

338. Rousseeuw, P. J., Hubert, M. (1997) Recent Developments in PROGRESS, in L_1-Statistical Procedures and Related Topics, edited by Y. Dodge. Institute of Mathematical Statistics Lecture Notes-Monograph Series, Volume 31, Hayward, California, 201-214.

339. Rousseeuw, P. J., Leroy, A. M. (1987) Robust Regression and Outlier Detection, John Wiley and Sons.

340. Roy, B. (1996) Multicriteria Methodology for Decision Aiding, Kluwer Academic Publishers.

341. Ruspini, E. H., Bonissone, P. P., Pedrycz, W. (eds.) (1998) Handbook of Fuzzy Computation, IOP Publishing.

342. Ryan, T. P. (1997) Modern Regression Methods, Wiley.

343. Saaty, R. W. (1987) The analytic hierarchy process – what it is and how it is used, Mathematical Modelling 9:3-5 161-176.

344. Saaty, T. L. (1980) The Analytic Hierarchy Process, McGraw-Hill.

345. Sadjadi, F. A. (ed.) (1996) Selected Papers on Sensor and Data Fusion, SPIE Optical Engineering Press.

346. Sander, W. (2002) Associative aggregation operators, in T. Calvo, G. Mayor, R. Mesiar (eds.) Aggregation Operators, Physica-Verlag, 124-158.

347. Schaefer, P. A., Mitchell, H. B. (1999) A generalized OWA operator, Int. J. of Intel. Syst. 14 123-143.

348. Schmeidler, D. (1986) Integral representation without additivity, Proceedings of the American Mathematical Society, 97 253-261.

349. Schweizer, B., Sklar, A. (1960) Statistical metric spaces, Pacific J. Math. 10 313-334.

350. Schweizer, B., Sklar, A. (1983) Probabilistic Metric Spaces, Elsevier-North-Holland.

351. Scott, D. (1964) Measurement structures and linear inequalities, Journal of Mathematical Psychology 1 233-247.

352. Searle, S. R. (1971) Linear Models, John Wiley and Sons.

353. Searle, S. R. (1982) Matrix Algebra Useful for Statistics, John Wiley and Sons.
354. Seber, G. A. F., Lee, A. J. (2003) Linear Regression, Wiley, 2nd Edition.
355. Shafer, G. (1976) A Mathematical Theory of Evidence, Princeton University Press, Princeton, New Jersey.
356. Shannon, C. E. (1948) A mathematical theory of communication, Bell System Tech. J. 27 379-423, 623-656. Reprinted in C. E. Shannon, W. Weaver, The Mathematical Theory of Communication, Univ. of Illinois Press, Urbana Ill., 1949.
357. Shapley, L. (1953) A value for $n$-person games, Annals of Mathematical Studies 28 307-317.
358. Shapley, L. S. (1971) Core of convex games, Int. J. of Game Theory 1 11-26
359. Shapley, L., Shubik, M. (1954) A method of evaluation of the distribution of power in a committee system, American Political Science Review 48 787-792.
360. Shortliffe, E. H., Buchanan, B. G. (1975) A model of inexact reasoning in medicine, Mathematical Biosciences 23 351-379.
361. Shudo, K. (2004) Performance comparison of Java/.NET Runtimes, http://www.shudo.net/jit/perf/
362. Simon, H. A. (1995) Artificial intelligence: an empirical science, Artificial Intelligence 77:1 95-127.
363. Simpson, T. (1756) A letter to the Right Honourable George Earl of Macclesfield, President of the Royal Society, on the advantage of taking the mean of a number of observations, in practical astronomy, Philosophical Transactions of the Royal Society of London for 1755, 49:1 82-93.
364. Šipoš, J. (1979) Integral with respect to premeasure, Math. Slovaca 29 141-145.
365. Šipoš, J. (1979) Non linear integral, Math. Slovaca 29:3 257-270.
366. Slany, W. (1996) Scheduling as a fuzzy multiple criteria optimization problem, Fuzzy Sets and Systems 78 197-222.
367. Smets, P. (1988) Belief functions, in P. Smets, E. H. Mamdani, D. Dubois, H. Prade (eds.) Non-Standard Logics for Automated Reasoning, Academic Press, 253-277.
368. Smets, P., Kennes, R. (1994) The transferable belief model, Artificial Intelligence 191-234.
369. Soler i Llopart, A. (1991) Encara sobre la data del *Blanquerna*, Studia Lulliana 31 113-123.
370. Soria-Frisch, A. (2003) Hybrid SOM and fuzzy integral frameworks for fuzzy classification, Proc. of the 12th IEEE Int. Conf. on Fuzzy Systems, 840-845.
371. Soria-Frisch, A. (2006) Unsupervised construction of fuzzy measures through self-organizing feature maps and its application in color image segmentation, Int. J. of Approx. Reasoning 41 23-42.
372. Srinivasan, V., Schocker, A. D. (1973) Linear programming techniques for multidimensional analysis of preferences, Psychometrica 38:3 337-369.

373. Srinivasan, V., Schocker, A. D. (1973) Estimating the weights for multiple attributes in a composite criterion using pairwise judgements, Psychometrica 38:4 473-493.

374. Stejić, Z., Takama, Y., Hirota, K. (2005) Mathematical aggregation operators in image retrieval: effect on retrieval performance and role in relevance feedback, Signal processing 85 297-324.

375. Stevens, S. S. (1946) On the theory of scales of measurement, Science 103 677-680.

376. Stevens, S. S. (1951) Mathematics, measurement and psychophysics, in S. S. Stevens (ed.) Handbook of Experimental Psychology, Wiley, 1-49.

377. Stevens, S. S. (1959) Measurement, psychophysics, and utility, in C. W. Churchman, P. Ratoosh (eds.) Measurement: Definitions and Theories, Wiley, 18-63.

378. Stigler, S. M. (1981) Gauss and the invention of least squares, Ann. Stat. 9 465-474.

379. Straffin, P. D. (1978) Homogeneity, independence and power indices, Public Choice 30 107-18.

380. Straffin, P. D. (1988) The Shapley-Shubik and Banzhaf power indices as probabilities, in A. E. Roth (ed.) The Shapley Value, Cambridge University Press, 71-81.

381. Stuart, A., Keith Ord, J. (1987) Kendall's Advanced Theory of Statistics (Originally by Sir M. Kendall), vol. I: Distribution Theory, Charles Griffin and Company Limited.

382. Sugeno, M. (1972) Fuzzy measures and fuzzy integrals (in Japanese), Trans. of the Soc. of Instrument and Control Engineers 8:2

383. Sugeno, M. (1973) Constructing fuzzy measures and grading similarity of patterns by fuzzy integrals (in Japanese), Trans. of the Soc. of Instrument and Control Engineers 9:3 361-368.

384. Sugeno, M. (1974) Theory of Fuzzy Integrals and its Applications, Ph. D. Dissertation, Tokyo Institute of Technology, Tokyo, Japan.

385. Sugeno, M., Murofushi, T. (1993) Fuzzy Measure (in Japanese), Tokyo, Nikkan Kogyo Shinbunsha.

386. Suppes, P., Krantz, D. H., Luce, R. D., Tversky, A. (1989) Foundations of Measurement, Vol. II, Geometrical, Threshold, and Probability Representations, Academic Press.

387. Taguchi, T. (1974) On Fechner's thesis and statistics with norm $p$, Annals of the Institute of Statistical Mathematics 26 175-193.

388. Tahani, H., Keller, J. M. (1990) Information fusion in computer vision using the fuzzy integral, IEEE Trans. on Systems, Man and Cybernetics 20:3 733-741.

389. Takahagi, E. (1999) On fuzzy integral representation in fuzzy switching functions, fuzzy rules and fuzzy control rules, Proc. of the 8th IFSA Conference, 289-293.

390. Takahagi, E. (2006) http://www.isc.senshu-u.ac.jp/~thc0456/EAHP/AHPweb.html

391. Tanaka, A., Murofushi, T. (1989) A learning model using fuzzy measure and the Choquet integral, Proc. 5th Fuzzy System Symp., Kobe, Japan, 213-217.

392. Torra, V. (1995) A New combination function in evidence theory, Int. J. of Intel. Syst. 10:12 1021-1033.

393. Torra, V. (1996) Negation functions based semantics for ordered linguistic labels, Int. J. of Intel. Syst. 11 975-988.
394. Torra, V. (1996) Weighted OWA operators for synthesis of information, Proc. of the 5th IEEE Int. Conf. on Fuzzy Systems 966-971.
395. Torra, V. (1997) The weighted OWA operator, Int. J. of Intel. Syst. 12 153-166.
396. Torra, V. (1998) On considering constraints of different importance in fuzzy constraint satisfaction problems, Int. J. of Unc., Fuzz. and Knowledge Based Systems 6:5 489-501.
397. Torra, V. (1998) On some relationships between the WOWA operator and the Choquet integral, Proc. of the IPMU 1998 Conference, Paris, France, 818-824.
398. Torra, V. (1999) On hierarchically S-decomposable fuzzy measures, Int. J. of Intel. Syst. 14:9 923-934.
399. Torra, V. (1999) On some relationships between hierarchies of quasi-arithmetic means and neural networks, Int. J. of Intel. Syst. 14:11 1089-1098.
400. Torra, V. (1999) On the learning of weights in some aggregation operators, Mathware and Soft Computing 6 249-265.
401. Torra, V. (2000) Learning weights for Weighted OWA operators, Proc. of the IEEE Int. Conf. on Industrial Electronics, Control and Instrumentation (IECON 2000), 2530-2535.
402. Torra, V. (2000) The WOWA operator and the interpolation function W*: Chen and Otto's interpolation method revisited, Fuzzy Sets and Systems 113:3 389-396.
403. Torra, V. (2001) Aggregation of linguistic labels when semantics is based on antonyms, Int. J. of Intel. Syst. 16 513-524.
404. Torra, V. (2001) Author's reply to [37], Fuzzy Sets and Systems 121 551.
405. Torra, V. (2001) Interpreting Sugeno $\lambda$-measures through Q-p-decomposable ones, Proc. of the EUROFUSE 2001 Workshop, Granada, Spain, 67-72.
406. Torra, V. (2002) Learning weights for the Quasi-Weighted Mean, IEEE Trans. on Fuzzy Systems 10:5 653-666.
407. Torra, V. (2003) La integral doble o *twofold* integral: Una generalització de les integrals de Choquet i Sugeno, Butlletí de l'Associació Catalana d'Intel·ligencia Artificial 29 13-19. Preliminary version in English: Twofold integral: A generalization of Choquet and Sugeno integral, IIIA Technical Report TR-2003-08.
408. Torra, V. (2004) OWA operators in data modeling and reidentification, IEEE Trans. on Fuzzy Systems 12:5 652-660.
409. Torra, V., Cortés, U. (1995) Towards an automatic consensus generator tool: EGAC, IEEE Trans. on Systems, Man and Cybernetics 25:5 888-894.
410. Torra, V., Godo, L. (2002) Continuous WOWA operators with application to defuzzification, in T. Calvo, G. Mayor, R. Mesiar (eds.) Aggregation Operators, Physica-Verlag, 159-176.
411. Torra, V., Narukawa, Y. (2004) On the interpretation of some fuzzy integrals, MDAI 2004, Lecture Notes in Artificial Intelligence 3131 316-326.

412. Torra, V., Narukawa, Y. (2004) Proc. 1st Int. Conf. Modeling Decisions for Artificial Intelligence, Lecture Notes in Artificial Intelligence 3131.
413. Torra, V., Narukawa, Y. (2005) On the meta-knowledge Choquet integral and related models, Int. J. of Intel. Syst. 20:10 1017-1036.
414. Torra, V., Narukawa, Y. (2006) The interpretation of fuzzy integrals and their application to fuzzy systems, Int. J. of Approx. Reasoning 41:1 43-58.
415. Torra, V., Narukawa, Y., Miyamoto, S. (2005) Proc. 2nd Int. Conf. Modeling Decisions for Artificial Intelligence, Lecture Notes in Artificial Intelligence 3558.
416. Torra, V., Narukawa, Y., Valls, A., Domingo-Ferrer, J. (2006) Proc. 3rd Int. Conf. Modeling Decisions for Artificial Intelligence, Lecture Notes in Artificial Intelligence 3885.
417. Traub, R. E. (1994) Reliability for the Social Sciences: theory and applications, SAGE publications.
418. Trillas, E. (1979) Sobre funciones de negación en la teoría de conjuntos difusos, Stochastica III-1 47-60.
419. Tsadiras, A. K., Margaritis, K. G. (1998) The MYCIN certainty factor handling function as uninorm operator and its use as a threshold function in artificial neurons, Fuzzy Sets and Systems 93 263-274.
420. Tsurumi, M., Tanino, T., Inuiguchi, M. (2001) A Shapley function on a class of cooperative fuzzy games, European Journal of Operational Research 129 596-618.
421. Tversky, A., Wakker, P. (1995) Risk attitudes and decision weights, Econometrica 63:6 1255-1280.
422. Vasilesco, F. (1937) La Notion de Capacité, Hermann (Paris).
423. Verkeyn, A., Botteldooren, D., De Baets, B., De Tré, G. (2003) Sugeno integrals for the modelling of noise annoyance aggregation, Proc. of the 10th IFSA Conference, Lecture Notes in Artificial Intelligence 2715 277-284.
424. Vitali, G. (1925) Sulla definizione di integrale delle funzioni di una variabile, Annali di Matematica Serie IV Tomo II 111-121
425. von Neumann, J., Morgenstern, O. (1944) Theory of Games and Economic Behavior, Princeton University Press.
426. Voting (2004) The History of Voting, http://www-gap.dcs.st-and.ac.uk/~history/HistTopics/Voting.html
427. Wang, Z., Klir, G. J. (1992) Fuzzy Measure Theory, Plenum Press.
428. Wang, Z., Leung, K.-S., Wong, M.-L., Fang, J., Xu, K. (2000) Nonlinear nonnegative multiregressions based on Choquet integrals, Int. J. of Approx. Reasoning 25 71-87.
429. Wang, D., Wang, X., Keller, J. M. (1997) Determining fuzzy integral densities using a genetic algorithm forpattern recognition, Proc. of the NAFIPS 1997 Conference, 263-267.
430. Weber, S. (1984) ⊥-decomposable measures and integrals for archimedean $t$-conorms ⊥, J. of Math. Analysis and Applications 101 114-138.
431. Weber, S. (1986) Two integrals and some modified versions – critical remarks, Fuzzy Sets and Systems 20 97-105.

432. Wei, Q., Yan, H., Ma, J., Fan, Z. (2000) A compromise weight for multi-criteria group decision making with individual preferences, The Journal of the Operational Research Society, 51:5 625-634.

433. Weintraub, E. R. (1992) Toward a history of game theory, Duke University Press (1992 annual supplement to History of Political Economy).

434. Weisstein, E. W., et al. (2006) Pythagorean means, from MathWorld–A Wolfram Web Resource.
http://mathworld.wolfram.com/PythagoreanMeans.html

435. White, F. (1987) Data fusion lexicon, prepared for Joint Directory Laboratories, Technical Panel for $C^3$ (JDLITPC$^3$), Data Fusion Subpanel (DFSP), Draft version, Naval Ocean Systems Center, 254-270.

436. Witten, I. H., Frank, E. (2000) Data Mining, Morgan Kaufmann.

437. World RPS Society, http://www.worldrps.com/

438. Xu, Z., Qingli, D. (2003) Approaches to obtaining the weights of the ordered weighted aggregation operators, J. of Southeast University (Natural Science Edition) 33:1 94-96.

439. Xu, Z. (2006) On generalized induced linguistic aggregation operators, Int. J. of General Systems 35:1 17-28.

440. Xu, Z. S., Da, Q. L. (2002) The ordered weighted geometric averaging operator, Int. J. of Intel. Syst. 17 709-716.

441. Yager, R. R. (1981) A new methodology for ordinal multiple aspect decisions based on fuzzy sets, Decision Sciences 12 589-600.

442. Yager, R. R. (1988) On ordered weighted averaging aggregation operators in multi-criteria decision making, IEEE Trans. on Systems, Man and Cybernetics 18 183-190.

443. Yager, R. R. (1992) Applications and extensions of OWA aggregations, Int. J. of Man-Machine Studies 37 103-122.

444. Yager, R. R. (1993) Families of OWA operators, Fuzzy Sets and Systems 59 125-148.

445. Yager, R. R. (1994) Misrepresentations and challenges: A response to Elkan, IEEE Expert August 41-42.

446. Yager, R. R. (1996) On the inclusion of variance in decision making under uncertainty, Int. J. of Unc., Fuzz. and Knowledge Based Systems 4:5 401-419.

447. Yager, R. R. (1996) Quantifier guided aggregation using OWA operators, Int. J. of Intel. Syst. 11 49-73.

448. Yager, R. R. (1999) On the entropy of fuzzy measures, Technical Report #MII-1917R, Machine Intelligence Institute, Iona College, New Rochelle, NY.

449. Yager, R. R. (1999) Nonmonotonic OWA operators, Soft Computing 3 187-196,

450. Yager, R. R. (2001) Induced aggregation operators, Fuzzy Sets and Systems 137 59-69.

451. Yager, R. R. (2002) Using importances in group preference aggregation to block strategic manipulation, in T. Calvo, G. Mayor, R. Mesiar (eds.) Aggregation Operators, Physica-Verlag, 177-191.

452. Yager, R. R. (2004) Generalized OWA aggregation operators, Fuzzy Optimization and Decision Making 3 93-107.

453. Yager, R. R. (2004) OWA Aggregation over a continuous interval argument with applications to decision making, IEEE Trans. on Systems, Man and Cybernetics 34:5 1952-1963.

454. Yager, R. R., Filev, D. P. (1994) Parameterized and-like and or-like OWA operators, Int. J. of General Systems 22 297-316.

455. Yager, R. R., Filev, D. P. (1999) Induced ordered weighted averaging operators, IEEE Trans. on Systems, Man and Cybernetics 29 141-150.

456. Yager, R. R., Kacprzyk, J. (eds.) (1997) The Ordered Weighted Averaging Operators: Theory and Applications, Springer.

457. Yager, R. R., Rybalov, A. (1996) Uninorm aggregation operators, Fuzzy Sets and Systems 80 111-120. Also as Tech. Report #MII-1407, Machine Intelligence Institute, Iona College, New Rochelle, N. Y., 1994.

458. Yoneda, M., Fukami, S., Grabisch, M. (1994) Human factor and fuzzy science (in Japanese), in K. Asai (ed.) Fuzzy Science, Kaibundo, 93 - 122.

459. Zadeh, L. A. (1965) Fuzzy sets, Inform. Control. 8 338-353.

460. Zadeh, L. A. (1978) Fuzzy sets as a basis for a theory of possibility, Fuzzy Sets and Systems 1 3-28.

461. Zadeh, L. A. (1983) A computational approach to fuzzy quantifiers in natural languages, Comp. & Maths Appl. 9:1 149-184.

462. Zadeh, L. A. (2002) From computing with numbers to computing with words – From manipulation of measurements to manipulations of perceptions, Int. J. Appl. Math. Comput. Sci. 12:3 307-324.

463. Zadeh, L. A., Kacprzyk, J. (1999) Computing with Words in Information/Intelligent Systems 1: Foundations, Physica-Verlag.

464. Zhuang, Z. (2002) Selected fables, New World Press.

# Index

# Cognitive Technologies

T. Frühwirth, S. Abdennadher:
**Essentials of Constraint Programming.**
IX, 144 pages. 2003

J. W. Lloyd:
**Logic for Learning.**
Learning Comprehensive Theories from Structured Data.
X, 256 pages. 2003

S. K. Pal, L. Polkowski, A. Skowron (Eds.):
**Rough-Neural Computing.**
Techniques for Computing with Words.
XXV, 734 pages. 2004

H. Prendinger, M. Ishizuka (Eds.):
**Life-Like Characters.**
Tools, Affective Functions, and Applications.
IX, 477 pages. 2004

H. Helbig:
**Knowledge Representation and
the Semantics of Natural Language.**
XVIII, 646 pages. 2006

P. M. Nugues:
**An Introduction to Language Processing
with Perl and Prolog.**
An Outline of Theories, Implementation,
and Application with Special Consideration
of English, French, and German.
XX, 513 pages. 2006

W. Wahlster (Ed.):
**SmartKom: Foundations of Multimodal Dialogue Systems.**
XVIII, 644 pages. 2006

B. Goertzel, C. Pennachin (Eds.):
**Artificial General Intelligence.**
XVI, 509 pages. 2007

O. Stock, M. Zancanaro (Eds.):
**PEACH - Intelligent Interfaces for Museum Visits.**
XVIII, 316 pages. 2007

V. Torra, Y. Narukawa:
**Modeling Decisions.**
Information Fusion and Aggregation Operators
XIV, 284 pages. 2007